THE THEORY OF PIEZOELECTRIC SHELLS AND PLATES

Nellya N. Rogacheva

CRC PRESS

Boca Raton London New York Washington, D.C.

Library of Congress Cataloging-in-Publication Data

Rogacheva, Nellya N.
The theory of piezoelectric shells and plates / Nellya N. Rogacheva.
p. cm.
Includes bibliographical references and index.
ISBN 0-8493-4459-X
1. Piezoelectric devices—Mathematical models. 2. Piezoelectric ceramics.
3. Shells (Engineering)—Mathematical models. 4. Plates
(Engineering—Mathematical models. I. Title
TK7872.P54R64 1998
621.381′4—dc20 93-11621

Direct all inquiries to CRC Press LLC, 2000 N.W. Corporate Blvd., Boca Raton, Florida 33431.

International Standard Book Number 0-8493-4459-X
Library of Congress Card Number 93-11621
Printed in the United States of America 4 5 6 7 8 9 0
Printed on acid-free paper

FOREWORD

This book contains a comprehensive treatment of problems arising in the construction and substantiation of applied theories of piezoelectrical shells.

Two tendencies have obviously appeared in modern mechanics due to computerization: i.e., a rapid development of numerical methods and their applications; and a noticeable lag in the field of theoretical investigations that give a qualitative analysis of the problem before it is solved and result in simple engineering theories. This is clearly manifested in electromagnetic elasticity, which is a relatively new field of mechanics. Thus, the design of electroelastic thin-walled elements is commonly based on three-dimensional theories, which causes unnecessary complications which can be explained by the inconsistency of the applied electroelastic shell theories and the lack of estimates for the introduced errors.

The work of my disciple, N. N. Rogacheva, is aimed at eliminating this discrepancy. She gives a mathematical substantiation for the transition from three-dimensional electroelasticity equations to two-dimensional differential equations without using any hypotheses, investigates the applicability of traditional hypotheses, and estimates their possible errors. Specifically, she gives examples of their inapplicability.

Much attention is given to the approximate methods for solving the differential equations of the theory of piezoelectrical shells and plates and computing the coefficient of electromechanical coupling. Some of the problems are solved for the first time.

A new problem is raised and solved concerning the elastic and electrical phenomena at the edge of a thin-walled element. It is shown that they can be more essential than the edge phenomena of elasticity theory.

The book will undoubtedly engage the interest of scientists and engineers dealing with problems of electroelasticity.

A. L. Gol'denveizer
Professor, Foreign Honorary Member of the American
Academy of Arts and Sciences

INTRODUCTION

Piezoeffect has important uses in modern engineering because it expresses the connection between the electrical and mechanical fields.

The piezoeffect was first applied during World War I by P. Langevin, a French physicist, in a device that used quartz mosaic for hydroacoustic measurements and investigations.

A new stage in the development of piezoeffect devices evolved due to the discovery of piezoelectric properties in a polarized ceramic, barium titanate. Piezoelectric crystals were soon replaced by piezoceramics that were much cheaper, had greater piezoelectric activity than quartz, and could be shaped into practically any form and size [12, 13, 37, 54, 62].

Piezoceramic transformers, frequency-band filters, acoustic radiators and receivers, ultrasonic delay lines, and gyro piezoelectric elements do not exhaust the list of piezoelectric devices. The reader can find detailed information on these devices in other sources [6, 7, 8, 54, 63, 66, 69, 95].

The reliability of piezoelectric devices greatly depends on design techniques. Even in the simplest linear formulations that do not allow for geometric and physical nonlinearities or temperature and magnetic effects, the electroelastic state of a piezoelement is described by the solution of a related electroelastic problem which becomes more complicated as the element form becomes more complex. It is expedient to develop approximate methods for designing actual thin-walled elements using the smallness of one of the sizes, as done in the theory of elastic plates and shells. This approach seems to be of special importance because, in the three-dimensional theory, the analytical solutions can only be obtained for the simplest boundary problems [5, 49, 50, 96, 102].

Although a lot of publications [e.g., 32–34, 48, 58–61, 97–99, 101–104] are devoted to thin-walled piezoelectric elements, a need is still felt for consistent theories of approximate techniques for solving dynamic problems by estimating the introduced errors.

By analogy with the theory of elastic shells and plates, various methods for passing from three-dimensional to two-dimensional problems are used for electroelastic shells and plates. The three-dimensional electroelasticity relations are replaced by an approximate system of differential equations containing two independent variables. Simultaneously, the three-dimensional boundary conditions are replaced by two-dimensional ones. This transition becomes possible due to the small thickness of the shells and the qualitative behavior resulting from this property.

By tradition, we will call the methods for reducing three-dimensional equations to two-dimensional ones, the reduction methods. We also agree to treat the elastic shells whose equations contain only mechanical quantities as nonelectrical shells. Problems with only mechanical unknown quantities will be called mechanical problems and those where all unknown quantities are electrical will be called electrical problems.

When constructing approximate theories, the known exact solutions for three-dimensional problems can be used to check the correctness of the two-dimensional theories. Today, the exact three-dimensional solutions have only been obtained for such simple cases as an infinite layer, an infinite cylindrical shell, and a hollow sphere [4, 49, 50, 56, 101–103].

The numerical solutions to three-dimensional problems obtained by the method of finite elements [4, 39–41, 100, 115, 116] can be used as standard solutions.

A three-dimensional problem can be reduced to a two-dimensional one by using certain hypotheses. The known Kirchhoff-Love theory was obtained in this way. Hypotheses were used for constructing the theory of electroelastic shells in [1–3, 8–11, 19, 24, 56, 58–61]. Kirchhoff's hypotheses were adopted for the mechanical quantities, while the electrical quantities were found using different hypotheses. There were attempts to construct refined shell theories [46, 47, 73] where, by analogy with Reisner's theory of nonelectrical shells [65, 77], the transverse shear was taken into account.

Another reduction method, based on expanding the required functions in power series in the thickness coordinate, stems from the works by Cauchi and Poisson. The number of terms kept in the series depends on the desired accuracy. Specifically, this method was used in a number of studies [2, 34, 59, 101].

The small thickness of a shell is most fully used in the asymptotic methods that allow us to investigate the qualitative behavior of the stressed-strained state, pass from the three-dimensional electroelastic problems to two-dimensional shell equations without using any hypotheses, and estimate the errors of the resulting theories [20, 44, 52, 78–82, 89]. The asymptotic reduction method was thoroughly tested in the construction of the theory of nonelectrical shells, when the Kirchhoff-Love hypotheses were substantiated, the boundary stressed-strained state was investigated, and asymptotic estimation of shell theory errors was carried out [22, 43].

This book is devoted to the linear theory of thin piezoelectrical shells with arbitrary form and differently directed preliminary polarization. We assume that the reader is acquainted with classical shell theories of Kirchhoff-Love type.

The book has two parts. In the first part we use some hypotheses to derive the generalized equations of piezoelectrical shells. Note that originally these theories were constructed by the asymptotic method without using any hypotheses. We preferred using the hypotheses because the method is physically obvious and allows a comparison with the known Kirchhoff-Love hypotheses for non-electrical shells. The hypotheses are dependent on the electrical conditions on the shell faces and on the direction of the preliminary polarization. We show that electroelastic shell problems can be subdivided into two classes. In the first class the problems are connected electroelastically and in the second, they are partitioned into separate mechanical and electrical parts.

We derive the boundary conditions for the theory of electroelastic shells on the basis of Saint Venant's principle generalized to electroelasticity. This principle is as important for the electroelasticity theory as is the Saint Venant principle for the elasticity theory.

We pay much attention to an approximate method of shell design consisting of partitioning the elecroelastic state into the principal, say, membrane electroelastic state and simple boundary effects.

A separate section deals with the dynamic behavior of piesoelectrical shells. We show that the entire spectrum of the vibrations of electroelastic shells can be broken into subspectra having different physical natures. For every vibration type we obtain approximate equations and respective boundary conditions. A similar investigation is carried out for forced vibrations.

We solve a number of dynamic problems using approximate theories for computing the dynamical electroelastic state of a shell. Special attention is payed to computing the electromechanic coupling coefficient.

For simplicity, we try to use the asymptotic analysis as little as possible in the first part, without affecting the physical meaning of the investigated problems.

In the second part of the book, the theory of piezoelectrical shells is constructed by investigating the limiting transition from three-dimensional electroelastic problems to two-dimensional ones. This allows us to substantiate the hypotheses adopted in the first part, define the applicability of the theory of electroelastic shells, and construct asymptotically consistent and refined theories. We also test the applicability of the Kirchhoff-Love hypotheses used in the theory of nonelectrical shells.

We study the essentially three-dimensional state of the boundary layer and obtain approximate equations of the boundary layer; i.e., the equations of the plane and antiplane electroelastic problems. The Saint Venant principle is generalized to the case of electroelasticity.

By investigating the interaction between the electroelastic state described by shell theory and the boundary layer at the edge, provided that the three-dimensional boundary conditions are satisfied, we get the boundary conditions for the theory of piezoelectrical shells and the boundary layer. We also solve some problems for the boundary layer.

TABLE OF CONTENTS

THE THEORY OF PIEZOELECTRIC SHELLS AND PLATES

Part I. Statics and Dynamics of Piezoelectric Shells and Plates

Chapter 1

THREE-DIMENSIONAL EQUATIONS OF ELECTROELASTICITY

1 PRELIMINARY INFORMATION

We will investigate thin-walled bodies, i.e., shells made of polarized ceramics. A polarized ceramic is an electroelastic material in which the link between deformation fields and internal electric fields is rather strong. The microstructure of the material and the piezoeffect mechanism were considered in [12, 13, 37, 62]; therefore, we will only describe the effect itself.

Elementary dipoles inside small ceramic regions (domains) spontaneously orient themselves parallel to each other; therefore, the domains are dipoles. Since the domains are chaotically arranged, no piezoeffect occurs in a ceramic. In a strong electric field, however, the domains become oriented and preserve their orientation when the electric field is removed. Thus we obtain polarized ceramic possessing piezoelectric properties.

The behavior of polarized ceramics can be illustrated by a simple example. In Figure 1, the solid line presents a rod polarized along its length; the dashed line presents the same rod deformed by an electric field applied at the rod's ends.

The electroelastic state is defined as a linear problem. We assume that (a) displacements are small compared to the body thickness and (b) the deformations, the mechanical stresses, and the electric field are directly proportional. We neglect temperature and magnetic effects.

Our assumptions are reasonable because most devices employ the linear behavior of piezomaterials and piezoelements and because it will be easier to turn to the nonlinear theory after the simpler linear formulations have been studied. We will take the linearized piezoelectric equations described in [8] as the original equations. A formulation for three-dimensional electroelastic problems can be found in [26, 57, 67, 73].

2 BASIC EQUATIONS

We place a thin-walled electroelastic body in the three-orthogonal coordinate system. We then choose curvilinear coordinates α_1 and α_2 so that they coincide with the curvatures of the middle surface of the shell and a linear coordinate γ along the normal to the middle surface.

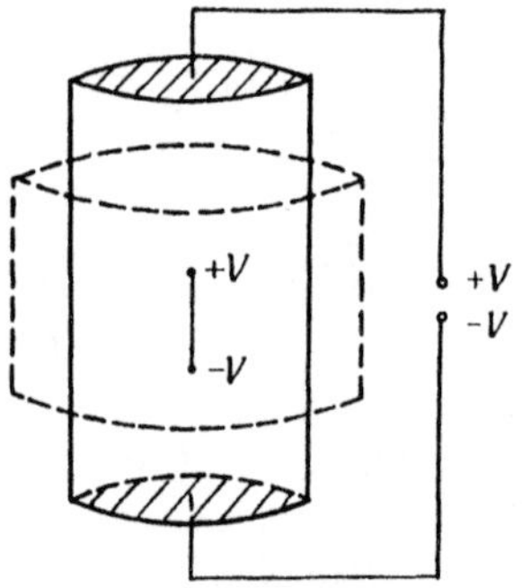

FIGURE 1. Strained piezoelement in electric field

A complete system of equations in electroelasticity theory consists of the equilibrium equations, the geometrical relationships between the components of the deformation tensor and those of the displacement vector, the constitutive relations, and the electrostatic equations. The equilibrium equations and geometrical formulas are taken from linear elasticity theory [105].

Let an electroelastic body be deformed by some forces. As a result its point M with coordinates $\alpha_1, \alpha_2, \gamma$ is displaced. The displacement vector components v_1, v_2, and v_3 are projected on the coordinate lines α_1, α_2, and γ, i.e.,

$$v_i = v_i(\alpha_1, \alpha_2, \gamma), \qquad v_3 = v_3(\alpha_1, \alpha_2, \gamma). \tag{2.1}$$

Every equation with subscripts i and j contains two equations: one for $i = 1$ and $j = 2$, and the other for $i = 2$ and $j = 1$. We will use Greek letters for subscripts that assume the values 1, 2, or 3 and Latin letters for subscripts that assume the values 1 or 2.

We say that displacements v_i and v_3 are positive if their projections on the coordinate axes are positive. In the linear theory, a deformed state is described by the deformation tensor $e_{\rho\mu}$; the deformations are related to the displacements v_ρ:

$$e_{ii} = \frac{1}{H_i}\frac{\partial v_i}{\partial \alpha_i} + \frac{1}{H_iH_j}\frac{\partial H_i}{\partial \alpha_j}v_j + \frac{1}{H_iH_3}\frac{\partial H_i}{\partial \gamma}v_3$$

$$e_{33} = \frac{1}{H_3}\frac{\partial v_3}{\partial \gamma} + \frac{1}{H_3H_1}\frac{\partial H_3}{\partial \alpha_1}v_1 + \frac{1}{H_2H_3}\frac{\partial H_3}{\partial \alpha_2}v_2$$

$$e_{ij} = \frac{H_i}{H_j}\frac{\partial}{\partial \alpha_j}\left(\frac{v_i}{H_i}\right) + \frac{H_j}{H_i}\frac{\partial}{\partial \alpha_i}\left(\frac{v_j}{H_j}\right)$$

$$e_{3i} = \frac{H_3}{H_i}\frac{\partial}{\partial \alpha_i}\left(\frac{v_3}{H_3}\right) + \frac{H_i}{H_3}\frac{\partial}{\partial \gamma}\left(\frac{v_i}{H_i}\right). \tag{2.2}$$

The components e_{11}, e_{22}, and e_{33} of the deformation tensor are relative elongations along the three coordinate axes at the point with coordinates $\alpha_1, \alpha_2, \gamma$. The components e_{ij} and e_{i3} are relative shifts in the planes $\alpha_i\alpha_j$ and $\alpha_i\gamma$. The H_i in formulas 2.2 are Lame coefficients. In our three-orthogonal coordinate system, the Lame coefficients are related to the coefficients of the first quadratic form A_i and the principal radius of curvatures of the middle surface R_i, i.e.,

$$H_i = A_i\left(1 + \frac{\gamma}{R_i}\right), \qquad H_3 = 1. \tag{2.3}$$

From the theory of surfaces we know that the coefficients of the first quadratic form are related to the curvature radius the coordinate lines via Codazzi-Gauss formulas

$$\frac{\partial}{\partial \alpha_j}\left(\frac{A_i}{R_i}\right) = \frac{1}{R_j}\frac{\partial A_i}{\partial \alpha_j}$$

$$\frac{1}{R_1 R_2} = -\frac{1}{A_1 A_2}\left[\frac{\partial}{\partial \alpha_1}\left(\frac{1}{A_1}\frac{\partial A_2}{\partial \alpha_1}\right) + \frac{\partial}{\partial \alpha_2}\left(\frac{1}{A_2}\frac{\partial A_1}{\partial \alpha_2}\right)\right]. \tag{2.4}$$

Using formulas 2.3 and 2.4 for the derivatives of H_i with respect to α_i and γ, we get

$$\frac{\partial H_i}{\partial \alpha_j} = \left(1 + \frac{\gamma}{R_j}\right)\frac{\partial A_i}{\partial \alpha_j},$$

$$\frac{\partial H_i}{\partial \gamma} = \frac{A_i}{R_i}. \tag{2.5}$$

We substitute the coefficients 2.3 into the right-hand sides of equations 2.2 and use 2.5 to obtain geometrical formulas that are more suitable for a theory of shells, i.e.,

$$e_{ii} = \frac{e_i}{a_i} \qquad e_{33} = \frac{\partial v_3}{\partial \gamma}$$

$$e_{ij} = e_{ji} = \frac{1}{a_j}m_i + \frac{1}{a_i}m_j \qquad e_{3i} = e_{i3} = \frac{\partial v_i}{\partial \gamma} + \frac{1}{a_i}g_i. \tag{2.6}$$

To be more concise, we use

$$e_i = \frac{1}{A_i}\frac{\partial v_i}{\partial \alpha_i} + k_i v_j + \frac{v_3}{R_i} \qquad m_i = \frac{1}{A_j}\frac{\partial v_i}{\partial \alpha_j} - k_j v_j \tag{2.7}$$

$$g_i = \frac{1}{A_i}\frac{\partial v_3}{\partial \alpha_i} - \frac{v_i}{R_i} \qquad k_i = \frac{1}{A_i A_j}\frac{\partial A_i}{\partial \alpha_j} \qquad a_i = 1 + \frac{\gamma}{R_i}. \tag{2.8}$$

The stresses state that the point M with coordinates $\alpha_1, \alpha_2, \gamma$ is fully described by the stress tensor $\sigma_{\rho\mu}(\rho, \mu = 1, 2, 3)$. Figure 2 shows stresses with positive direction. The components of the stresses should meet the differential equilibrium equations, which assume the form

$$\frac{\partial}{\partial \alpha_i}(H_j H_3 \sigma_{ii}) + \frac{\partial}{\partial \alpha_j}(H_i H_3 \sigma_{ij}) + \frac{\partial}{\partial \gamma}(H_i H_j \sigma_{i3}) - H_3 \frac{\partial H_j}{\partial \alpha_i}\sigma_{jj}$$
$$- H_j \frac{\partial H_3}{\partial \alpha_i}\sigma_{33} + H_3 \frac{\partial H_i}{\partial \alpha_j}\sigma_{ji} + H_j \frac{\partial H_i}{\partial \gamma}\sigma_{3i} + H_i H_j H_3 q_i = 0$$
$$\frac{\partial}{\partial \gamma}(H_1 H_2 \sigma_{33}) + \frac{\partial}{\partial \alpha_1}(H_3 H_2 \sigma_{31}) + \frac{\partial}{\partial \alpha_2}(H_3 H_1 \sigma_{32}) - H_2 \frac{\partial H_1}{\partial \gamma}\sigma_{11}$$
$$- H_1 \frac{\partial H_2}{\partial \gamma}\sigma_{22} + H_2 \frac{\partial H_3}{\partial \alpha_1}\sigma_{13} + H_1 \frac{\partial H_3}{\partial \alpha_2}\sigma_{23} + H_1 H_2 H_3 q_3 = 0 \quad (2.9)$$

in our three-orthogonal coordinate system. Here q_1, q_2, and q_3 are the components of the vector of the bulk forces. In order to obtain the differential equations of motion, we introduce the inertial terms

$$\rho \frac{\partial^2 v_i}{\partial t^2} \quad \text{and} \quad \rho \frac{\partial^2 v_3}{\partial t^2}$$

into the right-hand sides, where ρ is the density of the material and t is the time.

The relation between the elastic field tensors (strain $e_{\rho\mu}$ and stress $\sigma_{\rho\mu}$) and the electric field vectors (induction $\mathcal{D}$ with components $\mathcal{D}_1, \mathcal{D}_2$, and $\mathcal{D}_3$ and strength $\mathcal{E}$ with components $\mathcal{E}_1, \mathcal{E}_2$, and $\mathcal{E}_3$) can be represented by four equally important systems of piezoeffect equations:

$$\begin{cases} \sigma_{ij} = c^E_{ijkl} e_{kl} - \underline{e}_{kij} \mathcal{E}_k \\ D_i = \underline{e}_{ikl} e_{kl} + \varepsilon^S_{ij} \mathcal{E}_k \end{cases}$$

$$\begin{cases} e_{ij} = s^E_{ijkl} \sigma_{kl} + d_{kij} \mathcal{E}_k \\ D_i = d_{ikl} \sigma_{kl} + \varepsilon^T_{ik} \mathcal{E}_k \end{cases}$$

$$\begin{cases} e_{ij} = s^D_{ijkl} \sigma_{kl} + g_{kij} \mathcal{D}_k \\ E_i = g_{ikl} \sigma_{kl} + \beta^T_{ik} \mathcal{D}_k \end{cases}$$

$$\begin{cases} \sigma_{ij} = c^D_{ijkl} e_{kl} - h_{klij} \mathcal{D}_k, \\ E_i = -h_{ikl} e_{kl} + \beta^S_{ij} \mathcal{D}_k \end{cases}$$

where s^E_{ijkl} and s^D_{ijkl} are elastic compliances at constant electric field and induction, respectively; c^E_{ijkl} and c^D_{ijkl} are elasticity coefficients under the same conditions; $\varepsilon^s_{ij}, \varepsilon^T_{ij}$, and $\beta^s_{ij}, \beta^T_{ij}$ are dielectric permittivities and "nonpermittivities" for constant strain and elastic stress; $d_{kij}, \underline{e}_{kij}, g_{kij}$, and h_{kij} are piezoelectric constants that are also called a piezoelectric modulus.

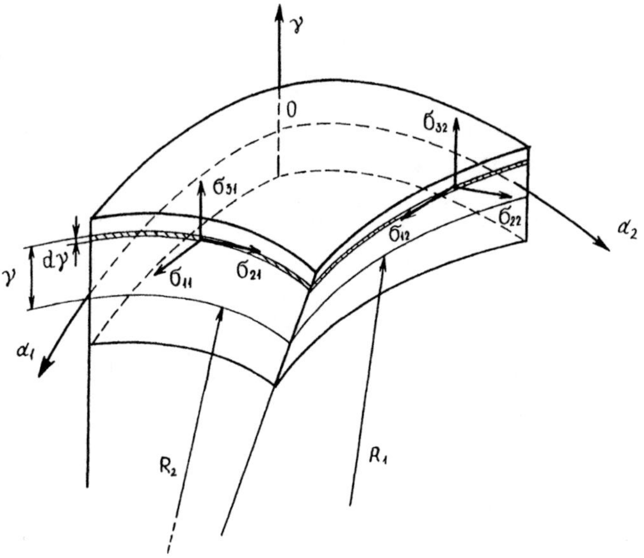

FIGURE 2. A shell element with applied stresses

We use a brief notation of tensor theory, where the pairs of subscripts that can be interchanged are replaced by the scheme $11 \rightarrow 1, 22 \rightarrow 2, 33 \rightarrow 3, 32 \rightarrow 4, 31 \rightarrow 5, 21 \rightarrow 6$. By using this notation, we can write the equations as square matrices. For example, the second system in matrix form looks like

$$
\begin{vmatrix} e_1 \\ e_2 \\ e_3 \\ e_{23} \\ e_{13} \\ e_{12} \\ \mathcal{D}_1 \\ \mathcal{D}_2 \\ \mathcal{D}_3 \end{vmatrix}
=
\left|\begin{array}{cccccc|ccc}
s_{11}^E & s_{12}^E & s_{13}^E & 0 & 0 & 0 & 0 & 0 & d_{13} \\
s_{12}^E & s_{11}^E & s_{13}^E & 0 & 0 & 0 & 0 & 0 & d_{13} \\
s_{13}^E & s_{13}^E & s_{44}^E & 0 & 0 & 0 & 0 & 0 & d_{44} \\
0 & 0 & 0 & s_{44}^E & 0 & 0 & 0 & d_{15} & 0 \\
0 & 0 & 0 & 0 & s_{44}^E & 0 & d_{15} & 0 & 0 \\
0 & 0 & 0 & 0 & 0 & s_{66}^E & 0 & 0 & 0 \\
\hline
0 & 0 & 0 & 0 & d_{15} & 0 & \varepsilon_{11}^T & 0 & 0 \\
0 & 0 & 0 & d_{15} & 0 & 0 & 0 & \varepsilon_{11}^T & 0 \\
d_{31} & d_{31} & d_{33} & 0 & 0 & 0 & 0 & 0 & \varepsilon_{33}^T
\end{array}\right|
\cdot
\begin{vmatrix} \sigma_{11} \\ \sigma_{22} \\ \sigma_{33} \\ \sigma_{23} \\ \sigma_{13} \\ \sigma_{12} \\ \mathcal{E}_1 \\ \mathcal{E}_2 \\ \mathcal{E}_3 \end{vmatrix}
$$

The relations between the stress tensors, deformation tensors, and electrical quantities for piezoceramics polarized along the γ-lines can be written as piezo-effect equations:

$$\begin{aligned} e_{ii} &= s_{11}^E \sigma_{ii} + s_{12}^E \sigma_{jj} + s_{13}^E \sigma_{33} + d_{31}\mathcal{E}_3 \\ e_{33} &= s_{13}^E(\sigma_{11} + \sigma_{22}) + s_{33}^E \sigma_{33} + d_{33}\mathcal{E}_3 \\ e_{ij} &= s_{66}^E \sigma_{ij} \\ e_{i3} &= s_{44}^E \sigma_{i3} + d_{15}\mathcal{E}_i \\ \mathcal{D}_i &= \varepsilon_{11}^T \mathcal{E}_i + d_{15}\sigma_{i3} \\ \mathcal{D}_3 &= \varepsilon_{33}^T \mathcal{E}_3 + d_{31}(\sigma_{11} + \sigma_{22}) + d_{33}\sigma_{33}. \end{aligned} \tag{2.10}$$

Table 1 gives physical constants for the piezoceramics PZT-4 and PZT-5 that will be needed for our computation [8].

We use formulas 2.3 and 2.5 to transform the equilibrium equations 2.9 and the relations 2.10 to a more convenient form, vis.

$$\begin{aligned} &\frac{1}{A_i}\frac{\partial \tau_{ii}}{\partial \alpha_i} + \frac{1}{A_j}\frac{\partial \tau_{ij}}{\partial \alpha_j} + k_j(\tau_{ii} - \tau_{jj}) + k_i(\tau_{ij} + \tau_{ij}) + \frac{1}{a_i}\frac{\partial}{\partial \gamma}(a_i^2 \tau_{i3}) + a_1 a_2 q_1 = 0, \\ &-\frac{\tau_{11}}{R_1} - \frac{\tau_{22}}{R_2} + \frac{1}{A_1}\frac{\partial \tau_{13}}{\partial \alpha_1} + \frac{1}{A_2}\frac{\partial \tau_{23}}{\partial \alpha_2} + k_2\tau_{13} + k_1\tau_{23} + \frac{\partial \tau_{33}}{\partial \gamma} + a_1 a_2 q_3 = 0. \end{aligned} \tag{2.11}$$

Here the asymmetric stress tensor $\tau_{\rho\mu}$ is related to the symmetric stress tensor $\sigma_{\rho\mu}$, i.e.,

$$\tau_{ii} = a_j\sigma_{ii}, \qquad \tau_{ij} = a_i\sigma_{ij}, \qquad \tau_{i3} = \tau_{3i} = a_j\sigma_{i3}, \qquad \tau_{33} = a_1 a_2 \sigma_{33}. \tag{2.12}$$

We will need this notation when considering forces and moments in shell theory.

We change the components $\mathcal{D}_\rho, \mathcal{E}_\rho$ for the quantities D_ρ and E_ρ according to the formulas

$$E_i = a_i\mathcal{E}_i, \qquad E_3 = a_1 a_2\mathcal{E}_3, \qquad D_i = a_j\mathcal{D}_i, \qquad D_3 = a_1 a_2\mathcal{D}_3 \tag{2.13}$$

which will simplify some of the equations.

The polarization that makes the domains perpendicular to the middle surface along the γ-lines is usually termed thickness polarization. We solve the first two equations 2.10 for σ_{11} and σ_{22}, introduce the asymmetric stress tensor 2.12

TABLE 1
Some physical properties of piezoceramics PZT-4, PZT-5[8]

Quantity	PZT-4	PZT-5
s_{11}^E, 10^{-12} m^2/H	12.3	16.4
s_{12}^E	−4.05	−5.74
s_{13}^E	−5.31	−7.22
s_{33}^E	15.5	18.8
s_{44}^E	39.0	47.5
s_{66}^E	32.7	44.3
d_{33}, 10^{-12} C/N	289	374
d_{31}	−123	−172
d_{15}	496	584
$\varepsilon_{11}^T/\varepsilon_0$	1475	1730
$\varepsilon_{33}^T/\varepsilon_0$	1300	1700
ρ, 10^3 kg/m^3	7.5	7.75
$\varepsilon_0 = 8.85 \times 10^{-12}$ F/m		

and the electric vector components E_ρ and D_ρ 2.13. Formulas 2.10 will then be written as

$$\begin{aligned}
\tau_{ii} &= \frac{1}{s_{11}^E(1-\nu^2)}\left(\frac{a_j}{a_i}e_i + \nu e_j\right) - \frac{s_{13}^E}{s_{11}^E}\frac{\nu}{1-\nu}\frac{\tau_{33}}{a_i} - \frac{d_{31}}{s_{11}^E}\frac{1}{1-\nu}\frac{E_3}{a_i} \\
\tau_{ij} &= \frac{1}{s_{66}^E}\left(\frac{a_i}{a_j}m_i + m_j\right) \\
\frac{\partial v_3}{\partial \gamma} &= s_{13}^E\left(\frac{\tau_{11}}{a_2} + \frac{\tau_{22}}{a_1}\right) + s_{33}^E\frac{\tau_{33}}{a_1 a_2} + d_{33}\frac{E_3}{a_1 a_2} \\
\frac{\partial v_i}{\partial \gamma} + \frac{g_i}{a_i} &= s_{44}^E\frac{\tau_{i3}}{a_j} + d_{15}\frac{E_i}{a_i} \qquad D_i = \varepsilon_{11}^T\frac{a_j}{a_i}E_i + d_{15}\tau_{i3} \\
D_3 &= \varepsilon_{33}^T E_3 + d_{31}(\tau_{11}a_1 + \tau_{22}a_2) + d_{33}\tau_{33} \\
\nu &= -\frac{s_{12}^E}{s_{11}^E}.
\end{aligned} \tag{2.14}$$

We write the piezoeffect equations for a piezoceramics with preliminary tangential polarization. We imply a polarization for which the domains are arranged along one family of coordinate lines in parallel with the middle surface, i.e.,

$$e_{11} = s_{11}^E\sigma_{11} + s_{12}^E\sigma_{22} + s_{12}^E\sigma_{33} + d_{31}\mathcal{E}_2$$

$$
\begin{aligned}
e_{22} &= s_{13}^E\sigma_{11} + s_{33}^E\sigma_{22} + s_{13}^E\sigma_{33} + d_{33}\mathcal{E}_2 \\
e_{33} &= s_{12}^E\sigma_{11} + s_{13}^E\sigma_{22} + s_{11}^E\sigma_{33} + d_{31}\mathcal{E}_2 \\
e_{12} &= s_{44}^E\sigma_{21} + d_{15}\mathcal{E}_1 \\
e_{21} &= s_{44}^E\sigma_{21} + d_{15}\mathcal{E}_1 \\
e_{23} &= s_{44}^E\sigma_{23} + d_{15}\mathcal{E}_3 \\
e_{32} &= s_{44}^E\sigma_{32} + d_{15}\mathcal{E}_3 \\
e_{13} &= s_{66}^E\sigma_{13} \\
e_{31} &= s_{66}^E\sigma_{31} \\
\mathcal{D}_1 &= \varepsilon_{11}^T\mathcal{E}_1 + d_{15}\sigma_{12} \\
\mathcal{D}_2 &= \varepsilon_{33}^T\mathcal{E}_2 + d_{31}(\sigma_{11} + \sigma_{22}) + d_{33}\sigma_{22} \\
\mathcal{D}_3 &= \varepsilon_{11}^T\mathcal{E}_3 + d_{15}\sigma_{23}.
\end{aligned} \tag{2.15}
$$

Here we assume that the polarization orients the domains along the family of α_2-lines.

We solve 2.15 for σ_{11} and σ_{22}, taking into account equations 2.12 and 2.13, to get the required piezoeffect equations:

$$
\begin{aligned}
\tau_{ii} &= n_{ii}\frac{a_j}{a_i}e_i + n_{ij}e_j - p_i\frac{\tau_{33}}{a_i} - c_i\frac{a_j}{a_2}E_2 \\
\tau_{ij} &= \frac{1}{s_{44}^E}\left(\frac{a_i}{a_j}m_i + m_j\right) - \frac{d_{15}}{s_{44}^E}\frac{a_i}{a_i}E_1 \\
\frac{\partial v_3}{\partial\gamma} &= s_{12}^E\frac{\tau_{11}}{a_2} + s_{13}^E\frac{\tau_{22}}{a_1} + s_{11}^E\frac{\tau_{33}}{a_1a_2} + \frac{d_{31}}{a_2}E_2 \\
\frac{\partial v_1}{\partial\gamma} + \frac{g_1}{a_1} &= s_{66}^E\frac{\tau_{13}}{a_2} \\
\frac{\partial v_2}{\partial\gamma} + \frac{g_2}{a_2} &= s_{44}^E\frac{\tau_{23}}{a_1} + \frac{d_{15}}{a_1a_2}E_3 \\
D_1 &= \varepsilon_{11}^T\frac{a_2}{a_1}E_1 + d_{15}\tau_{23} \\
D_2 &= \varepsilon_{33}^T\frac{a_2}{a_1}E_2 + d_{31}\left(\frac{a_2}{a}\tau_{11} + \frac{1}{a_2}\tau_{33}\right) + d_{33}\tau_{22} \\
D_3 &= \varepsilon_{11}^T E_3 + d_{15}a_2\tau_{23}
\end{aligned} \tag{2.16}
$$

where we introduced n_{ij}, p_i, and c_i by the formulas

$$
\begin{aligned}
n_{11} &= \frac{s_{33}^E}{\delta}, & n_{12} &= -\frac{s_{13}^E}{\delta}, & n_{22} &= \frac{s_{11}^E}{\delta} \\
P_1 &= \frac{s_{12}^E s_{33}^E - (s_{13}^E)^2}{\delta}, & P_2 &= \frac{s_{11}^E s_{13}^E - s_{12}^E s_{13}^E}{\delta} \\
c_1 &= \frac{d_{31} s_{33}^E - d_{33} s_{13}^E}{\delta}, & c_2 &= \frac{d_{33} s_{11}^E - d_{31} s_{13}^E}{\delta} \\
\delta &= s_{11}^E s_{33}^E - (s_{13}^E)^2.
\end{aligned}
\tag{2.17}
$$

The behavior of the electromagnetic field is described by Maxwell's equations [18, 45]. It is shown in [53] that for a wide class of electroelastic dynamic problems, where the deformation wave length is much less than that of the electromagnetic wave of the same frequency, Maxwell's equations can be replaced by electrostatic equations in vector form:

$$
div\,\mathcal{D} = 0, \qquad \mathcal{E} = -grad\psi. \tag{2.18}
$$

Or they can be replaced in scalar form when changing to D and E by formulas 2.13:

$$
\begin{aligned}
&\frac{1}{A_1}\frac{\partial D_1}{\partial \alpha_1} + \frac{1}{A_2}\frac{\partial D_2}{\partial \alpha_2} + k_1 D_2 + k_2 D_1 + \frac{\partial D_3}{\partial \gamma} = 0 \\
&E_i = -\frac{1}{A_i}\frac{\partial \psi}{\partial \alpha_i}, \qquad E_3 = -a_1 a_2 \frac{\partial \psi}{\partial \gamma}.
\end{aligned}
\tag{2.19}
$$

The equations describing a piezoelectric body include the equilibrium equations 2.11, the strain-displacement formulas 2.7, the piezoeffect equations 2.14 or 2.16, and the electrostatic equations 2.18 or 2.19.

3 BOUNDARY CONDITIONS IN ELECTROELASTICITY THEORY

The equilibrium equations 2.11, geometrical relations 2.6 or 2.7, piezoeffect equations 2.14 or 2.16, and the equations of electrostatics 2.18 or 2.19 constitute a complete system, where the number of equations is equal to the number of unknowns. To solve problems, we use this system of differential equations supplemented by additional boundary conditions for the sought-for electrical and mechanical quantities. In electroelasticity theory, mechanical boundary conditions are formulated just like in the classical elasticity theory: the stresses, displacements, or mixed boundary conditions are specified at the boundary of the body.

We assume that the surface load vector with components $q_1^\pm$, $q_2^\pm$, and $q_3^\pm$ is given at every point of the shell faces. In terms of the nonsymmetrical stress tensor, the boundary conditions on the shell faces will be written as

$$\sigma_{i3} = \frac{\tau_{i3}}{a_j} \,|_{\gamma=\pm h} = \pm q_i^\pm, \qquad \sigma_{33} = \frac{\tau_{33}}{a_1 a_2} \,|_{\gamma=\pm h} = \pm q_3^\pm. \tag{3.1}$$

Here, and below, every formula containing the $\pm$ sign, unites two formulas: one which is obtained by deleting all the minus signs and the other by deleting all the plus signs.

We should add the electrical boundary conditions to the mechanical ones. The former are formulated depending on the presence and location of the electrodes and the way the electroelastic states are excited in the body.

We will assume that the thickness of electrodes covering a part of the body surface are infinitely small compared to the smallest size of body.

Different versions of boundary conditions for electrical quantities can be found in [26].

We write some types of electrical boundary conditions on the shell faces. We can define the electrical potential

$$\psi \,|_{\gamma=\pm h} = \pm V \tag{3.2}$$

on the electrode-covered surfaces. If the electrodes are short-circuited, the electrical conditions assume the form

$$\psi \,|_{\gamma=\pm h} = 0. \tag{3.3}$$

If the shell faces are not electrode-covered and are in contact with a medium with low permittivity—say, vacuum or air—then the component of the electric induction vector, which is normal to the surface, is zero:

$$D_3 \,|_{\gamma=\pm h} = 0. \tag{3.4}$$

If the electrodes are shorted by a contour with known complex conductivity $Y = Y_0 + iY_1$, the electrical condition becomes

$$\int_\Omega \dot{D}_3 d\Omega = 2VY, \tag{3.5}$$

where the integral is evaluated over the surface Ω of one of the electrodes and the point denotes the time derivative.

4 GENERAL THEOREMS ON ELECTROELASTICITY

Consider a piezoceramic body of volume V bounded by the surface Ω in the coordinates x^ρ(here, as before, the Greek indices assume the values 1, 2, or 3 and the Latin indices assume the values 1 or 2). For simplicity, we will only deal with the static case [82].

We write the equation for energy

$$\begin{aligned}
B &= \int_V A dv \\
B &= \int_V q_\rho v^\rho dv + \int_\Omega (\sigma^{\rho\mu} n_\rho v_\mu - \psi n_\rho \mathcal{D}^\rho) d\Omega \\
A &= \sigma^{\rho\mu} e_{\rho\mu} + \mathcal{E}_\rho \mathcal{D}^\rho .
\end{aligned} \tag{4.1}$$

Here q_ρ are the components of the bulk forces; v^ρ are the components of the displacement vector; $\sigma^{\rho\mu}$ is the stress tensor; $e_{\rho\mu}$ is the strain tensor; $\mathcal{D}^\rho$ are the contravariant components of the electric induction vector; $\mathcal{E}_\rho$ are the covariant components of the electric strength vector; V is a volume body; and Ω is the body surface. The components of the tensor n_ρ are found from the vector relation

$$\mathbf{n} = n_\rho \mathbf{e}^\rho ,$$

where $\mathbf{n}$ is a unit vector of the normal external to the surface Ω and $\mathbf{e}^\rho$ are base vectors.

The electroelastic potential will be written as in elasticity theory, i.e.,

$$W = \frac{1}{2}(\sigma^{\rho\mu} e_{\rho\mu} + \mathcal{E}_\rho \mathcal{D}^\rho).$$

We write the constitutive relations 2.10 in the Cartesian coordinates x_1, x_2, and x_3 for a piezoceramic polarized along the x_3-lines:

$$\begin{aligned}
e_{ii} &= s_{11}^E \sigma_{ii} + s_{12}^E \sigma_{jj} + s_{13}^E \sigma_{33} + d_{31} \mathcal{E}_3 \\
e_{33} &= s_{13}^E (\sigma_{11} + \sigma_{22}) + s_{33}^E \sigma_{33} + d_{33} \mathcal{E}_3 \\
e_{ij} &= s_{66}^E \sigma_{ij}, \\
e_{i3} &= s_{44}^E \sigma_{i3} + d_{15} \mathcal{E}_i \\
\mathcal{D}_i &= \varepsilon_{11}^T \mathcal{E}_i + d_{15} \sigma_{i3} \\
\mathcal{D}_3^T &= \varepsilon_{33}^T \mathcal{E}_3 + d_{31}(\sigma_{11} + \sigma_{22}) + d_{33}\sigma_{33} .
\end{aligned} \tag{4.2}$$

Consider two electroelastic states. One state will be marked by superscript 1 and the other by superscript 2. We write the expressions for the work done by the electrical and mechanical forces of the first electroelastic state on the generalized displacements of the second state, and the work done by the forces of the second state on the generalized displacements of the first state. By generalized displacements, we mean the set of the quantities v_1, v_2, v_3, ψ. After simple mathematics using the constitutive relations 4.2, we can show that the work of the forces of the first electroelastic state on the generalized displacements of the second state is equal to the work of the forces of the second state on the generalized displacements of the first state. This proves the Clapeyron theorem generalized to the electroelastic case:

$$\sigma_{\rho\mu}^{(1)} e^{\rho\mu(2)} + \mathcal{E}_{\rho}^{(1)} \mathcal{D}^{\rho(2)} = \sigma_{\rho\mu}^{(2)} e^{\rho\mu(1)} + \mathcal{E}_{\rho}^{(2)} \mathcal{D}^{\rho(1)}. \tag{4.3}$$

In electroelasticity, the theorem on uniqueness will hold for appropriate boundary conditions if the integrand A in 4.1 is positive. We can show this by transforming the integrand and allowing for 4.2 to get a positive-definite quadratic form

$$\begin{aligned} A = a\Bigg[&\frac{1+2\nu_1}{4}(\sigma_{11}+\sigma_{22})^2 + \frac{1-2\nu_1}{4}(\sigma_{11}-\sigma_{22})^2 \\ &+\frac{1+2\nu_2}{4}(\sigma_{11}+\tilde{\sigma}_{33})^2 + \frac{1-2\nu_2}{4}(\sigma_{11}-\sigma_{33})^2 \\ &+\frac{1+2\nu_2}{4}(\sigma_{22}+\tilde{\sigma}_{33})^2 + \frac{1-2\nu_2}{4}(\sigma_{22}-\tilde{\sigma}_{33})^2\Bigg] \\ &+ e(\sigma_{13}^2+\sigma_{23}^2) + s_{66}^E\sigma_{12}^2 + \frac{1}{\varepsilon_{11}^T}(\mathcal{D}_1^2+\mathcal{D}_2^2) + \frac{1}{\varepsilon_{33}^T}\mathcal{D}_3^2 \end{aligned}$$

$$a = s_{11}^E - \frac{d_{31}^2}{\varepsilon_{33}^2}, \qquad b = s_{12}^E - \frac{d_{31}^2}{\varepsilon_{33}^T}, \qquad c = s_{13}^E - \frac{d_{31}d_{33}}{\varepsilon_{33}^T}$$

$$d = s_{33}^E - \frac{d_{33}^2}{\varepsilon_{33}^T}, \qquad e = s_{44}^E - \frac{d_{15}^2}{\varepsilon_{11}^T}, \qquad \nu_1 = \frac{b}{a}$$

$$\nu_2 = \frac{c}{\sqrt{ad}}, \qquad \tilde{\sigma}_{33} = \sqrt{\frac{d}{a}}\sigma_{33}. \tag{4.4}$$

The constant factors before the parentheses in 4.4 are positive numbers and can be calculated using physical constants that can be found, say, in [8].

When the general theorems are fulfilled, problems on electroelasticity can be solved by variational methods. The variational methods and extremal estimates used in elasticity theory can be generalized to electroelasticity.

Chapter 2

EQUATIONS OF THE THEORY OF PIEZOCERAMIC SHELLS

5 TWO-DIMENSIONAL EQUILIBRIUM EQUATIONS

We use the surfaces α_1 = const, and α_2 = const, $\alpha_1 + d\alpha_1$ = const, $\alpha_2 + d\alpha_2$ = const to single out an infinitesimally small element of the shell and choose positive stresses as shown in Figure 2. Consider the element's side lying in the plane $\alpha_2 + d\alpha_2$ = const (Figure 2). The length of the arc ab is

$$dl_2 = H_2 d\alpha_2$$

and the area of the hatched element bounded by the curves γ = const and $\gamma + d\gamma$ = const is equal to

$$dS_2 = dl_2 \cdot d\gamma = H_2 d\alpha_2 d\gamma.$$

Here H_1 and H_2 are Lame coefficients.

The resultants of the stresses acting on the element's side will be

$$\begin{aligned} F_{11} &= \int_{-h}^{+h} \sigma_{11} H_2 d\alpha_2 d\gamma = A_2 d\alpha_2 \int_{-h}^{+h} \tau_{11} d\gamma \\ F_{21} &= \int_{-h}^{+h} \sigma_{21} H_2 d\alpha_2 d\gamma = A_2 d\alpha_2 \int_{-h}^{+h} \tau_{21} d\gamma \\ F_{31} &= -\int_{-h}^{+h} \sigma_{31} H_2 d\alpha_2 d\gamma = -A_2 d\alpha_2 \int_{-h}^{+h} \tau_{31} d\gamma \end{aligned}$$

where we have taken into account formulas 2.3 and 2.12. Then the force per unit length of the α_2-line will be

$$\begin{aligned} T_1 &= \frac{F_{11}}{dl_2} = \int_{-h}^{+h} \tau_{11} d\gamma \\ S_{21} &= \frac{F_{21}}{dl_2} = \int_{-h}^{+h} \tau_{21} d\gamma \\ N_1 &= \frac{F_{31}}{dl_2} = -\int_{-h}^{+h} \tau_{31} d\gamma. \end{aligned} \tag{5.1}$$

In a similar way, we calculate the resultants of the moments acting on the element's side and get the formulas for the bending moment G_1 and the twisting

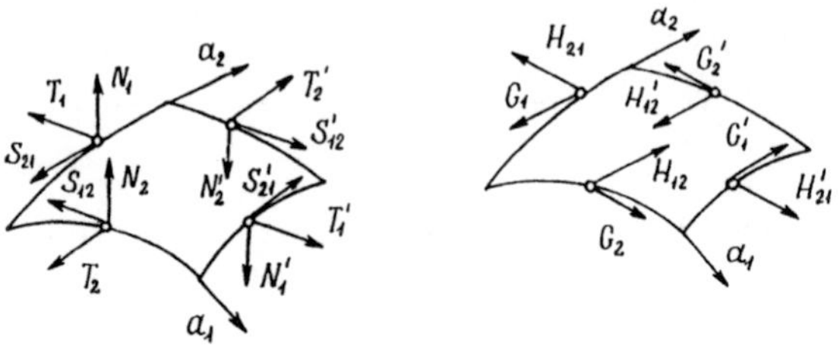

FIGURE 3. Forces and moments acting on an element of shell surface

moment H_{21}

$$G_1 = -\int_{-h}^{+h} \tau_{11}\gamma d\gamma \qquad H_{21} = \int_{-h}^{+h} \tau_{21}\gamma d\gamma. \tag{5.2}$$

The positive directions of forces and moments are shown in Figure 3. In order to get the equilibrium equations in the terms of forces and moments, we equate to zero the principal vector and the principal moment of all the forces acting on an element of the shell's middle surface. The necessary calculations can be found in any course on shell theory; and here we will only write the equilibrium equations

$$\frac{1}{A_iA_j}\frac{\partial}{\partial\alpha_i}(A_jT_i) + k_iS_{ji} + \frac{1}{A_iA_j}\frac{\partial(A_iS_{ij})}{\partial\alpha_j} - k_jT_j - \frac{N_i}{R_i} + X_i = 0$$
$$\frac{T_1}{R_1} + \frac{T_2}{R_2} + \frac{1}{A_1A_2}\left[\frac{\partial}{\partial\alpha_1}(A_2N_1) + \frac{\partial}{\partial\alpha_2}(A_1N_2)\right] + Z = 0$$
$$\frac{1}{A_iA_j}\frac{\partial}{\partial\alpha_i}(A_jH_{ji}) + k_iG_i - \frac{1}{A_iA_j}\frac{\partial}{\partial\alpha_j}(A_iG_j) + k_jH_{ij} + N_j = 0$$
$$S_{21} - S_{12} + \frac{H_{21}}{R_1} + \frac{H_{12}}{R_2} = 0. \tag{5.3}$$

The components of the surface load are expressed in terms of the components of the body forces. The components of the loads are applied to the surfaces of the shell [21], i.e.,

$$X_i = \left(1 + \frac{h^2}{R_iR_j}\right)\left(q_i^+ + q_i^-\right) + h\left(\frac{1}{R_i} + \frac{1}{R_j}\right)\left(q_i^+ - q_i^-\right) + \int_{-h}^{+h} a_ia_jq_id\gamma$$
$$Z = -\left(1 + \frac{h^2}{R_1R_2}\right)\left(q_3^+ + q_3^-\right) - h\left(\frac{1}{R_1} + \frac{1}{R_2}\right)\left(q_3^+ - q_3^-\right)$$
$$- \int_{-h}^{+h} a_1a_2q_3d\gamma.$$

6 HYPOTHESES OF THE THEORY OF NONELECTRIC SHELLS AND THE SAINT VENANT PRINCIPLE

In the previous section, we wrote the equilibrium equations of the theory of shells. The equations directly follow from statics laws and no additional assumptions are required.

Three-dimensional constitutive relations (piezoeffect equations), strain-displacement formulas, and electrostatic equations should also be transformed so that the sought-for quantities—i.e., the vector components of the induction, electric field strength, stresses, strains, and displacements—were referred to the middle surface of the shell.

In the theory of nonelectric shells of the Kirchhoff-Love type, some physically obvious assumptions on the dependence of the displacements and stresses on the thickness coordinate and on the smallness of some stresses relative to other stresses are a usual practice. Let us formulate the Kirchhoff-Love hypotheses.

First Hypothesis

In the constitutive relations, the normal stresses σ_{33} acting on the surface elements parallel to the middle surface can be neglected as compared to other stresses.

Second Hypothesis

A rectilinear element that was normal to the middle surface before the deformation remains rectilinear and perpendicular to the strained surface and is not extended after deformation.

In order for the Kirchhoff-Love theory of nonelectric shells to be applicable, the shell faces should not be fixed; therefore, we can only define the stresses on the shell faces. The two-dimensional layer theory, where the displacements on the faces are given, is based on other hypotheses and described by a theory fundamentally different from the Kirchhoff-Love theory.

We assume that, of all mechanical conditions, only stresses are given on the faces of a piezoelectric shell. We will investigate the kinds of electrical conditions enumerated in Section 3. We will show that the type of two-dimensional theory depends on the kind of electrical conditions on the shell faces and that the electroelastic shell theories for every kind of electrical conditions on the surfaces are constructed separately.

The Saint Venant principle is important in deriving boundary conditions in the theory of nonelectric shells. It implies that if the stresses are arbitrarily specified on the shell edge, the nonself-balanced edge load generates a deeply penetrating stressed-strained state. This should be allowed for in the boundary conditions of the shell theory, while the part of the edge load self-balanced with respect to thickness brings about a stressed-strained state that quickly decreases with the distance from the edge. This has to be calculated by the three-dimensional theory.

Originally, the Saint Venant principle was derived for an elastic cylinder edge—loaded at the end face—and was later generalized to the case of arbitrary elastic, nonlinear, and viscoelastic bodies. Modern mathematics and mechanics employ the Saint Venant principle in the generalized form. A review on the topic can be found in [35].

When writing the boundary conditions in the theory of electroelastic shells, we will use the generalized Saint Venant principle for the electroelastic case. As applied to piezoceramic shells, it consists in that (1) for shells with electrode-covered faces, the Saint Venant conditions are used as in elasticity theory and (2) for shells without electrode coverings of the faces the known mechanical Saint Venant conditions are used with an electrical condition. In other words, the shell theory should allow for the part of the component of the electric induction vector that is nonself-balanced over thickness and normal to the edge surface if the latter has no electrodes, or it should allow for the nonself-balanced part of the electrical potential if the edge is electrode-covered.

7 SHELLS WITH THICKNESS POLARIZATION (ELECTRODE-COVERED FACES)

Shells with thickness polarization and electrode-covered faces are best studied due to their behavior, which is very much like that of nonelectric shells described by the Kirchhoff-Love type of theory [26, 73, 79, 101].

Unless specifically mentioned, the approximate theories should not differentiate between the components of the nonsymmetrical and symmetrical tensors, i.e., a_i can be approximately taken to be equal to one:

$$a_i = 1 + \frac{\gamma}{R_i} \simeq 1. \tag{7.1}$$

We write three-dimensional equations of the electroelastic medium as the *Piezoeffect equations*

$$\sigma_{ii} = \frac{1}{s_{11}^E(1-\nu^2)}(e_i + \nu e_j) - \frac{s_{13}^E}{s_{11}^E}\frac{1}{1-\nu}\sigma_{33} - \frac{d_{31}}{s_{11}^E}\frac{1}{1-\nu}E_3 \tag{7.2}$$

$$\sigma_{ij} = \frac{1}{s_{66}^E}(m_i + m_j) \tag{7.3}$$

$$\frac{\partial v_3}{\partial \gamma} = s_{13}^E(\sigma_{11} + \sigma_{22}) + s_{33}^E\sigma_{33} + d_{33}E_3 \tag{7.4}$$

$$\frac{\partial v_i}{\partial \gamma} + g_i = s_{44}^E \sigma_{i3} + d_{15} E_i$$

$$D_i = \varepsilon_{11}^T E_i + d_{15}\sigma_{i3} \tag{7.5}$$

$$D_3 = \varepsilon_{33}^T E_3 + d_{31}(\sigma_{11} + \sigma_{22}) + d_{33}\sigma_{33} \tag{7.6}$$

$$\nu = -s_{12}^E / s_{11}^E$$

Equations of forced electrostatics

$$\frac{1}{A_1}\frac{\partial D_1}{\partial \alpha_1} + \frac{1}{A_2}\frac{\partial D_2}{\partial \alpha_2} + k_1 D_2 + k_2 D_1 + \frac{\partial D_3}{\partial \gamma} = 0 \tag{7.7}$$

$$E_i = -\frac{1}{A_i}\frac{\partial \psi}{\partial \alpha_i}, \qquad E_3 = -\frac{\partial \psi}{\partial \gamma}. \tag{7.8}$$

Strain-displacement formulas

$$e_i = \frac{1}{A_i}\frac{\partial v_i}{\partial \alpha_i} + k_i v_j + \frac{v_3}{R_i}$$

$$m_i = \frac{1}{A_j}\frac{\partial v_i}{\partial \alpha_j} - k_j v_j$$

$$g_i = \frac{1}{A_i}\frac{\partial v_3}{\partial \alpha_i} - \frac{v_i}{R_i}. \tag{7.9}$$

We assume that the electrical potential

$$\psi\, |_{\gamma = \pm h} = \pm V \tag{7.10}$$

is defined on the electrode-covered surfaces.

The theory of shells with thickness polarization can be based on hypotheses similar to those proposed by Kirchhoff, which are generalized to the electroelastic case as follows.

First Hypothesis

In the electroelasticity relations 7.2 and 7.6, the stresses σ_{33} can be neglected as compared to the principal stresses σ_{11}, σ_{22}. Therefore, we can replace 7.2 and 7.6 by the approximate formulas

$$\sigma_{ii} = \frac{1}{s_{11}^E(1-\nu^2)}(e_i + \nu e_j) - \frac{d_{31}}{s_{11}^E}\frac{1}{1-\nu}E_3 \tag{7.11}$$

$$D_3 = \varepsilon_{33}^T E_3 + d_{31}(\sigma_{11} + \sigma_{22}). \tag{7.12}$$

This hypothesis corresponds to Kirchhoff's first hypothesis for nonelectric shells.

Second Hypothesis

The second hypothesis coincides with the second Kirchhoff-Love hypothesis. We assume that the rectilinear element that was normal to the middle surface before deformation remains straight and perpendicular to the strained surface after deformation, i.e.,

$$\frac{\partial v_i}{\partial \gamma} + g_i = 0. \tag{7.13}$$

We can also neglect the elongation of the normal element in the relations for strains:

$$v_3 = -w, \qquad v_i = u_i - \gamma g_i, \qquad g_i = -\frac{1}{A_i}\frac{\partial w}{\partial \alpha_i}. \tag{7.14}$$

The second formula in 7.14 was obtained by integrating equation 7.13 with respect to γ. In 7.14, w and u_i are the deflection and the tangential displacements of the middle surface points, respectively:

$$u_i = v_i \,|_{\gamma=0}, \qquad w = -v_3 \,|_{\gamma=0} .$$

Third Hypothesis

We assume that the electrical potential varies with thickness by the square law

$$\psi = \psi^{(0)} + \gamma\psi^{(1)} + \gamma^2\psi^{(2)}. \tag{7.15}$$

Hence, using the last relation in 7.8 and the condition 7.10, we get

$$E_3 = E_3^{(0)} + \gamma E_3^{(1)}, \qquad E_3^{(0)} = -\frac{V}{h}. \tag{7.16}$$

The quantities with superscripts in parentheses do not depend on γ. (The numbers in the superscripts denote the degree of γ.)

Fourth Hypothesis

The component D_3 of the electric induction vector is constant with respect to thickness, i.e.,

$$D_3 = D_3^{(0)}. \tag{7.17}$$

Using our hypotheses, we get two-dimensional equations.
By substituting 7.14 into 7.9 we obtain

$$e_i = \varepsilon_i + \gamma\kappa_i, \qquad m_i = \omega_j + \gamma\tau^{(i)}, \qquad g_i^{(0)} = \gamma_i. \tag{7.18}$$

Here, ε_i, ω_j are tangential strains of the shell's middle surface; and $\tau^{(i)}$ and $\gamma^{(i)}$ are flexural strains [15, 21, 68, 106] that can be expressed in terms of the middle surface displacements as

$$\begin{aligned}
\varepsilon_i &= \frac{1}{A_i}\frac{\partial u_i}{\partial \alpha_i} + k_i u_j - \frac{w}{R_i} \\
\omega_i &= \frac{1}{A_i}\frac{\partial u_j}{\partial \alpha_i} - k_i u_i \\
\omega &= \omega_1 + \omega_2 \\
\gamma_i &= -\frac{1}{A_i}\frac{\partial w}{\partial \alpha_i} - \frac{u_i}{R_i} \\
\kappa_i &= -\frac{1}{A_i}\frac{\partial \gamma_i}{\partial \alpha_i} - k_i \gamma_j \\
\tau^{(i)} &= -\frac{1}{A_i}\frac{\partial \gamma_j}{\partial \alpha_i} + k_i \gamma_i \\
\tau &= \tau^{(1)} + \frac{\omega}{2R_1} = -\tau^{(2)} + \frac{\omega}{2R_2}.
\end{aligned} \tag{7.19}$$

Taking into account formulas 7.16 and 7.18, we will find that the principal stresses vary with thickness by the linear law

$$\sigma_{ii} = \sigma_{ii}^{(0)} + \gamma\sigma_{ii}^{(1)}, \qquad \sigma_{ij} = \sigma_{ij}^{(0)} + \gamma\sigma_{ij}^{(1)} \tag{7.20}$$

where

$$\begin{aligned}
\sigma_{ii}^{(0)} &= \frac{1}{s_{11}^E(1-\nu^2)}(\varepsilon_i + \nu\varepsilon_j) - \frac{d_{31}}{s_{11}^E}\frac{1}{1-\nu}E_3^{(0)} \\
\sigma_{ii}^{(1)} &= \frac{1}{s_{11}^E(1-\nu^2)}(\kappa_i + \nu\kappa_j) - \frac{d_{31}}{s_{11}^E}\frac{1}{1-\nu}E_3^{(1)} \\
\sigma_{ij}^{(0)} &= \frac{1}{s_{66}^E}\omega \\
\sigma_{ij}^{(1)} &= \frac{2}{s_{66}^E}\left(\tau - \frac{\omega}{2R_j}\right).
\end{aligned} \tag{7.21}$$

The formulas relating the forces and moments with stresses have the form

$$\begin{aligned}
T_i &= \int_{-h}^{+h} \sigma_{ii} d\gamma = 2h\sigma_{ii}^{(0)} \\
G_i &= -\int_{-h}^{+h} \sigma_{ii}\gamma d\gamma = -\frac{2h^3}{3}\sigma_{ii}^{(1)} \\
S_{ij} &= \int_{-h}^{+h} \sigma_{ij} d\gamma = 2h\sigma_{ij}^{(0)}
\end{aligned}$$

$$H_{ij} = \int_{-h}^{+h} \sigma_{ij}\gamma d\gamma = \frac{2h^3}{3}\sigma_{ij}^{(1)}. \tag{7.22}$$

Here we have taken equations 7.20 into account.

Allowing for 7.21, we rewrite 7.22 as

$$T_i = \frac{2h}{s_{11}^E(1-\nu^2)}(\varepsilon_i + \nu\varepsilon_j) + \frac{2d_{31}}{s_{11}^E(1-\nu)}V, \qquad S = S_{ij} = \frac{2h}{s_{66}^E}\omega \tag{7.23}$$

$$\begin{aligned} G_i &= -\frac{2h^3}{3s_{11}^E(1-\nu^2)}(\kappa_i + \nu\kappa_j) + \frac{2h^3}{3(1-\nu)}\frac{d_{31}}{s_{11}^E}E_3^{(1)} \\ H_{ij} &= \frac{4h^3}{3s_{66}^E}\left(\tau - \frac{\omega}{2R_j}\right). \end{aligned} \tag{7.24}$$

Based on equations 7.6 and 7.8 and using equations 7.16, 7.17, 7.20, and 7.22, we equate the coefficients at the same powers γ to get

$$D_3^{(0)} = -\varepsilon_{33}^T\frac{V}{h} + \frac{d_{31}}{2h}(T_1 + T_2) \tag{7.25}$$

$$\varepsilon_{33}^T E_3^{(1)} = \frac{3d_{31}}{2h^3}(G_1 + G_2). \tag{7.26}$$

We use equation 7.26 to eliminate $E_3^{(1)}$ from equations 7.24 and, after some algebra, get the following relations for the electroelasticity for bending moments:

$$\begin{aligned} G_i &= -\frac{2h^3B}{3}(\kappa_i + \sigma\kappa_j) \\ B &= \frac{2-(1-\nu)k_p^2}{2s_{11}^E(1-\nu^2)(1-k_p^2)} \\ \sigma &= \frac{2\nu + (1-\nu)k_p^2}{2-(1-\nu)k_p^2} \\ k_{31}^2 &= \frac{d_{31}^2}{s_{11}^E\varepsilon_{33}^T}, \qquad k_p^2 = \frac{2k_{31}^2}{(1-\nu)} \end{aligned} \tag{7.27}$$

In our case, the problem is divided into two simpler ones: mechanical and electrical. The mechanical problem coincides with that of the theory of nonelectrical shells within the coefficients. It is described by a system of differential equations that contains the equilibrium equations 5.3, strain-displacement formulas 7.19, and electroelasticity equations (equations 7.23, 7.24 and 7.27). The electrical quantity $E_3^{(0)}$ in the electroelasticity relations is expressed through the electric potential found from the conditions 7.10. The other electrical quantities can be found after solving the mechanical problem; we find $E_3^{(1)}$ and $D_3^{(0)}$ from equations 7.26 and 7.25. We use equations 7.15, 7.8, 7.10 and 7.16 to find

$$\psi^{(1)} = -E_3^{(0)}, \qquad \psi^{(2)} = -\frac{1}{2}E_3^{(1)}, \qquad \psi^{(0)} = -h^2\psi^{(2)}$$

$$E_i^{(\rho)} = -\frac{1}{A_i}\frac{\partial\psi^{(\rho)}}{\partial\alpha_i} \qquad (i = 1, 2; \rho = 1, 2, 3)$$

$$E_i = E_i^{(0)} + \gamma E_i^{(1)} + \gamma^2 E_i^{(2)}, \tag{7.28}$$

$$D_i^{(0)} = \varepsilon_{11}^T E_i^{(0)} + d_{15}\left[\frac{3}{4h}N_i - \frac{1}{4}(q_i^+ - q_i^-)\right]$$

$$D_i^{(1)} = \varepsilon_{11}^T E_i^{(1)} + d_{15}\frac{1}{2h}(q_i^+ + q_i^-)$$

$$D_i^{(2)} = \varepsilon_{11}^T E_i^{(2)} + d_{15}\left[-\frac{3h}{4}N_i + \frac{3h^2}{4}(q_i^+ - q_i^-)\right]. \tag{7.29}$$

For shells with thickness polarization and electrode-covered faces, the boundary conditions are purely mechanical and have a typical form. We write the boundary conditions for the edge $\alpha_i = \alpha_{i0}$ restricting ourselves to

1. *Free edge* $\alpha_i = \alpha_{i0}$

$$T_i = 0, \qquad S_{ji} = 0, \qquad G_i = 0, \qquad N_i - \frac{1}{A_j}\frac{\partial H_{ji}}{\partial\alpha_j} = 0. \tag{7.30}$$

2. *Hinge-supported edge* $\alpha_i = \alpha_{i0}$

$$u_j = 0, \qquad w = 0, \qquad G_i = 0, \qquad T_i = 0. \tag{7.31}$$

3. *Rigidly fixed edge* $\alpha_i = \alpha_{i0}$

$$u_i = 0, \qquad u_j = 0, \qquad w = 0, \qquad \gamma_i = 0. \tag{7.32}$$

We note that the second hypothesis does not imply that the elongation of the normal element to the middle surface is small. The relative elongation of the normal element is comparable to that along the coordinate lines ε_1 and ε_2 of the middle surface. When the mechanical problem is solved, the relative elongation can be found by the formula

$$\varepsilon_3 = \frac{\partial v_3}{\partial\gamma} = \frac{s_{13}^E}{2h}(T_1 + T_2) - \frac{d_{31}}{h}V.$$

The stresses σ_{i3} and σ_{33} can be neglected in a similar way only in the electroelasticity relations. We can see from the three-dimensional equilibrium equations that if σ_{ii} and σ_{ij} vary with thickness by a linear law, then σ_{i3} and σ_{33} vary with respect to γ by square and cubic laws, respectively. Just as in the theory of nonelectrical shells, σ_{i3} can be found after the two-dimensional problem has been solved from the formulas

$$\begin{aligned}
\sigma_{i3} &= \sigma_{i3}^{(0)} + \gamma\sigma_{i3}^{(1)} + \gamma^2\sigma_{i3}^{(2)} \\
\sigma_{i3}^{(0)} &= -\frac{3}{4h}N_i - \frac{1}{4}(q_i^+ - q_i^-) \\
\sigma_{i3}^{(1)} &= \frac{1}{2h}(q_i^+ + q_i^-) \\
\sigma_{i3}^{(2)} &= +\frac{3h}{4}N_i + \frac{3h^2}{4}(q_i^+ - q_i^-)
\end{aligned} \tag{7.33}$$

where the right-hand sides contain only known quantities, i.e., the components of the surface load (equations 3.1) and the transverse forces N_i. The stresses σ_{33} are found from the three-dimensional equilibrium equations and conditions 3.1 on the faces:

$$\begin{aligned}
\sigma_{33} &= \sigma_{33}^{(0)} + \gamma\sigma_{33}^{(1)} + \gamma^2\sigma_{33}^{(2)} + \gamma^3\sigma_{33}^{(3)} \\
\sigma_{33}^{(0)} &= -h^2\sigma_{33}^{(2)} + \frac{1}{2}(q_3^+ - q_3^-) \\
\sigma_{33}^{(1)} &= -h^2\sigma_{33}^{(3)} + \frac{1}{2h}(q_3^+ + q_3^-) \\
\sigma_{33}^{(2)} &= -\frac{1}{2}\left(\frac{1}{A_1A_2}\frac{\partial(A_2\sigma_{13}^{(1)})}{\partial\alpha_1} + \frac{1}{A_1A_2}\frac{\partial(A_1\sigma_{23}^{(1)})}{\partial\alpha_2} + \frac{\sigma_{11}^{(1)}}{R_1} + \frac{\sigma_{22}^{(1)}}{R_2}\right) \\
\sigma_{33}^{(3)} &= -\frac{1}{3}\left(\frac{1}{A_1A_2}\frac{\partial(A_2\sigma_{13}^{(2)})}{\partial\alpha_1} + \frac{1}{A_1A_2}\frac{\partial(A_1\sigma_{23}^{(2)})}{\partial\alpha_2}\right).
\end{aligned} \tag{7.34}$$

Consider different electrical conditions on the faces of an electrode-covered shell. Since the electrodes are equipotential surfaces, the potential difference is a constant quantity independent of the coordinates. In some cases, the constant V is known. For example, V is known in equations 7.10, 7.16, and 7.23 when an electric power is supplied to the electrodes. If the electrodes are short-circuited, the potential difference at the electrodes is zero and we can use $V = 0$ in equations 7.16 and 7.23.

If the shell's faces are completely covered by the electrodes and shorted by a contour with known complex conductivity (equation 3.5), we should add an equation integral to the system of differential equations of the shell theory. Recalling our hypotheses, we can write the later in two ways, i.e., for forces or strains:

$$\frac{d_{31}}{2h}\int_\Omega (T_1 + T_2)d\Omega = i\frac{2YV}{\omega} + \varepsilon_{33}^T\Omega\frac{V}{h}$$

$$\int_\Omega (\varepsilon_1 + \varepsilon_2)d\Omega = \frac{s_{11}^E \varepsilon_{33}^T}{d_{31}} \frac{\Omega(1-\nu)}{h} \left[i \frac{2hY}{\varepsilon_{33}^T \omega \Omega} + 1 - \frac{2k_{31}^2}{1-\nu} \right] V$$

$$k_{31}^2 = \frac{d_{31}^2}{s_{11}^E \varepsilon_{33}^T}. \tag{7.35}$$

Here i is an imaginary unit, Ω is the square of one of the electrodes, ω is the angular frequency of vibrations, operating by the law $e^{-i\omega t}$. If the electrodes are disconnected, we should put $Y = 0$ in equation 7.35.

8 SHELLS WITH THICKNESS POLARIZATION (FACES WITHOUT ELECTRODES)

Let the electrical conditions

$$D_3 \mid_{\gamma=\pm h} = 0 \tag{8.1}$$

hold on the faces that are not covered with electrodes and are in contact with vacuum or air. Consider two cases depending on the electrical conditions on the edges.

CASE ONE

The shell's edges may have electrodes, but no electric power is supplied to them. Here the Kirchhoff-Love hypotheses for mechanical quantities and the third hypothesis for electrical quantities given in Section 7 remain true.

The *fourth hypothesis* implies that we can neglect D_3 in equation 7.6 as compared to $\varepsilon_{33}^T E_3$ to get

$$\varepsilon_{33}^T E_3 = -d_{31}(\sigma_{11} + \sigma_{22}). \tag{8.2}$$

This assumption can be made because D_3 is equal to zero on the faces (equation 8.1), and on the interval $[-h, +h]$ where γ varies, D_3 does not assume values comparable to $\varepsilon_{33}^T E_3$. We now use the introduced hypotheses to construct a theory for shells that do not have electrodes on their faces. By the second hypothesis, the tangential displacements and principal stresses vary with thickness by a linear law. By the third hypothesis, the electrical potential varies with respect to the thickness coordinate γ by the square law (equation 7.15), while E_3 varies by the linear law (the first equation 7.16):

$$E_3 = E_3^{(0)} + \gamma E_3^{(1)}. \tag{8.3}$$

We can use equations 7.20, 8.2, and 8.3 to express $E_3^{(0)}$ and $E_3^{(1)}$ in terms of $\sigma_{ii}^{(0)}$ and $\sigma_{ii}^{(1)}$ as

$$E_3^{(0)} = -\frac{d_{31}}{\varepsilon_{33}^T}(\sigma_{11}^{(0)} + \sigma_{22}^{(0)}), \qquad E_3^{(1)} = -\frac{d_{31}}{\varepsilon_{33}^T}(\sigma_{11}^{(1)} + \sigma_{22}^{(1)}). \tag{8.4}$$

We substitute $E_3^{(0)}$ and $E_3^{(1)}$ from equations 8.4 into equations 7.21 and get the relationships

$$T_i = 2hB(\varepsilon_i + \sigma\varepsilon_j)$$
$$S = S_{ij} = \frac{2h}{s_{66}^E}\omega$$
$$H_{ij} = \frac{4h^3}{3s_{66}^E}(\tau - \frac{\omega}{2R_j})$$
$$G_i = -\frac{2H^3B}{3}(\kappa_i + \sigma\kappa_j). \tag{8.5}$$

The numbers B and σ are found from 7.27.

We turn to forces and moments in 8.4:

$$E_3^{(0)} = -\frac{d_{31}}{2h\varepsilon_{31}^T}(T_1 + T_2), \qquad E_3^{(1)} = \frac{3d_{31}}{2h^3\varepsilon_{33}^T}(G_1 + G_2). \tag{8.6}$$

As in Section 7, the laws of variation with respect to the thickness coordinate γ and the relations 7.28, 7.29, 7.33, and 7.34 remain valid for the sought-for quantities $\sigma_{i3}, \sigma_{33}, E_i$, and D_i. Since D_i changes according to a square law, it follows from equation 7.7, that D_3 changes by a cubic law. We substitute D_3 in the form of a polynomial in γ into equation 7.7 and equate the coefficients at similar powers of γ to get

$$D_3^{(n)} = -\frac{1}{nA_1A_2}\left[\frac{\partial}{\partial\alpha_1}\left(A_2D_1^{(n-1)}\right) + \frac{\partial}{\partial\alpha_2}\left(A_1D_2^{(n-1)}\right)\right] \tag{8.7}$$
$$n = 1, 2, 3.$$

The boundary conditions 8.1 will take the form

$$D_3^{(0)} + h^2D_3^{(2)} = 0, \qquad D_3^{(1)} + h^2D_3^{(3)} = 0. \tag{8.8}$$

By substituting the values of $D_3^{(1)}$ and $D_3^{(3)}$ found from equation 8.7 into the second relation in 8.8, we obtain

$$\frac{\partial}{\partial\alpha_1}A_2\left(D_1^{(0)} + \frac{h^2}{3}D_1^{(2)}\right) + \frac{\partial}{\partial\alpha_2}A_1\left(D_2^{(0)} + \frac{h^2}{3}D_2^{(2)}\right) = 0. \tag{8.9}$$

We use equations 7.28, 7.29 and 8.6 to transform equation 8.9 to the form

$$\left(\frac{\partial}{\partial\alpha_1}\frac{A_2}{A_1}\frac{\partial}{\partial\alpha_1} + \frac{\partial}{\partial\alpha_2}\frac{A_1}{A_2}\frac{\partial}{\partial\alpha_2}\right)\left[\psi^{(0)} - \frac{d_{31}}{4h\varepsilon_{33}^T}(G_1 + G_2)\right]$$
$$+ \frac{d_{15}}{2h\varepsilon_{11}^T}\left(\frac{\partial}{\partial\alpha_1}A_2N_1 + \frac{\partial}{\partial\alpha_2}A_1N_2\right) = 0. \tag{8.10}$$

Since we have obtained a differential equation for finding $\psi^{(0)}$, the system of differential equations in the theory of piezoceramic shells has the tenth order and not the eighth order as in the previous section. Therefore, five boundary conditions should hold at each edge: four mechanical conditions and one electrical condition. The mechanical boundary conditions are like 7.30 to 7.32 in the theory of nonelectrical shells. Let us find the electrical boundary condition. According to the Saint Venant principle generalized to electroelasticity, the integral condition

$$\int_{-h}^{+h} \psi d\gamma = 2hV \tag{8.11}$$

should hold at the electrode-covered edge with a given value of the electrical potential. At the edge $\alpha_i = \alpha_{i0}$ without electrodes

$$\int_{-h}^{+h} D_i d\gamma = 2h\left(D_i^{(0)} + \frac{h^2}{3}D_i^{(2)}\right) = 0. \tag{8.12}$$

We transform the condition 8.11. By the third hypothesis, the potential is a quadratic function of γ. We substitute 7.15 into 8.11 and integrate with respect to γ to get

$$\psi^{(0)} + \frac{h^2}{3}\psi^{(2)} = V. \tag{8.13}$$

Instead of $\psi^{(2)}$, we substitute its expression in terms of the bending moments into equation 8.13 and use equations 7.28 and 8.6 to get the condition for the edge:

$$\psi^{(0)} - \frac{d_{31}}{4h\varepsilon_{33}^T}(G_1 + G_2) = V. \tag{8.14}$$

Following the same lines, we can show that for the edge $\alpha_i = \alpha_{i0}$ without electrodes, we will have the condition

$$\frac{1}{A_i}\frac{\partial\psi^{(0)}}{\partial\alpha_i} - \frac{d_{31}}{4h\varepsilon_{33}^T}\frac{1}{A_i}\frac{\partial}{\partial\alpha_i}(G_1 + G_2) = 0. \tag{8.15}$$

If the electrodes are connected by an external contour with complex conductivity Y, the following condition is obeyed (for $\alpha_i = \alpha_{i0}$):

$$-i\omega\int_{-h}^{+h} d\gamma\int_{li} D_i dl_i = 2VY$$

$$-i\omega h\int_{li}\left(D_i^{(0)} + \frac{h^2}{3}D_i^{(2)}\right) dl_i = VY. \tag{8.16}$$

Here l_i is the edge line on the middle surface.

Whether the problem can be divided into mechanical and electrical problems depends on the electrical conditions on the edges. If the edges have no electrodes or the edges have electrodes but no electric power supplied to them, the mechanical problem with elasticity relations 8.5 can be solved separately. Then the boundary conditions match those in the previous section, and the largest electrical quantities $E_3^{(0)}$ and $E_3^{(1)}$ are calculated after the mechanical problem has been solved using equation 8.6. In order to find other electrical quantities, we should integrate equation 8.10 and then find $\psi^{(1)}$, $\psi^{(2)}$, $E_i^{(0)}$, $E_i^{(1)}$, $E_i^{(2)}$, $D_i^{(0)}$, $D_i^{(1)}$, $D_i^{(2)}$, $D_3^{(0)}$, $D_3^{(1)}$, $D_3^{(2)}$, and $D_3^{(3)}$ using equations 7.29, 8.7, and 8.8.

If the edges of the shell have electrodes and their electrical potential is given, the electroelastic problem cannot be divided into two.

CASE TWO

An electrical potential

$$\psi\, |_{\alpha_i=\alpha_{i1}} = +V, \qquad \psi\, |_{\alpha_i=\alpha_{i2}} = -V \tag{8.17}$$

is given on the electrode-covered faces of the edges. For simplicity, we will assume that the mechanical load is absent.

In order to construct the shell theory consisting of the previous four hypotheses, we should reconsider the second *Kirchhoff's hypothesis* and replace equation 7.13 with an approximate formula

$$\frac{\partial v_i}{\partial \gamma} + g_i = d_{15}E_i. \tag{8.18}$$

We add the *fifth hypothesis* in that σ_{i3} is neglected in equation 7.5:

$$D_i = \varepsilon_{11}^T E_i. \tag{8.19}$$

The *sixth hypothesis* will discard $D_3^{(1)}$ for $n = 1$ in equation 8.7:

$$\frac{\partial}{\partial \alpha_1} A_2 D_1^{(0)} + \frac{\partial}{\partial \alpha_2} A_1 D_2^{(0)} = 0. \tag{8.20}$$

Using these hypotheses after transformations, we get the equations of the electroelastic shells theory:

$$\frac{\partial}{\partial \alpha_1}\frac{A_2}{A_1}\frac{\partial \psi^{(0)}}{\partial \alpha_1} + \frac{\partial}{\partial \alpha_2}\frac{A_1}{A_2}\frac{\partial \psi^{(0)}}{\partial \alpha_2} = 0 \tag{8.21}$$

$$T_i = 2hB(\varepsilon_i + \sigma\varepsilon_j), \qquad S = \frac{2h}{s_{66}^E}\omega$$

$$
\begin{aligned}
G_i &= -\frac{2h^3 B}{3}\left[(\kappa_i + \sigma\kappa_j) + d_{15}\left(\frac{1}{A_i}\frac{\partial E_i^{(0)}}{\partial\alpha_i} + \sigma\left(\frac{1}{A_j}\frac{\partial E_j^{(0)}}{\partial\alpha_j} + k_j E_i^{(0)}\right)\right)\right] \\
H &= \frac{4h^3}{3s_{66}^E}\left[\tau + d_{15}\left(\frac{1}{A_1}\frac{\partial E_2^{(0)}}{\partial\alpha_1} - k_1 E_1^{(0)}\right)\right] \\
&= \frac{4h^3}{3s_{66}^E}\left[\tau + d_{15}\left(\frac{1}{A_2}\frac{\partial E_1^{(0)}}{\partial\alpha_2} - k_2 E_2^{(0)}\right)\right] \\
E_i^{(0)} &= -\frac{1}{A_i}\frac{\partial\psi^{(0)}}{\partial\alpha_i}, \qquad (8.22)
\end{aligned}
$$

$$
\begin{aligned}
D_i^{(0)} &= \varepsilon_{11}^T E_i^{(0)} \\
\psi^{(1)} &= -E_3^{(0)} = \frac{d_{31}}{2h\varepsilon_{33}^T}(T_1 + T_2), \\
\psi^{(2)} &= -\frac{3d_{31}}{4h^3\varepsilon_{33}^T}(G_1 + G_2) \\
nD_3^{(n)} &= -\frac{1}{A_1A_2}\left[\frac{\partial}{\partial\alpha_1}A_2D_1^{(n-1)} + \frac{\partial}{\partial\alpha_2}A_1D_2^{(n-1)}\right], \qquad n = 2, 3 \\
D_3^{(0)} &= -h^2 D_3^{(2)}, \\
D_3^{(1)} &= -h^2 D_3^{(3)}. \qquad (8.23)
\end{aligned}
$$

The mechanical boundary conditions are similar to those of the Kirchhoff theory with the exception of the boundary conditions at the free edge: $\alpha_1 = \alpha_{10}$. In this case, the boundary conditions of shell theory can be found after the boundary-layer problem has been solved.

The problem can be divided into mechanical and electrical ones; but, in contrast to the first case, we should first solve the electrical problem, i.e., integrate equation 8.21 with the boundary conditions

$$
\psi^{(0)}|_{\alpha_i=\alpha_{i1}} = +V, \qquad \psi^{(0)}|_{\alpha_i=\alpha_{i2}} = -V
$$

and then solve the mechanical problem. Having solved both problems, we can find the other quantities using 8.23.

9 SHELLS WITH TANGENTIAL POLARIZATION (FACES WITHOUT ELECTRODES)

We will assume that the shell is polarized along the α_2-lines and that the electrical conditions

$$
D_3\,|_{\gamma=\pm h} = 0 \qquad (9.1)
$$

and mechanical conditions 3.1 hold on the faces of the shell.

We write three-dimensional constitutive relations

$$\sigma_{ii} = n_{ii}e_i + n_{ij}e_j - p_i\sigma_{33} - c_iE_2, \qquad \sigma_{ij} = \frac{1}{s_{44}^E}(m_i + m_j) - \frac{d_{15}}{s_{44}^E}E_1; \quad (9.2)$$

$$\frac{\partial v_3}{\partial \gamma} = s_{12}^E\sigma_{11} + s_{13}^E\sigma_{22} + s_{11}^E\sigma_{33} + d_{31}E_2$$

$$\frac{\partial v_1}{\partial \gamma} + g_1 = s_{66}^E\sigma_{13}, \qquad \frac{\partial v_2}{\partial \gamma} + g_2 = s_{44}^E\sigma_{23} + d_{15}E_3, \quad (9.3)$$

$$D_3 = \varepsilon_{11}^T E_3 + d_{15}\sigma_{23},$$
$$D_1 = \varepsilon_{11}^T E_1 + d_{15}\sigma_{21},$$
$$D_2 = \varepsilon_{33}^T E_2 + d_{31}(\sigma_{11} + \sigma_{33}) + d_{33}\sigma_{22}$$
$$\frac{\partial}{\partial \alpha_1}A_2D_1 + \frac{\partial}{\partial \alpha_2}A_1D_2 + \frac{\partial D_3}{\partial \gamma} = 0$$
$$E_3 = -\frac{\partial \psi}{\partial \gamma},$$
$$E_i = -\frac{1}{A_i}\frac{\partial \psi}{\partial \alpha_i}. \quad (9.4)$$

The constants n_{ij}, c_i, and p_i are found from 2.17.

In order to construct the theory of shells, we use the following simplifying hypotheses [80].

First Hypothesis

Neglecting the stresses σ_{33} as compared to σ_{ii} in the constitutive relations 9.2 and 9.4, we obtain

$$\sigma_{ii} = n_{ii}e_i + n_{ij}e_j - c_iE_2, \qquad D_2 = \varepsilon_{33}^T E_2 + d_{31}\sigma_{11} + d_{33}\sigma_{22}. \quad (9.5)$$

Second Hypothesis

It coincides with the Kirchhoff-Love second hypothesis (Section 6) and leads to the relations 7.14.

Third Hypothesis

We assume that the components E_i do not depend on γ; therefore, ψ does not change with thickness, i.e.,

$$\psi = \psi^{(0)}, \qquad E_i = E_i^{(0)} = -\frac{1}{A_i}\frac{\partial \psi^{(0)}}{\partial \alpha_i}. \quad (9.6)$$

Fourth Hypothesis

We neglect $D_3^{(1)}$ in the last formula 9.4 as compared to $D_i^{(0)}$. Using these hypotheses and the usual scheme, we find the electroelasticity relations

$$\begin{aligned} T_i &= 2h(n_{ii}\varepsilon_i + n_{ij}\varepsilon_j) - 2hc_iE_2^{(0)} \\ S = S_{ij} &= \frac{2h}{s_{44}^E}(\omega - d_{15}E_1^{(0)}) \\ G_i &= -\frac{2h^3}{3}(n_{ii}\kappa_i + n_{ij}\kappa_j) \\ H = H_{ij} &= \frac{4h^3}{3s_{44}^E}\tau, \end{aligned} \tag{9.7}$$

$$D_1^{(0)} = \varepsilon_{11}^T E_1^{(0)} + \frac{d_{15}}{2h}S_{12}, \qquad D_2^{(0)} = \varepsilon_{33}^T E_2^{(0)} + \frac{d_{31}}{2h}T_1 + \frac{d_{33}}{2h}T_2$$

$$\frac{1}{A_1}\frac{\partial D_1^{(0)}}{\partial \alpha_1} + \frac{1}{A_2}\frac{\partial D_2^{(0)}}{\partial \alpha_2} + k_2D_1^{(0)} + k_1D_2^{(0)} = 0. \tag{9.8}$$

If we add the equilibrium equation 5.3 and strain-displacement relations 7.19 to 9.7 and 9.8, we will get a complete system of tenth-order differential equations for the sought-for quantities.

Having solved the problem, we can, if necessary, define less important electrical and mechanical quantities. The stresses σ_{i3} and σ_{33} are calculated by equations 7.33 and 7.34. Since σ_{ii} and σ_{ij} vary with respect to γ by a linear law, 9.4 implies that D_i also varies by a linear law. According to equation 7.7, D_3 changes by a square law. We write the formulas for the coefficients D_i and D_3 in the polynomial in γ, taking into account the conditions 9.1:

$$D_1^{(1)} = \frac{3d_{15}}{2h^3}H_{12}, \qquad D_2^{(1)} = -\frac{3}{2h^3}(d_{31}G_1 + d_{33}G_2)$$

$$D_3^{(0)} = -h^2D_3^{(2)} = \frac{h^2}{2A_1A_2}\left[\frac{\partial(A_2D_1^{(1)})}{\partial\alpha_1} + \frac{\partial(A_1D_2^{(1)})}{\partial\alpha_2}\right]$$

$$D_3^{(1)} = 0.$$

The coefficients E_3 and ψ in the polynomial in γ can be found using the corresponding formulas in equations 9.4:

$$E_3^{(0)} = \frac{1}{\varepsilon_{11}^T}D_3^{(0)} + \frac{d_{15}}{4\varepsilon_{11}^T}\left[\frac{3N_2}{h} + (q_2^+ - q_2^-)\right]$$

$$E_3^{(1)} = -\frac{d_{15}}{2h\varepsilon_{11}^T}(q_2^+ + q_2^-)$$

$$
\begin{aligned}
E_3^{(2)} &= \frac{1}{\varepsilon_{11}^T} D_3^{(2)} - \frac{3d_{15}}{4\varepsilon_{11}^T h^2}\left(\frac{N_2}{h} + q_2^+ - q_2^-\right) \\
\psi^{(k+1)} &= -\frac{1}{k+1} E_3^{(k)}, \qquad k = 0, 1, 2.
\end{aligned}
$$

In our case, the system of equations of the shell theory is of order ten; therefore, four mechanical and one electrical condition should be satisfied at each edge. Using the Saint Venant principle, generalized to the electroelasticity case, we find the boundary conditions

$$D_i^{(0)} = 0 \tag{9.9}$$

at the edge without electrodes with $\alpha_i = \alpha_{i0}$, and

$$\psi^{(0)} \mid_{\alpha_i=\alpha_{i1}} = +V, \qquad \psi^{(0)} \mid_{\alpha_i=\alpha_{i2}} = -V \tag{9.10}$$

at the electrode-covered edges with $\alpha_i = \alpha_{i1}$ and $\alpha_i = \alpha_{i2}$. The mechanical conditions have the form in equations 7.30 to 7.32.

10 SHELLS WITH TANGENTIAL POLARIZATION (ELECTRODE-COVERED FACES)

As before we will assume that the shell is polarized along the α_2-lines.

The electroelastic state considered in this section does not resemble the stressed and strained state of nonelectrical shells [80]. We will show that this electroelastic state cannot arise due to mechanical loads. Consider two cases depending on the electrical conditions on the faces.

CASE ONE

Suppose that an electrical potential

$$\psi \mid_{\gamma=\pm h} = \pm V \tag{10.1}$$

is given on the electrodes. Under the effect of the electric field, the domains oriented along the α_2-lines tend to become normal to the middle plane. Therefore, the largest displacement is $v_2^{(1)}$, and the largest stresses are $\sigma_{ii}^{(1)}$ and $\sigma_{ij}^{(1)}$, which are used to determine the moments. We do not need to solve the problem to obtain the formulas

$$
\begin{aligned}
v_2^{(1)} &= d_{15} E_3^{(0)} \\
E_3^{(0)} &= -\frac{V}{h} \\
G_i &= -\frac{2h^3}{3} n_{i1} k_1 d_{15} E_3^{(0)}
\end{aligned}
$$

$$H_{ij} = -\frac{2h^2}{3s_{44}^E} k_2 d_{15} E_3^{(0)}. \tag{10.2}$$

To be more exact in defining the needed quantities, we construct a theory of shells based on the following hypotheses.

First Hypothesis

We can neglect the quantities σ_{33}, E_1, and E_2 as compared to the principle quantities σ_{ii} and σ_{ij} in the electroelasticity relations 9.2 and write the latter in a simplified form

$$\sigma_{ii} = n_{ii}e_{ii} + n_{ij}e_{jj}, \qquad \sigma_{ij} = \frac{1}{s_{44}^E}(m_i + m_j).$$

Second Hypothesis

The electrical potential varies with the thickness coordinate as

$$\psi = \gamma\psi^{(1)}$$

We find from 10.1 that

$$\psi = \gamma\frac{V}{h}.$$

It is clear from these formulas that E_3 is a constant quantity, i.e.,

$$E_3 = -\frac{\partial\psi}{\partial\gamma} = -\frac{V}{h}. \tag{10.3}$$

Third Hypothesis

We replace the relations 9.3 by the approximate formulas

$$\frac{\partial v_1}{\partial\gamma} + g_1 = 0, \qquad \frac{\partial v_2}{\partial\gamma} + g_2 = d_{15}E_3, \qquad \frac{\partial v_3}{\partial\gamma} = 0 \tag{10.4}$$

from which we get the three-dimensional strains e_i, m_i and the stresses σ_{ii}, σ_{ij}:

$$\begin{aligned} e_1 &= \varepsilon_1 + \gamma(\kappa_1 + k_1 d_{15}E_3), & e_2 &= \varepsilon_2 + \gamma\kappa_2 \\ m_1 &= \omega_2 + \gamma(\tau - k_2 d_{15}E_3), & m_2 &= \omega_1 + \gamma\tau \\ \sigma_{ii} &= \sigma_{ii}^{(0)} + \gamma\sigma_{ii}^{(1)}, & \sigma_{ij} &= \sigma_{ij}^{(0)} + \gamma\sigma_{ij}^{(1)}. \end{aligned} \tag{10.5}$$

We use the formulas 5.1 to pass from stresses to forces and moments and get from 10.3 to 10.5 the electroelasticity relations

$$
\begin{aligned}
T_i &= 2h(n_{ii}\varepsilon_i + n_{ij}\varepsilon_j),\\
S = S_{ij} &= \frac{2h}{s_{44}^E}\omega\\
G_i &= -\frac{2h^3}{3}(n_{ii}\kappa_i + n_{ij}\kappa_j + n_{i1}k_1 d_{15}E_3)\\
H = H_{ij} &= \frac{4h^3}{3s_{44}^E}\left(\tau - \frac{1}{2}k_2 d_{15}E_3\right),\\
E_3 &= -\frac{V}{h}.
\end{aligned}
\tag{10.6}
$$

To define the electrical quantities, we need one more hypothesis.

Fourth Hypothesis

We will neglect the stresses σ_{23} and σ_{33} and the components $\varepsilon_{11}^T E_1$ and $\varepsilon_{33}^T E_2$ of the electric field intensity in 9.4 as compared to the components D_3 and D_2 of the electric induction. Then

$$
D_3 = \varepsilon_{11}^T E_3, \qquad D_1 = d_{15}\sigma_{21}, \qquad D_2 = d_{31}\sigma_{11} + d_{33}\sigma_{22}.
$$

Since the component E_3 is constant and σ_{ij} varies with γ by a linear law, a simple transformation of 9.4 will give

$$
\begin{aligned}
D_3 &= D_3^{(0)} = -\varepsilon_{11}^T \frac{V}{h}\\
D_1 &= D_1^{(0)} + \gamma D_1^{(1)}, \qquad D_2 = D_2^{(0)} + \gamma D_2^{(1)},
\end{aligned}
\tag{10.7}
$$

where

$$
\begin{aligned}
D_1^{(0)} &= \frac{d_{15}}{2h}S_{21}, & D_1^{(1)} &= \frac{3d_{15}}{2h^3}H_{ij}\\
D_2^{(0)} &= \frac{1}{2h}(d_{31}T_1 + d_{33}T_2), & D_2^{(1)} &= -\frac{3}{2h^3}(d_{31}G_1 + d_{33}G_2).
\end{aligned}
\tag{10.8}
$$

Electroelasticity relations 10.6 for forces and moments do not contain unknown electrical quantities; therefore, just like in the case of a shell with thickness polarization and electrode-covered faces, the problem can be divided into mechanical and electrical ones. The first coincides to constant factors with the theory of nonelectrical shells, and the second gives the electrical quantities directly from equations 10.7 and 10.8 after the mechanical problem has been solved.

Let us write the boundary conditions. Since the directions α_1 and α_2 are not equivalent for the tangential polarization, we will consider the edges $\alpha_1 = \alpha_{10}$

and $\alpha_2 = \alpha_{20}$, separately. At the hinge-supported and free edges $\alpha_1 = \alpha_{10}$ and $\alpha_2 = \alpha_{20}$, the boundary conditions have the usual form 7.30 and 7.31. At the rigidly fixed edge $\alpha_2 = \alpha_{20}$, the condition for the rotation angle γ_2 should be changed to take into account the angle that is constant over the entire shell and is due to the electrical load

$$u_1 = 0 \qquad u_2 = 0, \qquad w = 0, \qquad \gamma_2 + \frac{d_{15}}{h} V = 0. \tag{10.9}$$

If the geodesic curvature $k_1 = (1/A_1A_2)(\partial A_1/\partial \alpha_2)$ of the α_1-line is equal to zero at the rigidly fixed edge $\alpha_1 = \alpha_{10}$, the boundary conditions have the usual form 7.32. If $k_1 \neq 0$, the boundary conditions of the shell theory cannot be found without calculating the three-dimensional electroelastic state localized at the edge (boundary layer). We will discuss this problem later using an asymptotic approach.

CASE TWO

Consider an electroelastic state of a shell with electrode-covered faces connected by an external contour with conductivity Y. We supply electric power to the electrodes

$$-i\omega \int_{\Omega} D_3 d\Omega = 2VY \tag{10.10}$$

where V is an unknown constant.

For this electroelastic state, the first hypothesis of this section and the second Kirchhoff's hypothesis remain valid. By assuming these hypotheses we arrive at the laws for variation with thickness of the displacements and stresses that were used in the Kirchhoff theory.

Third Hypothesis

The electrical potential varies with thickness by a cubic law

$$\psi = \psi^{(0)} + \gamma\psi^{(1)} + \gamma^2\psi^{(2)} + \gamma^3\psi^{(3)}. \tag{10.11}$$

Fourth Hypothesis

In the relations 9.4, we can neglect E_i for D_i.

Using our hypotheses, we find that E_3 and D_3 vary with respect to γ by a square law; D_i varies by a linear law:

$$\begin{aligned} E_3 &= E_3^{(0)} + \gamma E_3^{(1)} + \gamma^2 E_3^{(2)} \\ D_3 &= D_3^{(0)} + \gamma D_3^{(1)} + \gamma^2 D_3^{(2)} \\ D_i &= D_i^{(0)} + \gamma D_i^{(1)}. \end{aligned} \tag{10.12}$$

The system obtained with the help of the hypotheses breaks into two subsystems. One is mechanical and coincides with the Kirchhoff theory, and the other is electrical.

Let us write the elasticity relations

$$T_i = 2h(n_{ii}\varepsilon_i + n_{ij}\varepsilon_j), \qquad G_i = -\frac{2h^3}{3}(n_{ii}\kappa_i + n_{ij}\kappa_j)$$
$$S_{ij} = \frac{2h}{s_{44}^E}\omega, \qquad H_{ij} = \frac{4h^3}{3s_{44}^E}\tau. \tag{10.13}$$

After the mechanical problem has been solved, we can find the electrical quantities from the equations

$$\begin{aligned}
D_1^{(0)} &= \frac{d_{15}}{2h}S_{21},\\
D_1^{(1)} &= \frac{3d_{15}}{2h^3}H_{ij}\\
D_2^{(0)} &= \frac{1}{2h}(d_{31}T_1 + d_{33}T_2),\\
D_2^{(1)} &= -\frac{3}{2h^3}(d_{31}G_1 + d_{33}G_2)\\
D_3^{(1)} &= -\frac{1}{A_1A_2}\left[\frac{\partial}{\partial\alpha_1}A_2D_1^{(0)} + \frac{\partial}{\partial\alpha_2}A_1D_2^{(0)}\right]\\
2D_3^{(2)} &= -\frac{1}{A_1A_2}\left[\frac{\partial}{\partial\alpha_1}A_2D_1^{(1)} + \frac{\partial}{\partial\alpha_2}A_1D_2^{(1)}\right]\\
D_3^{(0)} &= \varepsilon_{11}^T E_3^{(0)} + d_{15}\sigma_{23}^{(0)}\\
\psi &= \psi^{(0)} \pm h\psi^{(1)} + h^2\psi^{(2)} \pm h^3\psi^{(3)} = \pm V\\
E_3^{(0)} &= -\psi^{(1)} = h^2\psi^{(3)} - \frac{V}{h}\\
E_3^{(1)} &= -2\psi^{(2)} = \frac{1}{\varepsilon_{11}^T}D_3^{(1)} - \frac{d_{15}}{2h\varepsilon_{11}^T}(q_2^+ + q_2^-)\\
E_3^{(2)} &= -3\psi^{(3)}\\
&= -\frac{d_{15}}{\varepsilon_{11}^T}\frac{3}{4h^2}\left(q_2^+ - q_2^- + \frac{N_2}{h}\right) + \frac{1}{\varepsilon_{11}^T}D_3^{(2)}\\
&\quad -i\omega\int_\Omega (D_3^{(0)} + hD_3^{(1)} + h^2D_3^{(2)})d\Omega = 2VY.
\end{aligned} \tag{10.14}$$

11 FREE SHELLS WITH THICKNESS POLARIZATION

We know that a stressed-strained state appears in nonelectrical shells with all edges unfixed (free shells) under the effect of an external load. In this case, the extending strains and shears are small compared to the flexural strains; the largest stresses are defined by the moments.

A similar stressed state is typical for free piezoceramic shells. Nevertheless, although the elasticity relations for free nonelectrical shells have a form similar to that for shells with fixed edges, we observe quite a different situation in the theory of piezoelectrical shells.

Let us formulate hypotheses for the theory of free electroelastic shells with electrode-covered faces. Note that this theory can be used to describe slowly-varying electroelastic states at a distance from the edges. Besides, in contrast to the previous cases, we cannot assume that a_i in 2.8 is equal to unity, even in the roughest approximation, because the principal terms in some equations cancel each other while the less significant terms may become principal.

First Hypothesis

We assume that the principal stresses in the constitutive relations 2.14 vary with respect to γ as

$$\tau_{ii} = \gamma \tau_{ii}^{(1)}, \qquad \tau_{ij} = \gamma \tau_{ij}^{(1)}. \tag{11.1}$$

Second Hypothesis

We can neglect the stresses τ_{33} compared to the principal stresses τ_{ii} in the constitutive relations

$$\tau_{ii} = \frac{1}{s_{11}^E(1-\nu^2)}\left(\frac{a_j}{a_i}e_i + \nu e_j\right) - \frac{d_{31}}{s_{11}^E}\frac{1}{1-\nu}\frac{E_3}{a_i}$$

$$D_3 = \varepsilon_{33}^T E_3 + d_{31}(a_1\tau_{11} + a_2\tau_{22}). \tag{11.2}$$

Third Hypothesis

As a result of the strain, an element perpendicular to the middle surface remains perpendicular, i.e.,

$$\frac{\partial v_i}{\partial \gamma} + \frac{g_i}{a_i} = 0 \tag{11.3}$$

and its elongation is

$$e_3 = \frac{\partial v_3}{\partial \gamma} = d_{33}E_3. \tag{11.4}$$

Fourth Hypothesis

The electrical potential is a quadratic function of γ

$$\psi = \psi^{(0)} + \gamma\psi^{(1)} + \gamma^2\psi^{(2)} \tag{11.5}$$

where the second term is much greater than the other terms. From equation 11.5 we get

$$E_3 = -\frac{\partial \psi}{\partial \gamma} = E_3^{(0)} + \gamma E_3^{(1)}. \tag{11.6}$$

Fifth Hypothesis

The quantity D_3 is independent of γ:

$$D_3 = D_3^{(0)}. \tag{11.7}$$

It follows from equations 11.3 and 11.4 that the displacements v_i and v_3 are linear functions of γ:

$$v_i = v_i^{(0)} - \gamma g_i^{(0)}, \qquad v_3 = v_3^{(0)} + \gamma d_{33} E_3^{(0)}$$

whence

$$e_i = e_i^{(0)} + \gamma e_i^{(1)}, \qquad m_i = m_i^{(0)} + \gamma m_i^{(1)}. \tag{11.8}$$

We represent the sought-for quantities as functions of γ (equations 11.1, 11.6, and 11.8) and substitute them into the first equation of 11.2 and the second equation of 2.14. Equating the coefficients at the same powers of γ, we get

$$\frac{1}{s_{11}^E(1-\nu^2)}(e_i^{(0)} + \nu e_j^{(0)}) = \frac{d_{31}}{s_{11}^E}\frac{1}{1-\nu}E_3^{(0)}$$
$$m_i^{(0)} + m_j^{(0)} = 0$$

$$\begin{aligned}
\tau_{ii}^{(1)} &= \frac{1}{s_{11}^E(1-\nu^2)}\left[e_i^{(1)} + \nu e_j^{(1)} + \left(\frac{1}{R_j} - \frac{1}{R_i}\right)e_i^{(0)}\right] \\
&\quad - \frac{d_{31}}{s_{11}^E}\frac{1}{1-\nu}\left(E_3^{(1)} - \frac{1}{R_i}E_3^{(0)}\right) \\
\tau_{ij}^{(1)} &= \frac{1}{s_{66}^E}\left[m_i^{(1)} + m_j^{(1)} + \left(\frac{1}{R_i} - \frac{1}{R_j}\right)m_i^{(0)}\right].
\end{aligned}$$

As we pass from stresses to moments these relations assume the form

$$\begin{aligned}
&\varepsilon_i = d_{31}E_3^{(0)}, \qquad \omega = 0 \\
&G_i = -\frac{2h^3 B}{3}(\kappa_i + \sigma\kappa_j) - \frac{2h^3 B}{3}\left[d_{33}\left(\frac{1}{R_i} + \frac{\sigma}{R_j}\right) + d_{31}\left(\frac{\sigma}{R_i} + \frac{1}{R_j}\right)\right]E_3^{(0)} \\
&H = H_{ij} = \frac{4h^3}{3s_{66}^E}\tau, \qquad E_3^{(0)} = -\frac{V}{h}.
\end{aligned} \tag{11.9}$$

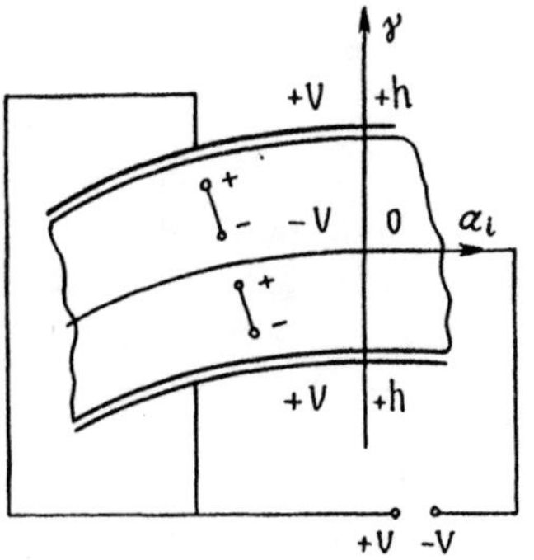

FIGURE 4. Section of a bimorphic shell element by the $\gamma\alpha_i$ - plane

The numbers B and σ are found from 7.27.

We note that, just like in the theory of nonelectrical shells, there are no elasticity relations for the forces T_i and S_{ij}, and they can only be found from the equilibrium equations. The formulas for other quantities coincide with equations 7.25, 7.26, 7.28, and 7.29.

12 BIMORPHIC SHELLS AND PLATES

Consider a shell that consists of two rigidly connected piezoceramic layers of the same thickness, h. There are electrode coverings on the middle surface of the two-layer shell and on its faces. Both piezoceramic layers are thickness-polarized in one direction. Electric power is supplied to the electrodes (Figure 4). Should the layers be disconnected, one of the layers would suffer elongation, and the other would suffer compression under the action of the electrical load. But since the layers are in contact, the shell bends.

Let us introduce two-dimensional equations for the bimorphic shell. We assume that the hypotheses we used in Section 7 when constructing the equations for one-layer shells with thickness polarization are valid for each of the layers. Then the electric potential will vary with the thickness coordinate γ by a square law

$$\psi_\pm = \psi_\pm^{(0)} + \gamma\psi_\pm^{(1)} + \gamma^2\psi_\pm^{(2)}. \tag{12.1}$$

Here, and below, the formulas with the $\pm$ sign unite two expressions: the upper sign holds for the layer $0 \le \gamma \le h$ and the lower sign holds for the layer $-h \le \gamma \le 0$. The electrical conditions on the electrodes are written as

$$\psi_\pm \mid_{\gamma=0} = -V, \qquad \psi_+ \mid_{\gamma=h} = V, \qquad \psi_- \mid_{\gamma=-h} = V. \tag{12.2}$$

By substituting 12.1 into 12.2, we find

$$\psi_\pm = -V \pm \gamma \left(\frac{2V}{h} - h\psi_\pm^{(2)} \right) + \gamma^2 \psi_\pm^{(2)}. \tag{12.3}$$

Taking into account the formulas of Section 7 and meeting the conditions for equal displacements on the contact surface of the layers, we get the following formulas for the stresses and electrical quantities:

$$\begin{aligned}
\sigma_{ii\pm}^{(0)} &= \frac{1}{s_{11}^E(1-\nu^2)}(e_i^{(0)} + \nu e_j^{(0)}) - \frac{d_{31}}{s_{11}^E(1-\nu)} E_{3\pm}^{(0)} \\
\sigma_{ii}^{(1)} &= \sigma_{ii+}^{(1)} = \sigma_{ii-}^{(1)} = \frac{1}{s_{11}^E(1-\nu^2)}(e_i^{(1)} + \nu e_j^{(1)}) - \frac{d_{31}}{s_{11}^E(1-\nu)} E_3^{(1)} \\
E_{3+}^{(0)} &= -E_{3-}^{(0)} = -\left(\frac{2V}{h} - h\psi_\pm^{(2)} \right) \\
E_{3\pm}^{(1)} &= -2\psi_\pm^{(2)} = -\frac{d_{31}}{\varepsilon_{33}^T} \left(\sigma_{11}^{(1)} + \sigma_{22}^{(1)} \right) \\
D_{3\pm}^{(0)} &= \varepsilon_{33}^T E_{3\pm}^{(0)} + d_{31} \left(\sigma_{11\pm}^{(0)} + \sigma_{22\pm}^{(0)} \right).
\end{aligned} \tag{12.4}$$

The formulas for $\sigma_{ij}^{(0)}$ and $\sigma_{ij}^{(1)}$ are the same as in Section 7. We pass from the stresses to the forces and moments

$$T_i = \frac{2h}{s_{11}^E(1-\nu^2)}(\varepsilon_i + \nu\varepsilon_j), \qquad S_{ij} = \frac{2h}{s_{66}^E}\omega \tag{12.5}$$

$$\begin{aligned}
G_i &= -\frac{2h^3 B_{11}}{3}(\kappa_i + \nu_* \kappa_j) - \frac{2hd_{31}}{s_{11}^E(1-\nu)} V \\
H_{ij} &= \frac{4h^3}{3s_{66}^E}\left(\tau - \frac{\omega}{2R_j} \right)
\end{aligned} \tag{12.6}$$

$$B_{11} = B\left(1 - \frac{3}{8}(1+\sigma)k_p^2 \right), \qquad \nu_* = \frac{B}{B_{11}}\left(\sigma - \frac{3}{8}(1+\sigma)k_p^2 \right). \tag{12.7}$$

The forces and moments are related to the stresses as

$$\begin{aligned}
\sigma_{ii\pm}^{(0)} &= \frac{T_i}{2h} - \frac{d_{31}}{s_{11}^E(1-\nu)} E_{3\pm}^{(0)}, & \sigma_{ij}^{(0)} &= \frac{S_{ij}}{2h} \\
\sigma_{ii}^{(1)} &= -\frac{3G_i}{2h^3} - \frac{3}{4h}\left(\sigma_{ii+}^{(0)} - \sigma_{ii-}^{(0)} \right), & \sigma_{ij}^{(1)} &= \frac{3H_{ij}}{3h^3}.
\end{aligned} \tag{12.8}$$

The electroelasticity relations 12.5 differ from 7.23 for one-layer shells with thickness polarization: in the latter, the electrical load enters the electroelasticity relations for tangential forces, while for bimorphic shells the electrical terms enter the electroelasticity relations for bending moments 12.5. After the mechanical problem has been solved, the greatest electrical quantities can be found by the formulas

$$\psi^{(0)} = -V, \qquad \psi_{\pm}^{(1)} = \pm\left(\frac{2V}{h} - h\psi^{(2)}\right) \tag{12.9}$$

$$\psi^{(2)} = \frac{d_{31}B}{2\varepsilon_{33}^{T}}(1+\sigma)(\kappa_1+\kappa_2), \qquad E_{3\pm}^{(0)} = -\psi_{\pm}^{(1)}, \qquad E_{3\pm}^{(1)} = -2\psi^{(2)}$$

$$D_{3\pm}^{(0)} = \pm\varepsilon_{33}^{T}\left(\frac{2V}{h} - h\psi^{(2)}\right) + \frac{d_{31}}{2h}(T_1+T_2). \tag{12.10}$$

Consider a special case of a bimorphic shell, i.e., a bimorphic plate, where a bending electroelastic state appears due to the described load. The electroelasticity relations for a bimorphic plate have the form 12.6. The greatest electrical quantities are found by the formulas 12.9 and an additional relation

$$D_{3\pm}^{(0)} = \pm\varepsilon_{33}^{T}\left(\frac{2V}{h} - h\psi^{(2)}\right).$$

13 THE THEORY OF PIEZOCERAMIC PLATES WITH THICKNESS POLARIZATION

The theory of piezoceramic plates can be constructed from the three-dimensional electroelasticity relations and the hypotheses used for the derivation of the equations of shell theory. We will not repeat our argument and will obtain the equations for the plates as a special case of the respective equations of the shell theory introduced in Sections 7 and 8.

For this purpose, we use the plate geometry and choose the zero curvatures in the shell theory equations, i.e.,

$$\frac{1}{R_1} = 0, \qquad \frac{1}{R_2} = 0 \tag{13.1}$$

and in all formulas of the shell theory we take

$$a_1 = \left(1 + \frac{h}{R_1}\right) = 1, \qquad a_2 = \left(1 + \frac{h}{R_2}\right) = 1. \tag{13.2}$$

Substituting 13.1 and 13.2 into the equations of the theory of piezoceramic shells, we get the equations of the theory of electroelastic plates that, just as in the case of the theory of elastic plates, can be divided into simpler subsystems: one describes the plane stressed-strained state and the other describes the bending stressed-strained state. We write the equations of the plate theory for different electrical conditions on the faces separately.

Suppose that a mechanical load

$$\sigma_{i3}|_{\gamma=\pm h} = \pm q_i^{\pm}, \qquad \sigma_{33}|_{\gamma=\pm h} = \pm q_3^{\pm} \tag{13.3}$$

is given on the plate's faces.

For *plates with electrode-covered faces,* the system of the plane problem includes:

Equilibrium Equations

$$\frac{1}{A_i}\frac{\partial T_i}{\partial \alpha_i} + \frac{1}{A_j}\frac{\partial S}{\partial \alpha_j} + k_j(T_i - T_j) + 2k_i S + 2h\rho\omega^2 u_i + X_i = 0. \tag{13.4}$$

Here, and below, A_i are coefficients of the first quadratic form on the plane, and X_i and Z are components of the surface load related to $q_i^{\pm}$ and $q_3^{\pm}$ through the formulas

$$X_i = q_i^{+} + q_i^{-}, \qquad Z = -(q_3^{+} + q_3^{-}). \tag{13.5}$$

Electroelasticity Relations

$$T_i = \frac{2h}{s_{11}^E(1-\nu^2)}(\varepsilon_i + \nu\varepsilon_j) - \frac{2hd_{31}}{s_{11}^E(1-\nu^2)}E_3^{(0)}$$

$$S = S_{12} = S_{21} = \frac{2h}{s_{66}^E}\omega. \tag{13.6}$$

Strain-Displacement Formulas

$$\varepsilon_i = \frac{1}{A_i}\frac{\partial u_i}{\partial \alpha_i} + k_i u_j, \qquad \omega = \frac{A_2}{A_1}\frac{\partial}{\partial \alpha_1}\left(\frac{u_2}{A_2}\right) + \frac{A_1}{A_2}\frac{\partial}{\partial \alpha_2}\left(\frac{u_1}{A_1}\right). \tag{13.7}$$

The system of equations 13.4, 13.6, and 13.7 differs from the respective equations of the theory of nonelectrical plates by the presence of the component $E_3^{(0)}$ of the electric field strength in the electroelasticity relations for the forces T_i. The value of $E_i^{(0)}$ depends on the electrical conditions on the faces of the plate.

If we are given the electrical potential

$$\psi|_{\gamma=\pm h} = \pm V \tag{13.8}$$

on the electrodes, then in the electroelasticity relations of the plane problem (equations 13.6) $E_3^{(0)}$ is a known quantity defined by

$$E_3^{(0)} = -\frac{V}{h}. \tag{13.9}$$

If the electrodes are connected by an external contour with complex conductivity Y and give off electrical energy, the resultant strain causes a potential difference $2V$. This is an unknown constant that can be found by integrating $D_3^{(0)}$ over the surface Ω of one of the electrodes (the point denotes the time derivative):

$$\int_\Omega \dot{D}_3^{(0)} d\Omega = 2VY. \tag{13.10}$$

This formula can be transformed into

$$\int_\Omega (\varepsilon_1 + \varepsilon_2) d\Omega = \frac{s_{11}^E \varepsilon_{33}^T (1-\nu)}{d_{31}} \frac{\Omega}{h} \left[i \frac{2Yh}{\omega \Omega \varepsilon_{33}^T} + 1 - \frac{2k_{31}^2}{1-\nu} \right] V. \tag{13.11}$$

If the electrodes are disconnected, we should set the conductivity Y equal to zero in 13.10 and 13.11. Then

$$\int_\Omega D_3^{(0)} d\Omega = 0, \int_\Omega (\varepsilon_1 + \varepsilon_2) d\Omega = \frac{s_{11}^E \varepsilon_{33}^T (1-\nu)}{d_{31}} \frac{\Omega}{h} \left[1 - \frac{2k_{31}^2}{1-\nu} \right] V. \tag{13.12}$$

If the electrodes are short-circuited, we should set the electrical potential V equal to zero in formulas 13.8 and 13.9:

$$V = 0 \tag{13.13}$$

Just as in the theory of nonelectrical plates, two mechanical boundary conditions should be met at every edge of the plate. For different fixation types, the conditions are as follows for:

A free edge $\alpha_i = \alpha_{i0}$

$$T_i = 0, \qquad S = 0 \tag{13.14}$$

A hinge-supported edge $\alpha_i = \alpha_{i0}$

$$u_j = 0, \qquad T_i = 0 \tag{13.15}$$

A rigidly fixed edge $\alpha_i = \alpha_{i0}$

$$u_1 = 0, \qquad u_2 = 0. \tag{13.16}$$

Having solved the plane problem, we can find the electrical quantities

$$\psi^{(1)} = -E_3^{(0)} = \frac{V}{h}, \qquad E_i^{(1)} = 0$$
$$D_3^{(0)} = \varepsilon_{33}^T E_3^{(0)} + \frac{d_{31}}{2h}(T_1 + T_2), \qquad D_i^{(1)} = \frac{d_{15}}{2h} X_i \tag{13.17}$$

where the electrical quantities vary with thickness of the plate γ as

$$\psi = \gamma \psi^{(1)}, \qquad E_i = \gamma E_i^{(1)}, \qquad E_3 = E_3^{(0)}$$
$$D_3 = D_3^{(0)}, \qquad D_i = \gamma D_i^{(1)}. \tag{13.18}$$

Equations for Bending Plates

Equilibrium equations

$$\frac{1}{A_1A_2}\frac{\partial}{\partial\alpha_1}(A_2N_1)+\frac{1}{A_1A_2}\frac{\partial}{\partial\alpha_2}(A_1N_2)+2h\rho\omega^2 w+Z=0$$
$$N_i=\frac{1}{A_i}\frac{\partial G_i}{\partial\alpha_i}-\frac{1}{A_j}\frac{\partial H}{\partial\alpha_j}+k_j(G_i-G_j)-2k_iH \tag{13.19}$$

Elasticity relations

$$G_i=-\frac{2h^3B}{3}(\kappa_i+\alpha\kappa_j),\qquad H_i=H_{12}=H_{21}=\frac{4h^3}{3s_{66}^E}\tau \tag{13.20}$$

where B and α are found from 7.27.

Strain-displacement formulas

$$\gamma_i=-\frac{1}{A_i}\frac{\partial w}{\partial\alpha_i},\qquad \kappa_i=-\frac{1}{A_i}\frac{\partial\gamma_i}{\partial\alpha_i}-k_i\gamma_j,\qquad \tau=-\frac{1}{A_i}\frac{\partial\gamma_j}{\partial\alpha_i}+k_i\gamma_i \tag{13.21}$$

In our case, the equations for bending plates completely coincide with those for bending nonelectrical plates, because the bending equations 13.19 to 13.21 do not contain electrical quantities.

When the bending problem is solved, we can find the electrical quantities:

$$\psi=\psi^{(0)}+\gamma^2\psi^{(2)},\qquad E_3=\gamma E_3^{(1)},\qquad E_i=E_i^{(0)}+\gamma^2E_i^{(2)}$$
$$D_i=D_i^{(0)}+\gamma^2D_i^{(2)},\qquad D_3=\gamma D_3^{(1)}+\gamma^3D_3^{(3)} \tag{13.22}$$

$$E_i^{(n)}=-\frac{1}{A_i}\frac{\partial\psi^{(n)}}{\partial\alpha_i}$$
$$(n+1)D_3^{(n+1)}=-\frac{1}{A_1A_2}\left[\frac{\partial}{\partial\alpha_1}(A_2D_1^{(n)})+\frac{\partial}{\partial\alpha_2}(A_1D_2^{(n)})\right],\qquad n=0,2$$
$$D_i^{(0)}=\varepsilon_{11}^TE_i^{(0)}-\frac{3d_{15}}{4h}N_i,$$
$$D_i^{(2)}=\varepsilon_{11}^TE_i^{(2)}+\frac{3d_{15}}{4h}N_i$$
$$\psi^{(0)}=-h^2\psi^{(2)},$$
$$\psi^{(2)}=-\frac{1}{2}E_3^{(1)}=-\frac{3d_{31}}{4h^3\varepsilon_{33}^T}(G_1+G_2). \tag{13.23}$$

The mechanical conditions at the edge of a plate are the same as in the theory of nonelectric plate for

A free edge $\alpha_i = \alpha_{i0}$

$$G_i = 0, \qquad N_i - \frac{1}{A_i}\frac{\partial H_{ji}}{\partial \alpha_i} = 0 \tag{13.24}$$

A hinge-supported edge $\alpha_i = \alpha_{i0}$

$$w = 0, \qquad G_i = 0 \tag{13.25}$$

A rigidly fixed edge $\alpha_i = \alpha_{i0}$

$$w = 0, \qquad \gamma_i = 0 \tag{13.26}$$

Since the equilibrium equations, strain-displacement formulas 13.4, 13.7, 13.19, and 13.21, mechanical boundary conditions 13.14 to 13.16, 13.24 to 13.26, and formulas for computing the three-dimensional stresses from forces and moments 7.22, 7.33, and 7.34 remain valid for all the electrical conditions on the plate faces considered below, we will only write the electroelasticity relations and formulas for computing the electrical quantities.

Plates with Faces Having No Electrodes

The edges of the plate may or may not have electrodes.

The elasticity relations for the plane problem have the form

$$T_i = 2hB(\varepsilon_i + \sigma\varepsilon_j), \qquad S = \frac{2h}{s_{66}^E}\omega. \tag{13.27}$$

Polynomials 13.18 in the thickness coordinate γ remain valid for the electrical quantities of the plane problem. The coefficients at different powers of γ are found from

$$E_3^{(0)} = -\psi^{(1)} = -\frac{d_{31}}{2h\varepsilon_{33}^T}(T_1 + T_2), \qquad E_i^{(1)} = -\frac{1}{A_i}\frac{\partial \psi^{(1)}}{\partial \alpha_i}$$
$$D_i^{(1)} = \varepsilon_{11}^T E_i^{(1)} + \frac{d_{15}}{2h}X_i, \qquad D_3^{(0)} = -h^2 D_3^{(2)}$$
$$2D_3^{(2)} = -\frac{1}{A_1 A_2}\left[\frac{\partial}{\partial \alpha_1}[A_2 D_1^{(1)}] + \frac{\partial}{\partial \alpha_2}[A_1 D_2^{(1)}]\right]. \tag{13.28}$$

The *problem for a bending plate* can be formulated differently depending on the electrical conditions at the edges. As in the theory of shells, we will deal with two cases.

CASE ONE

We assume that there is no electrical load at the electrodes. The elasticity relations will have the form

$$G_i = -\frac{2h^3B}{3}(\kappa_i + \sigma\kappa_j), \qquad H = \frac{4h^3}{3s_{66}^E} \tag{13.29}$$

Note that in this case the bending equations, just as in the plane problem, do not contain unknown electrical quantities. Therefore, the bending problem is a mechanical one.

If necessary, the electrical quantities can be found by equations 13.22 and the formulas

$$\begin{aligned}
E_i^{(n)} &= -\frac{1}{A_i}\frac{\partial\psi^{(n)}}{\partial\alpha_i},\\
(n+1)D_3^{(n+1)} &= -\frac{1}{A_1A_2}\left[\frac{\partial}{\partial\alpha_1}(A_2D_1^{(n)}) + \frac{\partial}{\partial\alpha_2}(A_1D_2^{(n)})\right], \qquad n = 0, 2\\
D_3^{(1)} &= -h^2D_3^{(3)}, \qquad D_i^{(0)} = \varepsilon_{11}^TE_i^{(0)} - \frac{3d_{15}}{4h}N_i,\\
D_i^{(2)} &= \varepsilon_{11}^TE_i^{(2)} + \frac{3d_{15}}{4h}N_i.
\end{aligned} \tag{13.30}$$

$$\begin{aligned}
&\left(\frac{\partial}{\partial\alpha_1}\frac{A_2}{A_1}\frac{\partial}{\partial\alpha_1} + \frac{\partial}{\partial\alpha_2}\frac{A_1}{A_2}\frac{\partial}{\partial\alpha_2}\right)\left[\psi^{(0)} - \frac{d_{15}}{4h\varepsilon_{33}^T}(G_1 + G_2)\right]\\
&\quad + \frac{d_{15}}{2h\varepsilon_{33}^T}\left(\frac{\partial}{\partial\alpha_1}A_2N_1 + \frac{\partial}{\partial\alpha_2}A_1N_2\right) = 0.
\end{aligned} \tag{13.31}$$

For $\psi^{(0)}$ we get a second-order differential equation 13.31. In order to find the arbitrary integration functions for this equation, we set one condition for every edge.

At the edge $\alpha_i = \alpha_{i0}$ which has no electrodes and is in contact with air or vacuum, the component of the electric induction vector normal to the edge should be equal to zero:

$$\begin{aligned}
\left(D_i^{(0)} + \frac{h^2}{3}D_i^{(2)}\right)\Big|_{\alpha_i=\alpha_{i0}} &= \frac{1}{A_i}\frac{\partial\psi^{(0)}}{\partial\alpha_i} - \frac{d_{31}}{4h\varepsilon_{33}^T}\frac{1}{A_i}\frac{\partial}{\partial\alpha_i}(G_1 + G_2) + \frac{d_{15}}{2h\varepsilon_{11}^T}N_i\Big|_{\alpha_i=\alpha_{i0}}\\
&= 0.
\end{aligned} \tag{13.32}$$

On two electrode-covered edges $\alpha_i = \alpha_{i1}$ and $\alpha_i = \alpha_{i2}$ connected by an external contour with complex conductivity Y, the condition

$$\int_l\left(\dot{D}_i^{(0)} + \frac{h^2}{3}\dot{D}_i^{(2)}\right)dl_i = -i\omega\int_l\left(D_i^{(0)} + \frac{h^2}{3}D_i^{(2)}\right)dl_i = 2VY \tag{13.33}$$

should hold, where l_i is the line of the electrode-covered edge $\alpha_i = \alpha_{ii}$ on the middle surface.

For disconnected electrodes, the condition 13.33 should be satisfied where $Y = 0$, while for short-circuited electrodes, the condition 13.13 should be met.

CASE TWO

We are given a potential difference

$$\psi|_{\alpha_i=\alpha_{i1}} = +V, \qquad \psi|_{\alpha_i=\alpha_{i2}} = -V \tag{13.34}$$

on the electrode-covered edges $\alpha_i = \alpha_{i1}$ and $\alpha_i = \alpha_{i2}$ of the plate.

For simplicity, we will assume that the mechanical load is absent.

Using the respective shell theory formulas, we get the following electro-elasticity relations and strain-displacement formulas:

$$\begin{aligned} G_i &= -\frac{2h^3B}{3}\left[\kappa_i + \sigma\kappa_j + d_{15}\left(\frac{1}{A_i}\frac{\partial E_i^{(0)}}{\partial \alpha_i} + k_iE_j^{(0)} + \sigma\left(\frac{1}{A_j}\frac{\partial E_j^{(0)}}{\partial \alpha_j} + k_jE_i^{(0)}\right)\right)\right] \\ H &= -\frac{4h^3}{3s_{66}^E}\left[\tau + d_{15}\left(\frac{1}{A_1}\frac{\partial E_2}{\partial \alpha_1} - k_1E_1\right)\right] \\ &= -\frac{4h^3}{3s_{66}^E}\left[\tau + d_{15}\left(\frac{1}{A_2}\frac{\partial E_1}{\partial \alpha_2} - k_2E_2\right)\right]. \end{aligned} \tag{13.35}$$

As compared to equations 13.29, in equations 13.35 we see additional terms containing electrical quantities. For defining $\psi^{(0)}$, we have the differential equation

$$\left(\frac{\partial}{\partial \alpha_1}\frac{A_2}{A_1}\frac{\partial}{\partial \alpha_1} + \frac{\partial}{\partial \alpha_2}\frac{A_1}{A_2}\frac{\partial}{\partial \alpha_2}\right)\psi^{(0)} = 0. \tag{13.36}$$

The electroelastic state is calculated in the following order. First, we solve the electrical problem that consists in integrating equation 13.36 with the boundary conditions

$$\psi^{(0)}|_{\alpha_i=\alpha_{i1}} = +V, \qquad \psi^{(0)}|_{\alpha_i=\alpha_{i2}} = -V$$

and find $\psi^{(0)}$. Then we solve the mechanical problem in the usual way with elasticity relations 13.35, where the electrical terms, E_i, are known. The remaining electrical quantities can be calculated by equations 13.30.

14 PLATES WITH TANGENTIAL POLARIZATION

As in the theory of shells with tangential polarization, we will assume, for definiteness, that the plate has been polarized along the α_2-lines.

The equilibrium equations and strain-displacement formulas for the plane and bending problems are given in the previous section. Here we will only write the electroelasticity and electrostatics equations obtained as a special case of the shell theories given in Sections 9 and 10.

PLATES WITH FACES NOT COVERED BY ELECTRODES

The Plane Problem

The electroelasticity and electrostatics equations have the form

$$T_i = 2h(n_{ii}\varepsilon_i + n_{ij}\varepsilon_j) - 2hc_i E_2^{(0)}, \qquad S = \frac{2h}{s_{66}^E}\omega$$
$$D_1^{(0)} = \varepsilon_{11}^T E_1^{(0)} + \frac{d_{15}}{2h}S, \qquad D_2^{(0)} = \varepsilon_{33}^T E_2^{(0)} + \frac{d_{31}}{2h}T_1 + \frac{d_{33}}{2h}T_2$$
$$\frac{\partial}{\partial \alpha_1}\left[A_2 D_1^{(0)}\right] + \frac{\partial}{\partial \alpha_2}\left[A_1 D_2^{(0)}\right] = 0, \qquad E_i^{(0)} = -\frac{1}{A_i}\frac{\partial \psi^{(0)}}{\partial \alpha_i}. \tag{14.1}$$

We note that the electrical quantities of the plane problem do not depend on the thickness coordinate

$$E_i = E_i^{(0)}, \qquad \psi = \psi^{(0)}, \qquad D_i = D_i^{(0)}.$$

In our case, the plane problem is a connected electroelastic one. This is because it is described by a system of equations that contains both mechanical and electrical quantities. The equations of the plane problem have the sixth order: therefore, three boundary conditions should be met at each edge: two are the mechanical conditions (equations 13.14 to 13.16) and the third is an electrical condition, which for the edge $\alpha_i = \alpha_{i0}$ without electrodes is written as

$$D_i^{(0)}\Big|_{\alpha_i=\alpha_{i0}} = 0. \tag{14.2}$$

For the electrode-covered edges $\alpha_i = \alpha_{i1}$ and $\alpha_i = \alpha_{i2}$ with specified electrical potential, there holds the conditions

$$\psi^{(0)}\Big|_{\alpha_i=\alpha_{i1}} = +V, \qquad \psi^{(0)}\Big|_{\alpha_i=\alpha_{i2}} = -V. \tag{14.3}$$

If the electrodes are short-circuited, we should set $V = 0$ in conditions 14.3.

If the electrodes are connected by an external contour with conductivity Y and give off electricity, the unknown constant V can be found by taking a curvilinear integral over the edge line $l_1(\alpha_1 = \alpha_{10})$

$$i\omega\left[\varepsilon_{11}^T \int_{l_1} \frac{1}{A_1}\frac{\partial \psi^{(0)}}{\partial \alpha_1} dl_1 - \frac{d_{15}}{2h}\int_{l_1} S\, dl_1\right] = 2VY \tag{14.4}$$

and over the edge line $l_2(\alpha_2 = \alpha_{20})$

$$i\omega\left[\varepsilon_{33}^T \int_{l_2} \frac{1}{A_2}\frac{\partial\psi^{(0)}}{\partial\alpha_2} dl_2 - \int_{l_2}\left[\frac{d_{31}}{2h}T_1 + \frac{d_{33}}{2h}T_2\right] dl_2\right] = 2VY. \qquad (14.5)$$

If the electrodes are disconnected, we should take $Y = 0$ in equations 14.4 and 14.5.

The Problem of a Bending Plate

The electroelasticity relations are of the form

$$G_i = -\frac{2h^3}{3}(n_{ii}\kappa_i + n_{ij}\kappa_j), \qquad H = \frac{4h^3}{3s_{44}^E}\tau. \qquad (14.6)$$

The equations of the bending problem contain only mechanical quantities. The boundary conditions coincide with 13.24 to 13.26.

Having solved the mechanical problem, we can compute the electrical quantities

$$D_i = \gamma D_i^{(1)}, \qquad E_3 = E_3^{(0)} + \gamma^2 E_3^{(2)}, \qquad D_3 = D_3^{(0)} + \gamma^2 D_3^{(2)}$$
$$\psi = \gamma\psi^{(1)} + \gamma^3\psi^{(3)}, \qquad E_i = \gamma E_i^{(1)} + \gamma^3 E_i^{(3)} \qquad (14.7)$$

$$D_1^{(1)} = \frac{3d_{15}}{2h^3}H,$$
$$D_2^{(1)} = -\frac{3}{2h^3}(d_{31}G_1 + d_{33}G_2)$$
$$D_3^{(0)} = -h^2 D_3^{(2)} = \frac{h^2}{2A_1A_2}\left[\frac{\partial}{\partial\alpha_1}(A_2D_1^{(1)}) + \frac{\partial}{\partial\alpha_2}(A_1D_2^{(1)})\right]$$
$$E_3^{(0)} = \frac{1}{\varepsilon_{11}^T D_3^{(0)}} + \frac{d_{15}}{4h\varepsilon_{11}^T}\left[3N_2 + h(q_2^+ - q_2^-)\right], \qquad \psi^{(1)} = -E_3^{(0)}$$
$$E_3^{(2)} = \frac{1}{\varepsilon_{11}^T}D_3^{(2)} - \frac{d_{15}}{4h^3\varepsilon_{11}^T}\left[N_2 + h(q_2^+ - q_2^-)\right], \qquad \psi^{(3)} = -\frac{1}{3}E_3^{(2)}. \qquad (14.8)$$

PLATES WITH ELECTRODE-COVERED FACES

The Plane Problem

The electroelasticity relations for forces have the form

$$T_i = 2h(n_{ii}\varepsilon_i + n_{ij}\varepsilon_j), \qquad S = \frac{2h}{s_{44}^E}\omega. \qquad (14.9)$$

The boundary conditions have the usual form.

The greatest electrical quantities are the components of the electric induction vector D_i defined by the formulas

$$D_i = D_i^{(0)}, \qquad D_1^{(0)} = \frac{d_{15}}{2h} S, \qquad D_2^{(0)} = \frac{d_{31}}{2h} T_1 + \frac{d_{33}}{2h} T_2. \tag{14.10}$$

The formulation of the bending problem greatly depends on the electrical conditions on the plate's faces.

CASE ONE

We assume that the electrical potential 13.8 is given on the electrodes.

The electroelasticity relations for moments have the form

$$\begin{aligned} G_i &= -\frac{2h^3}{3}(n_{ii}\kappa_i + n_{ij}\kappa_j) + n_{i_1} k_1 d_{15} E_3^{(0)} \\ H &= \frac{4h^3}{3s_4^E}\tau - \frac{2h^3}{3s_{44}^E} k_2 d_{15} E_3^{(0)} \end{aligned} \tag{14.11}$$

where $E_3^{(0)} = -V/h$.

The boundary conditions at the free and hinge-supported edges have the form 13.24 and 13.25.

At a rigidly fixed edge, $\alpha_2 = \alpha_{20}$ the boundary condition for the rotation angle γ_2, i.e.,

$$w = 0, \qquad \gamma_2 + \frac{d_{15}}{h} V = 0 \tag{14.12}$$

is nonhomogeneous.

The boundary conditions for a rigidly fixed edge, $\alpha_1 = \alpha_{10}$, need special consideration. Under the effect of the electrical load, the edge elements normal to the middle surface tend to rotate by the angle γ_2 to the plane of the edge and cause a three-dimensional electroelastic state near the rigidly fixed edge. This is described by the boundary layer. This electroelastic state is much more intensive than the respective state described by the plates theory and changes the boundary conditions of the plates theory. Only at the edge along which $k_1 = \frac{1}{A_1 A_2}\frac{\partial A_1}{\partial \alpha_2}$ vanishes (rectilinear edge) do the boundary conditions have a usual form 13.26.

The formulas for defining the electrical quantities assume the form

$$D_i = \gamma D_i^{(1)}, \qquad \psi = \gamma \psi^{(1)}, \qquad D_3 = D_3^{(0)}, \qquad E_3 = E_3^{(0)} \tag{14.13}$$

$$E_3^{(0)} = -\frac{V}{h}, \qquad D_1^{(1)} = \frac{3d_{15}}{2h^3} H, \qquad D_2^{(1)} = -\frac{3d_{31}}{2h^3} G_1 - \frac{3d_{33}}{2h^3} G_2. \tag{14.14}$$

CASE TWO

The electrical load is absent; the electroelastic state is due to the mechanical load. In this case the bending problem is broken into the mechanical and electrical problems.

The elasticity relations of the mechanical problem are

$$G_i = -\frac{2h^3}{3}(n_{ii}\kappa_i + n_{ij}\kappa_j), \qquad H = \frac{4h^3}{3s_{44}^E}\tau. \tag{14.15}$$

Having solved the mechanical problem with usual mechanical boundary conditions 13.24 to 13.26, we can calculate the electrical quantities. If the electrodes are connected by an external contour with conductivity Y, we get the following formulas for the electrical quantities:

$$\psi = \gamma\psi^{(1)} + \gamma^3\psi^{(3)}, \qquad E_3 = E_3^{(0)} + \gamma^2 E_3^{(2)} \tag{14.16}$$

$$D_3 = D_3^{(0)} + \gamma^2 D_3^{(2)}, \qquad D_i = \gamma D_i^{(1)}$$
$$D_1^{(1)} = d_{15}\frac{3}{2h^3}H, \qquad D_2^{(1)} = -d_{31}\frac{3}{2h^3}G_1 - d_{33}\frac{3}{2h^3}G_2 \tag{14.17}$$

$$2D_3^{(2)} = -\frac{1}{A_1A_2}\Big[\frac{\partial}{\partial\alpha_1}(A_2D_1^{(1)}) + \frac{\partial}{\partial\alpha_2}(A_1D_2^{(1)})\Big]$$
$$E_3^{(2)} = -3\psi^{(3)} = -\frac{d_{15}}{\varepsilon_{11}^T}\frac{3}{4h^2}\Big(q_2^+ - q_2^- + \frac{N_2}{h}\Big) + \frac{1}{\varepsilon_{11}^T}D_3^{(2)} \tag{14.18}$$

$$E_3^{(0)} = -\psi^{(1)} = h^2\psi^{(3)} - \frac{V}{h}$$
$$2V\Big(-\frac{iY}{\omega} + \varepsilon_{11}^T\frac{\Omega}{h}\Big) = \int_\Omega \Big[d_{15}\frac{N_2}{2h} + \frac{2h^2}{3}D_3^{(2)}\Big)d\Omega. \tag{14.19}$$

If the electrodes are disconnected, we should take $Y = 0$ in 14.19; and if the electrodes are short-circuited, we should take $V = 0$ in 14.18.

Chapter 3

THE METHOD OF PARTITIONING A STATIC ELECTROELASTIC STATE

15 PRELIMINARY REMARKS

When calculating nonelectrical shells, it is a common practice to use an approximate technique for finding the general solution, i.e., representation of the stressed-strained state as a sum of simpler stressed-strained states [21, 29, 68]. For example, for a large class of problems of the nonelectrical shells theory, the stressed-strained state can be represented as a sum of the membrane state or the pure moment state and simple edge effects. To describe these states, simpler systems of equations are used. These simpler systems are obtained from the complete equations of shell theory by neglecting, in the initial relations, the smaller quantities that for the considered stressed-strained state.

These questions have been well-studied in the theory of nonelectrical shells. We will briefly discuss the data necessary for our investigation, whose detailed treatment can be found in any course in classical shell theory.

An essential role in the construction of approximate theories is played by the stressed-strained state variability index that characterizes the variation rate of the sought-for quantities: forces, moments, strains, displacements, and electrical quantities, along the coordinates of the middle surface. The variability index s is introduced by extending the scale along the coordinates α_i,

$$\alpha_i = \eta^s R\xi_i, \qquad \eta = h/R. \tag{15.1}$$

Here η is a small parameter equal to the half-thickness of the shell h divided by the characteristic dimension R of the shell. New dimensionless variables are introduced so that their differentiations will not cause an essential increase or decrease of the needed functions.

Some papers employ a close notion of the characteristic size of the strain pattern instead of the stressed-strained state variability index. A null variability index corresponds to the characteristic size of the strain pattern compared to the characteristic size of the shell. A variability index equal to one corresponds to a characteristic size of the strain pattern which corresponds to the thickness of the shell. With the growth of the variability index, the error of the shell theory increases; and for the variability index equal to one, the error becomes so great that the shell theory gives no reliable description of the stressed-strained state.

The total stressed-strained state of a shell is composed of stressed-strained states with different variabilities. Approximate approaches to defining a stressed-strained state are based on distinguishing different variability states from the entire stressed-strained state set of the complete problem. We can make appropriate simplifications for a stressed-strained state with a certain variability in the initial system of equations. For example, in the membrane theory we can neglect the transverse forces and moments; and for a pure moment state we can neglect the extension-compression strains as compared to the bending strains.

The idea of the partition method is that the principal stressed-strained state and boundary effects essentially differ in their properties. The principal stressed-strained state has small variability and extends through the whole shell. The boundary effects are localized along some lines and damp down fast with the distance from those lines.

We know from the theory of nonelectrical shells [21, 29] that in order to break a stressed-strained state into the principal stressed-strained state and simple boundary effects we require that

1. All the edges of the shell are nonasymptotic
2. The shell geometry is specific: say, a cylindrical shell is not very long, a conical shell does not contain the vertex, and the shell does not contain plane regions
3. The principal stressed-strained state has small variability

We will consider two types of the principal electroelastic state: the membrane and pure moment states. For a membrane stressed-strained state to exist in a shell, its edges should be well fastened. The edge clampings should provide good rigidity of the shell against bending. When the shell edges are free or not well fastened, a pure moment state can appear in the shell.

The aim of this chapter is to generalize the method of partitioning developed in the theory of nonelectrical shells to the theory of electroelastic shells. The main difficulty of this task is to correctly define the place of the electrical quantities in the approximate equations and corresponding boundary conditions. Besides, the electroelastic shells may have electroelastic states that are not like the stressed-strained state of nonelectrical shells.

16 THE MEMBRANE THEORY OF SHELLS WITH THICKNESS POLARIZATION

Consider a shell with thickness polarization and faces covered with electrodes. We assume that all the conditions for the membrane theory of nonelectrical shells are valid and that the electrical load (i.e., the potential difference on the electrodes) also meets these conditions. It has small variability, because the electrodes are quasi-potential surfaces and the potential difference is independent of the coordinates of the shell's middle surface. The complete system of equations describing the electroelastic state of a shell with thickness polarization is given in Section 7.

In order to get the equations of the membrane theory, we neglect the transverse forces in the equilibrium equations for the forces in equation 5.3. Then, the simplified equilibrium equations for the forces, the electroelasticity relations for the forces, and the formulas for tangential strains and displacements will form the system for the membrane theory of shells. Let us write these equations.

The equilibrium equations for the forces are

$$\frac{1}{A_i}\frac{\partial T_i}{\partial \alpha_i} + \frac{1}{A_j}\frac{\partial S}{\partial \alpha_j} + k_j\left(T_i - T_j\right) + 2k_i S + X_i = 0, \qquad \frac{T_1}{R_1} + \frac{T_2}{R_2} + Z = 0. \tag{16.1}$$

The electroelasticity relations for the forces are

$$T_i = \frac{2h}{s_{11}^E(1-\nu^2)}(\varepsilon_i + \nu\varepsilon_j) - \frac{2hd_{31}}{s_{11}^E(1-\nu)}E_3^{(0)}, \qquad S = S_{12} = S_{21} = \frac{2h}{s_{66}^E}\omega. \tag{16.2}$$

The formulas for the tangential strains and displacements are

$$\varepsilon_i = \frac{1}{A_i}\frac{\partial u_i}{\partial \alpha_i} + k_i u_j - \frac{w}{R_i}, \qquad \omega = \frac{A_2}{A_1}\frac{\partial}{\partial \alpha_1}\frac{u_2}{A_2} + \frac{A_1}{A_2}\frac{\partial}{\partial \alpha_2}\frac{u_1}{A_1}. \tag{16.3}$$

If we are given the potential difference $2V$ on the electrodes, the electrical quantities of the membrane theory are found from

$$E_3^{(0)} = -\psi^{(1)} = -\frac{V}{h}, \qquad D_3^{(0)} = -\varepsilon_{33}^T\frac{V}{h} + \frac{d_{31}}{2h}\left(T_1 + T_2\right). \tag{16.4}$$

If the faces do not have electrodes, we find $E_3^{(0)}$ from

$$E_3^{(0)} = -\frac{d_{31}}{2h\varepsilon_{33}^T}\left(T_1 + T_2\right). \tag{16.5}$$

Using equation 16.5 to replace $E_3^{(0)}$ in 16.2 by the forces, we obtain

$$T_i = 2hB(\varepsilon_i + \sigma\varepsilon_j). \tag{16.6}$$

Note that for shells with thickness polarization without electrodes on the faces, the equations of the membrane theory coincide with the equations of the membrane theory for nonelectrical shells within constant factors of the elasticity relations. If the shell has electrodes on which the potential difference is specified, the system of membrane equations describing the shell is similar to that of thermoelastic shells. The analytical and numerical solutions to such problems are well described by the theory of nonelectrical shells and present no difficulty.

Having solved the membrane problem, we use the remaining equations of shell theory to find the less important quantities of the membrane theory: the bending strains, moments, and transverse forces. These equations have the form

$$\gamma_i = -\frac{1}{A_i}\frac{\partial w}{\partial \alpha_i} - \frac{u_i}{R_i}, \qquad \kappa_i = -\frac{1}{A_i}\frac{\partial \gamma_i}{\partial \alpha_i} - k_i\gamma_j$$

$$\tau = -\frac{1}{A_j}\frac{\partial \gamma_i}{\partial \alpha_j} + k_j\gamma_j + \frac{1}{R_i}\left(\frac{1}{A_j}\frac{\partial u_i}{\partial \alpha_j} - k_j u_j\right) \tag{16.7}$$

$$G_i = -\frac{2Bh^3}{3}(\kappa_i + \sigma\kappa_j), \qquad H_{ij} = \frac{4h^3}{3s_{66}^E}\left(\tau - \frac{\omega}{2R_j}\right) \tag{16.8}$$

$$N_i = \frac{1}{A_i}\frac{\partial G_i}{\partial \alpha_i} - \frac{1}{A_j}\frac{\partial H_{ij}}{\partial \alpha_j} + k_j(G_i - G_j) - k_i(H_{ij} + H_{ji}). \tag{16.9}$$

The computation of small quantities, like moments and transverse forces, presents no practical interest; however, they can help in estimating the error introduced into the membrane equations by neglecting the transverse forces in the force equilibrium equations. We will assume that the differentiation of the sought-for quantities leads to their η^{-s}-times growth (see 15.1). With the forces T_i having the same order as RZ, i.e.,

$$T_i \approx O(RZ) \tag{16.10}$$

we get the following estimates from 16.1 for the order of the shearing forces:

$$S \approx O(T_1) \approx O(RZ). \tag{16.11}$$

By analyzing the electroelasticity relations 16.2, we get the estimate

$$\varepsilon_i \approx O\left(\frac{s_{11}^E}{2h}T_i\right) \approx O\left(\frac{s_{11}^E}{2h}RZ\right) \tag{16.12}$$

for the order of the tangential strains. Formulas 16.3 for the tangential strains and displacements give the estimates

$$w \approx O(R\varepsilon_i) \approx O\left(\frac{s_{11}^E}{2h}R^2 Z\right), \qquad u_i \approx O(\eta^s \omega) \tag{16.13}$$

for the displacements.

Using 16.7 to 16.9, we can get by analogy

$$
\begin{aligned}
(\kappa_i, \tau) &\approx O\left(\eta^{-2s}\frac{w}{R^2}\right), \\
(G_i, H_{ij}) &\approx O\left(\eta^{2-2s}R^2 Z\right), \\
N_i &\approx O\left(\eta^{2-3s}RZ\right). \qquad (16.14)
\end{aligned}
$$

We will call formulas 16.10 to 16.14, describing the order of the needed quantities with respect to the load, the asymptotics of the membrane problem. When deriving formulas 16.10 to 16.14, we assumed, for simplicity, that the components of the surface load X_i and the electrical load cause an electroelastic state similar to that of the load Z normal to the surface.

Returning to the force equilibrium equations 5.3, we note that the transverse forces are $O(\eta^{2-4s})$ relative to the greatest terms of the force equations 5.3; therefore, by neglecting them we get an error

$$
\varepsilon = O\left(\eta^{2-4s}\right). \qquad (16.15)
$$

This formula shows that the membrane theory is only valid for the variability indexes smaller than 1/2.

17 THE MEMBRANE THEORY OF SHELLS WITH TANGENTIAL POLARIZATION

Consider a shell without electrodes on the faces (see Section 9).

By neglecting the transverse forces in the equilibrium equations, we get a membrane system of equations including the membrane equilibrium equations 16.1, formulas 16.3 for tangential strains and displacements, and the following electroelasticity relations and electrostatic equations:

$$
\begin{aligned}
T_i &= 2h(n_{ii}\varepsilon_i + n_{ij}\varepsilon_j) - 2hc_i E_2^{(0)} \\
S = S_{12} = S_{21} &= \frac{2h}{s_{44}^E}\left(\omega - d_{15}E_1^{(0)}\right) \\
D_1^{(0)} &= \varepsilon_{11}^T E_1^{(0)} + \frac{d_{15}}{2h} S \\
D_2^{(0)} &= \varepsilon_{33}^T E_2^{(0)} + \frac{d_{31}}{2h} T_1 + \frac{d_{33}}{2h} T_2
\end{aligned}
$$

$$
\frac{\partial}{\partial \alpha_1}\left(A_2 D_1^{(0)}\right) + \frac{\partial}{\partial \alpha_2}\left(A_1 D_2^{(0)}\right) = 0
$$

$$
E_i^{(0)} = -\frac{1}{A_i}\frac{\partial \psi^{(0)}}{\partial \alpha_i}. \qquad (17.1)
$$

Note that the membrane problem for shells with tangential polarization is a connected electroelastic problem. In contrast to the nonelectrical membrane problem whose equations are of the fourth order, the above system is of order six.

The estimate, obtained in the previous section for the asymptotics of the required quantities and the estimate for the error of the membrane equations 16.15 are true for the case we are considering. The asymptotics for the electrical quantities can be obtained in a similar way.

18 THE PRINCIPAL STRESSED STATE OF SHELLS WITH TANGENTIAL POLARIZATION AND ELECTRODE-COVERED FACES

Let an electroelastic shell be deformed under the action of the electrical load

$$\psi\,|_{\gamma=\pm h} = \pm V. \tag{18.1}$$

We showed in Section 10 that for this case the electroelasticity relations are of the form

$$T_i = 2h(n_{ii}e_i + n_{ij}e_j), \qquad S = S_{12} = S_{21} = \frac{2h}{s_{44}^E}\omega \tag{18.2}$$

$$G_i = -\frac{2h}{3}\left(n_{ii}\kappa_i + n_{ij}\kappa_j - n_{i1}k_1 d_{15}\frac{V}{h}\right)$$

$$H_{ij} = \frac{4h^3}{3s_{44}^E}\left(\tau + \frac{1}{2}k_2 d_{15}\frac{V}{h}\right). \tag{18.3}$$

This electroelastic state does not resemble the stressed-strained state of non-electrical shells because the moments and transverse forces in it are important whatever the fastening of the edges.

In order to construct the equations for the principal electroelastic state, we assume that the components of the bending strain can be neglected in the electroelasticity relations 18.3 for the moments and

$$G_i = \frac{2h^2}{3}n_{i1}k_1 d_{15}V, \qquad H_{ij} = +\frac{2h^2}{3}k_2 d_{15}V. \tag{18.4}$$

This simplification allows us to integrate the system in consecutive steps. The quantities determined at a previous step are treated as known at the next step.

Step One

Find the moments G_i and H_{ij} by equations 18.4.

Step Two

Find the transverse forces N_i from the equilibrium equations 16.9 for moments.

Step Three

Find the forces T_i and S by integrating the force equilibrium equations 5.3.

Step Four

Find the components of the tangential strains ε_i and ω from the electroelasticity relations 18.2.

Step Five

Find the displacements u_i and w by integrating the geometrical relations 16.3.

We estimate the relative order of the needed quantities at every step and get

$$
\begin{aligned}
&G_i \approx O(H_{ij}), && N_i \approx O\left(\eta^{-s}\frac{G_i}{R}\right),\\
&(S, T_i) \approx O\left(\eta^{-2s}\frac{G_i}{R}\right) && (\varepsilon_i, \omega) \approx O\left(\eta^{-2s}\frac{G_i}{2hRn_{11}}\right),\\
&w \approx O\left(\eta^{-2s}\frac{G_i}{2hn_{11}}\right) && u_i \approx O\left(\eta^{-s}\frac{G_i}{2hn_{11}}\right),\\
&(\kappa_i, \tau) \approx O\left(\eta^{-4s}\frac{G_i}{2hR^2n_{11}}\right) && h^3n_{11}\kappa_i \approx O\left(\eta^{2-4s}G_i\right),\\
&h^3n_{11}\tau \approx O\left(\eta^{2-4s}H_{ij}\right).
\end{aligned}
\tag{18.5}
$$

We see from 18.5 that the electrical load stimulates an electroelastic state in the shell where the stresses defined by the moments are η^{-1+2s} times greater than those defined by the forces.

Using the asymptotics 18.5, we can show that the error ε, introduced due to neglecting the components of the bending strain, is defined in the elasticity relations for the moments 16.15.

19 PURE MOMENT ELECTROELASTIC STATE OF A SHELL WITH THICKNESS POLARIZATION

In Section 11, we obtained the electroelasticity relations for the pure moment state of a shell. They are

$$
\varepsilon_i = -d_{31}\frac{V}{h}, \qquad \omega = 0 \tag{19.1}
$$

$$G_i = -\frac{2h^3B}{3}(\kappa_i + \sigma\kappa_j) + \frac{2h^2B}{3}\left[d_{33}\left(\frac{1}{R_i} + \frac{\sigma}{R_j}\right) + d_{31}\left(\frac{\sigma}{R_i} + \frac{1}{R_j}\right)\right]V$$

$$H_{ij} = \frac{4h^3}{3s_{66}^E}\tau. \qquad (19.2)$$

Note that instead of the electroelasticity relations for the forces we have gotten the approximate relations 19.1. Also, the electroelasticity relations for the moments 19.2 contain electrical terms that are essential only for the considered electroelastic state.

We add to equations 19.1 and 19.2 the equilibrium equations 5.3 and deformation formulas 16.3 and 16.7 to obtain a complete system. It is integrated just like the system for the pure moment electroelastic state of nonelectrical shells. The steps for defining the sought-for quantities are as follows.

Step One

Integrating equations 19.1 and 16.3 with respect to u_i and w.

Step Two

Finding the bending strains κ_i and τ from equation 16.7.

Step Three

Finding the moments G_i and H_{ij} from equation 19.2.

Step Four

Finding the transverse forces N_i from equation 16.9.

Step Five

Finding the forces T_i and S by integrating the force equilibrium equations 5.3.

By estimating the magnitudes of the needed quantities at every step, we can show that the error in the approximate equations 19.1 is ε, with ε defined by 16.15.

20 SIMPLE EDGE EFFECTS IN ELECTROELASTIC SHELLS WITH THICKNESS POLARIZATION

The electroelastic state of a shell with thickness polarization is very much like the stressed-strained state of nonelectrical shells; therefore, the equations describing simple edge effects for mechanical quantities are analogous to the simple edge effect equations of nonelectrical shells. We will construct the simple edge effect equations near the nonasymptotic distortion line of the electroelastic state, which is given by the equation $\alpha_1 = \alpha_{10} = \text{constant}$.

We will consider a shell with electrode-covered faces. By expanding the needed quantities into series in a small parameter, we can construct the theory of simple edge effects. A simple edge effect is characterized by the k-times growth of the sought-for functions as they are differentiated to the variable α_1, i.e., in the direction normal to the edge $\alpha_1 = \alpha_{10}$. Being differentiated to α_2, the functions either do not grow or grow by less than k times with

$$k = \sqrt{R/h} = \eta^{-1/2}. \tag{20.1}$$

We change the variables

$$k(\alpha_1 - \alpha_{10}) = R\xi_1 \tag{20.2}$$

where

$$\frac{\partial}{\partial \alpha_1} = k\frac{1}{R}\frac{\partial}{\partial \xi_1} \tag{20.3}$$

and assume that the differentiation to ξ_1 does not affect the order of the needed quantities. The forces, moments, strains, displacements, and electrical quantities will be given by the series

$$P = k^c \sum_{n=0}^{\infty} k^{-n} P, n. \tag{20.4}$$

Here, P is any of the enumerated quantities, and the number c is taken so that

$$\begin{aligned}
&c = 2 \text{ for } \kappa_1, E_3^{(1)}, && c = -2 \text{ for } u_2, T_2,\\
&c = 1 \text{ for } \tau, \kappa_2, && c = -3 \text{ for } T_1, N_1,\\
&c = 0 \text{ for } w, \varepsilon_1, \varepsilon_2, D_3^{(0)}, && c = -4 \text{ for } G_1, G_2, N_2,\\
&c = -1 \text{ for } u_1, \omega, && c = -5 \text{ for } H_{ij}.
\end{aligned} \tag{20.5}$$

We can verify the correctness of the choice of c by a comparison with the analytical solution of the simple edge effect problem for a nonelectrical shell and by the fact that the approximations obtained below are not contradictory.

The equations of shell theory contain the following functions describing the metric and geometry of the middle surface:

$$\frac{1}{A_i}, \qquad \frac{1}{R_i}, \qquad k_i = \frac{1}{A_i A_j}\frac{\partial A_i}{\partial \alpha_i}. \tag{20.6}$$

We expand them in the Taylor series in degrees of $\alpha_1 - \alpha_{10}$, take into account equation 20.2, and write the series in the form

$$Q = \sum_{n=0}^{\infty} \left(\frac{\xi}{k}\right)^n Q, n. \tag{20.7}$$

Here, Q is any of the functions 20.6; and Q, n are coefficients, respectively, specified as

$$\frac{1}{A_{i,n}}, \qquad \frac{1}{R_{i,n}}, \qquad k_{i,n}.$$

The general equations of the shell theory that interests us include the equilibrium equations 5.3, formulas for the strain-displacement component 7.19, electroelasticity equations 7.23 and 7.24, and formulas for the electrical quantities 7.25 and 7.26.

When constructing the equations for simple edge effect, we should take

$$E_3^{(0)} \equiv 0$$

in the electroelasticity relations, because the electrodes are equipotential surfaces and the potential difference on them cannot have the variability required for the edge effect.

We substitute expansions 20.4, 20.5, and 20.7 into the first equations 5.3, take into account 20.2 and 20.3 and equate the coefficients at the same powers of k in the resultant relations. For every power of k, we get an approximate relation

$$\frac{1}{A_{1,n}}\frac{1}{R}\frac{\partial T_{1,n}}{\partial \xi_1} + \frac{1}{A_{2,n}}\frac{\partial S_{,n-1}}{\partial \alpha_2} + k_{2,n}\left(T_{1,n-1} - T_{2,n}\right) + 2k_{1,n}S_{,n-1} - \frac{N_{1,n-2}}{R_{1,n}} = 0$$
$$n = 0, 1, 2, \ldots$$

The subscripts after commas denote the approximation numbers. The numbers with negative subscripts should be treated as equal to zero.

By treating all the other equations in our theory in the same way, we get the approximate equations. For the zeroth approximation, we get the system

$$\varepsilon_1 = \frac{1}{A_1}\frac{\partial u_1}{\partial \alpha_1} - \frac{w}{R_1}, \qquad \varepsilon_2 = -\frac{w}{R_2}$$

$$\omega = \frac{1}{A_1}\frac{\partial u_2}{\partial \alpha_1} + \frac{1}{A_2}\frac{\partial u_1}{\partial \alpha_2} - k_1 u_1$$

$$\kappa_1 = \frac{1}{A_1^2}\frac{\partial^2 w}{\partial \alpha_1^2}, \qquad \kappa_2 = \frac{k_2}{A_1}\frac{\partial w}{\partial \alpha_1} \tag{20.8}$$

$$\tau = \frac{1}{A_2}\frac{\partial}{\partial \alpha_2}\frac{1}{A_1}\frac{\partial w}{\partial \alpha_1},$$

$$T_2 = \frac{2h}{s_{11}^E(1-\nu^2)}(\varepsilon_2 + \nu\varepsilon_1), \qquad \varepsilon_1 + \nu\varepsilon_2 = 0, \qquad D_3^{(0)} = \frac{d_{31}}{2h}T_2 \tag{20.9}$$

$$S = \frac{2h}{s_{66}^E}\omega, \qquad G_1 = -\frac{2h^3 B}{3}\kappa_1,$$
$$G_2 = \sigma G_1, \qquad H_{ij} = \frac{4h^3}{3s_{66}^E}\omega \tag{20.10}$$

$$\frac{1}{A_1}\frac{\partial T_1}{\partial \alpha_1} - k_2 T_2 = 0,$$
$$\frac{\partial S}{\partial \alpha_1} + \frac{1}{A_2}\frac{\partial}{\partial \alpha_2}(A_1 T_2) = 0$$
$$\frac{T_2}{R_2} + \frac{1}{A_1}\frac{\partial N_1}{\partial \alpha_1} = 0,$$
$$N_1 = \frac{1}{A_1}\frac{\partial G_1}{\partial \alpha_1}$$
$$N_2 = \frac{1}{A_2}\frac{\partial G_2}{\partial \alpha_2} - \frac{1}{A_1}\frac{\partial H_{21}}{\partial \alpha_1} + k_1(G_2 - G_1)$$
$$E_3^{(1)} = -2\psi^{((2)} = \frac{3d_{31}}{2h^3\varepsilon_{33}^T}(G_1 + G_2). \tag{20.11}$$

In the formulas in 20.8 to 20.11, we passed from the coordinates ξ_1 to the coordinates α_1 using 20.2 and omitted the subscript 0 that denotes the number of the approximation. We stress that in the zeroth approximation theory for simple edge effect, the coefficients characterizing the metric and curvature of the middle surface are treated as constants with respect to α_1 near the edge $\alpha_1 = \alpha_{10}$.

After some transformations, we get the equation for the simple edge effect with respect to the displacement w and the formulas for the rest of the quantities we are seeking for

$$\frac{1}{A_1}\frac{\partial^4 w}{\partial \alpha_1^4} + 4g^4 w = 0, \qquad 4g^4 = \frac{3}{h^2 B s_{11}^E R_2^2}$$

$$u_1 = -\frac{h^2 B s_{11}^E R_2^2}{3A_1^3}\left(\frac{1}{R_1} + \frac{\nu}{R_2}\right)\frac{1}{A_1^3}\frac{\partial^3 w}{\partial \alpha_1^3}$$
$$u_2 = \frac{h^2 B}{3}\left[s_{11}^E \frac{A_1^2}{A_2}\frac{\partial}{\partial \alpha_2}\left(\frac{1}{R_1} + \frac{\nu}{R_2}\right)\frac{R_2^2}{A_1^2} - s_{66}^E \frac{A_1}{A_2}\frac{\partial}{\partial \alpha_2}\frac{R_2}{A_1}\right]\frac{1}{A_1^2}\frac{\partial^2 w}{\partial \alpha_1^2}$$
$$T_2 = -\frac{2h}{s_{11}^E R_2}w, \qquad D_3^{(0)} = -\frac{d_{31}}{s_{11}^E R_2}w$$
$$T_1 = \frac{2h^3 B}{3}k_2 R_2 \frac{1}{A_1^3}\frac{\partial^3 w}{\partial \alpha_1^3}, \qquad S = -\frac{2h^3 B}{3}\frac{1}{A_2}\frac{\partial}{\partial \alpha_2}\frac{R_2}{A_1^3}\frac{\partial^3 w}{\partial \alpha_1^3} \tag{20.12}$$

$$G_1 = -\frac{2h^3 B}{3}\frac{1}{A_1^2}\frac{\partial^2 w}{\partial \alpha_1^2}, \qquad G_2 = \sigma G_1$$

$$N_1 = -\frac{2h^3 B}{3}\frac{1}{A_1^3}\frac{\partial^3 w}{\partial \alpha_1^3}$$

$$N_2 = -\frac{2h^3}{3}\left[\sigma \frac{B}{A_2}\frac{\partial}{\partial \alpha_2} - B(1-\sigma)k_1 + \frac{2}{s_{66}^E}\left(\frac{1}{A_2}\frac{\partial}{\partial \alpha_2} + k_1\right)\right]\frac{1}{A_1^2}\frac{\partial^2 w}{\partial \alpha_1^2}$$

$$E_3^{(1)} = -\frac{d_{31}B(1+\sigma)}{\varepsilon_{33}^T}\frac{1}{A_1^2}\frac{\partial^2 w}{\partial \alpha_1^2}. \tag{20.13}$$

The general solution of the displacement equation has the form

$$w = [F_1 \cos A_1 g(\alpha_1 - \alpha_{10}) + F_2 \sin A_1 g(\alpha_1 - \alpha_{10})]\, e^{A_1 g(\alpha_1 - \alpha_{10})}$$
$$+ \left[F_3 \cos A_1 g(\alpha_1 - \alpha_{10}) + F_4 \sin A_1 g(\alpha_1 - \alpha_{(10)})\right] e^{-A_1 g(\alpha_1 - \alpha_{10})}. \tag{20.14}$$

On the right-hand side of formula 20.14, the first two terms are fast decreasing with the distance from the edge $\alpha_1 = \alpha_{10}$ inside the region $\alpha_1 < \alpha_{10}$, increases and the last two terms quickly decrease as the distance from the edge $\alpha_1 = \alpha_{10}$ inside the region $\alpha_1 > \alpha_{10}$ increases. It is obvious that the first two terms correspond to the edge effect in the region $\alpha_1 \leq \alpha_{10}$, and the last two terms give the edge effect in the region $\alpha_1 \geq \alpha_{10}$. Thus, we will assume that for the displacement w

$$w = [F_1 \cos A_1 g(\alpha_1 - \alpha_{10}) + F_2 \sin A_1 g(\alpha_1 - \alpha_{10})]\, e^{A_1 g(\alpha_1 - \alpha_{10})} \tag{20.15}$$

for $\alpha_1 \leq \alpha_{10}$, and

$$w = [F_3 \cos A_1 g(\alpha_1 - \alpha_{10}) + F_4 \sin A_1 g(\alpha_1 - \alpha_{10})]\, e^{-A_1 g(\alpha_1 - \alpha_{10})} \tag{20.16}$$

for $\alpha_1 \geq \alpha_{10}$.

By substituting solutions 20.15 and 20.16 into the approximate equations for displacements, forces, moments, and electrical quantities, we will get the analytical formulas for them.

Similar formulas can be obtained for subsequent approximations. We can find every next approximation of the simple edge effect after the preceding computation has been fulfilled. For example, the first approximation formulas have a structure similar to that of the zeroth approximation equations. They only differ by the presence of free terms that contain only zeroth approximation quantities.

The scheme may be used to obtain the equations of simple edge effect for other electrical conditions on the shell faces and other directions of preliminary polarization. We will give them without derivation since they are the same. Equations 20.8, 20.10, and 20.11 remain valid for a simple edge effect at the edge $\alpha_1 = \alpha_{10}$ in a shell with thickness polarization and no electrodes on the faces. However, formulas 20.9 should be replaced by

$$\varepsilon_1 + \sigma\varepsilon_2 = 0, \qquad T_2 = 2hB(\varepsilon_2 + \sigma\varepsilon_1). \tag{20.17}$$

Also, we should add the formulas for the electrical quantities:

$$E_3^{(0)} = -\psi^{(1)} = -\frac{d_{31}}{2h\varepsilon_{33}^T}T_2, \qquad \psi^{(0)} = \frac{(1+\sigma)d_{31} - 2d_{15}}{4h\varepsilon_{33}^T}G_1. \tag{20.18}$$

The final equation for simple edge effect and formulas expressing some of the required quantities in terms of the displacement w will be written as

$$\begin{aligned}
&\frac{1}{A_1}\frac{\partial^4 w}{\partial\alpha_1^4} + 4g^4 w = 0\\
&\qquad 4g^4 = \frac{3(1-\sigma^2)}{h^2R_2^2}\\
&\qquad T_2 = -2hB(1-\sigma^2)\frac{w}{R_2}\\
&\qquad u_1 = -\frac{h^2R_2^2}{3(1-\sigma^2)}\left(\frac{1}{R_1} + \frac{\sigma}{R_2}\right)\frac{1}{A_1^3}\frac{\partial^3 w}{\partial\alpha_1^3}\\
&\qquad u_2 = \frac{h^2}{3}\left[\frac{A_1^2}{A_2}\frac{\partial}{\partial\alpha_2}\left(\frac{1}{R_1} + \frac{\sigma}{R_2}\right)\frac{R_2^2}{A_1^2} - Bs_{66}^E\frac{A_1}{A_2}\frac{\partial}{\partial\alpha_2}\frac{R_2}{A_1}\right]\frac{1}{A_1^2}\frac{\partial^2 w}{\partial\alpha_1^2}\\
&\qquad \psi^{(0)} = -h^2B\frac{(1+\sigma)d_{31} - 2d_{15}}{6\varepsilon_{33}^T}\frac{1}{A_1^2}\frac{\partial^2 w}{\partial\alpha_1^2}.
\end{aligned} \tag{20.19}$$

The rest of the formulas coincide with formulas 20.13.

21 SIMPLE EDGE EFFECTS IN SHELLS WITH TANGENTIAL POLARIZATION

Since the directions α_1 and α_2 are not equally important in a shell pre-polarized along the α_2-lines, we consider the edges $\alpha_1 = \alpha_{10} =$ constant and $\alpha_2 = \alpha_{20} =$ constant separately.

Suppose that the shell faces are not covered with electrodes. Then, in the first approximation, the equations of simple edge effect near the edge $\alpha_1 = \alpha_{10}$

will be written in the form of 20.8 and 20.11 plus the formulas

$$n_{11}\varepsilon_1 + n_{12}\varepsilon_2 = 0, \qquad T_2 = 2h(n_{22}\varepsilon_2 + n_{12}\varepsilon_1)$$

$$G_1 = -\frac{2h^3}{3}n_{11}\kappa_1, \qquad G_2 = -\frac{2h^3}{3}n_{21}\kappa_1$$

$$H_{ij} = \frac{4h^3}{3s_{44}^E}\tau, \qquad D_2^{(0)} = \frac{d_{33}}{2h}T_2,$$

$$D_1^{(1)} = \frac{3d_{15}}{2h^3}H_{21} \qquad D_2^{(1)} = -\frac{3}{2h^3}(d_{31}G_1 + d_{33}G_2) \quad (21.1)$$

$$S^{(0)} = \frac{2h}{s_{44}^E}\left(\omega - d_{15}E_1^{(0)}\right),$$

$$E_i^{(0)} = -\frac{1}{A_i}\frac{\partial\psi^{(0)}}{\partial\alpha_i}$$

$$2h\varepsilon_{11}^T\frac{1}{A_1^2}\frac{\partial^2\psi^{(0)}}{\partial\alpha_1^2} = \frac{d_{15}}{A_1}\frac{\partial}{\partial\alpha_1}S^{(0)} + \frac{d_{33}}{A_1A_2}\frac{\partial}{\partial\alpha_2}A_1T_2^{(0)}$$

$$D_1^{(0)} = \varepsilon_{11}^TE_1^{(0)} + \frac{d_{15}}{2h}S^{(0)}. \quad (21.2)$$

After elementary transformations, we get

$$\frac{1}{A_1^4}\frac{\partial^4 w}{\partial\alpha_1^4} + 4g^4w = 0, \qquad 4g^4 = \frac{3}{n_{11}s_{33}^Eh^2R_2^2}$$

$$u_1 = -\frac{n_{11}s_{33}^Eh^2R_2^2}{3}\left(\frac{1}{R_1} + \frac{\nu_1}{R_2}\right)\frac{1}{A_1^3}\frac{\partial^3 w}{\partial\alpha_1^3}, \qquad \nu_1 = \frac{n_{12}}{n_{11}}$$

$$T_1 = k_2R_2\frac{2h^3n_{11}}{3}\frac{1}{A_1^3}\frac{\partial^3 w}{\partial\alpha_1^3}, \qquad S = -\frac{2h^3n_{11}}{3}\frac{1}{A_2}\frac{\partial}{\partial\alpha_2}\frac{R_2}{A_1^3}\frac{\partial^3 w}{\partial\alpha_1^3}$$

$$T_2 = -\frac{2h}{s_{33}^E}\frac{w}{R_1}, \qquad G_1 = -\frac{2h^3n_{11}}{3}\frac{1}{A_1^2}\frac{\partial^2 w}{\partial\alpha_1^2}$$

$$G_2 = -\frac{2h^3n_{21}}{3}\frac{1}{A_1^2}\frac{\partial^2 w}{\partial\alpha_1^2}, \qquad H_{ij} = \frac{2h^3}{3s_{44}^E}\frac{1}{A_2}\frac{\partial}{\partial\alpha_2}\frac{1}{A_1}\frac{\partial w}{\partial\alpha_1}$$

$$N_1 = -\frac{2h^3n_{11}}{3}\frac{1}{A_1^3}\frac{\partial^3 w}{\partial\alpha_1^3}$$

$$N_2 = -\frac{2h^3}{3}\left[n_{11}\frac{1}{A_2}\frac{\partial}{\partial\alpha_2} + \frac{1}{s_{44}^E}\left(\frac{1}{A_2}\frac{\partial}{\partial\alpha_2} + k_1\right) + k_1(n_{11} - n_{21})\right]\frac{1}{A_1^2}\frac{\partial^2 w}{\partial\alpha_1^2}$$

(21.3)

$$u_2 = \frac{h^2 n_{11} s_{33}^E}{3} \frac{A_1^2}{A_2} \frac{\partial}{\partial \alpha_2} \left(\frac{1}{R_1} + \frac{\nu_1}{R_2} \right) \frac{R_2^2}{A_1^4} \frac{\partial^2 w}{\partial \alpha_1^2}$$
$$+ \frac{h^2 n_{11}}{3} \left[\frac{d_{15}(d_{33} - d_{15})}{\varepsilon_{11}^T} - s_{44}^E \right] \frac{A_1}{A_2} \frac{\partial}{\partial \alpha_2} \frac{R_2}{A_1^3} \frac{\partial^2 w}{\partial \alpha_1^2}$$
$$\psi^{(0)} = \frac{h^2 n_{11}}{3\varepsilon_{11}^T} (d_{33} - d_{15}) \frac{A_1}{A_2} \frac{\partial}{\partial \alpha_2} \frac{R_2}{A_1^3} \frac{\partial^2 w}{\partial \alpha_1^2}. \tag{21.4}$$

The solution of the resultant equation is given by 20.15 and 20.16.

If the faces are covered with electrodes, we will have simple edge effect equations 21.1 combined with (for the edge $\alpha_1 = \alpha_{10}$)

$$S = \frac{2h}{s_{44}^E} \omega, \qquad D_1^{(0)} = \frac{d_{15}}{2h} S. \tag{21.5}$$

After transformations, we get 21.3 to which we should add the equations

$$u_2 = \frac{h^2 n_{11}}{3} \left[s_{33}^E \frac{A_1^2}{A_2} \frac{\partial}{\partial \alpha_2} \left(\frac{1}{R_1} + \frac{\nu_1}{R_2} \right) \frac{R_2^2}{A_1^2} - s_{44}^E \frac{1}{A_2} \frac{\partial}{\partial \alpha_2} \frac{R_2}{A_1} \right] \frac{1}{A_1^2} \frac{\partial^2 w}{\partial \alpha_1^2}$$
$$D_1^{(0)} = -\frac{h^2 n_{11} d_{15}}{3} \frac{1}{A_2} \frac{\partial}{\partial \alpha_2} \frac{R_2}{A_1^3} \frac{\partial^3 w}{\partial \alpha_1^3}. \tag{21.6}$$

At the edge $\alpha_2 = \alpha_{20}$ of a shell without electrodes on the faces. The zeroth approximation equations for simple boundary effect will be written as

$$\varepsilon_2 = \frac{1}{A_2} \frac{\partial u_2}{\partial \alpha_2} - \frac{w}{R_2}, \qquad \varepsilon_1 = -\frac{w}{R_1}$$
$$\omega = \frac{1}{A_1} \frac{\partial u_2}{\partial \alpha_2} + \frac{1}{A_2} \frac{\partial u_1}{\partial \alpha_2} - k_2 u_2$$
$$\kappa_1 = k_1 \frac{1}{A_2} \frac{\partial w}{\partial \alpha_2}, \qquad \kappa_2 = \frac{1}{A_2^2} \frac{\partial^2 w}{\partial \alpha_2^2}$$
$$\tau = \frac{1}{A_1} \frac{\partial}{\partial \alpha_1} \frac{1}{A_2} \frac{\partial w}{\partial \alpha_2} \tag{21.7}$$
$$n_{22}\varepsilon_2 + n_{21}\varepsilon_1 - c_2 E_2^{(0)} = 0$$

$$T_1 = 2h \left(n_{11}\varepsilon_1 + n_{12}\varepsilon_2 - c_1 E_2^{(0)} \right)$$
$$S = \frac{2h}{s_{44}^E} \left(\omega - d_{15} E_1^{(0)} \right), \qquad \psi^{(0)} = -\frac{d_{31} R_1}{2h\varepsilon_{33}^T} N_2$$
$$D_1^{(0)} = \varepsilon_{11}^T E_1^{(0)} + \frac{d_{15}}{2h} S, \qquad D_2^{(0)} = \varepsilon_{33}^T E_2^{(0)} + \frac{d_{31}}{2h} T_1 \tag{21.8}$$

$$
\begin{aligned}
G_1 &= -\frac{2h^3}{3} n_{12}\kappa_2, \\
G_2 &= -\frac{2h^3}{3} n_{22}\kappa_2 \\
\frac{1}{A_2}\frac{\partial T_2}{\partial \alpha_2} - k_1 T_1 &= 0 \\
\frac{\partial S}{\partial \alpha_2} + \frac{1}{A_1}\frac{\partial}{\partial \alpha_1} A_2 T_1 &= 0 \\
\frac{T_1}{R_1} + \frac{1}{A_2}\frac{\partial N_2}{\partial \alpha_2} &= 0 \\
N_2 &= \frac{1}{A_2}\frac{\partial G_2}{\partial \alpha_2} \\
N_1 &= \frac{1}{A_1}\frac{\partial G_1}{\partial \alpha_1} - \frac{1}{A_2}\frac{\partial H_{12}}{\partial \alpha_2} + k_2(G_1 - G_2) \\
D_1^{(1)} &= \frac{3d_{15}}{2h^3} H_{12}, \\
D_2^{(1)} &= -\frac{3}{2h^3}(d_{31}G_1 + d_{33}G_2).
\end{aligned}
\tag{21.9}
$$

We transform the formulas to a more convenient form:

$$
\begin{aligned}
\frac{1}{A_2^4}\frac{\partial^4 w}{\partial \alpha_2^4} + 4g^4 w &= 0, \\
4g^4 &= \frac{3p}{h^2 n_{22} R_1^2} \\
u_2 &= -\left(\frac{1}{R_1} - pd\frac{1}{R_2}\right)\frac{h^2 n_{22} R_1^2}{3p}\frac{1}{A_2^3}\frac{\partial^3 w}{\partial \alpha_2^3}, \\
T_1 &= -\frac{2hp}{R_1} w \\
E_2^{(0)} &= \frac{d_{31}p}{\varepsilon_{33}^T R_1} w, \\
\psi^{(0)} &= \frac{h^2 d_{31} n_{22}}{3\varepsilon_{33}^T}\frac{R_1}{A_2^3}\frac{\partial^3 w}{\partial \alpha_2^3} \\
\frac{1}{p} &= s_{11}^E\left(1 - \frac{d_{33}d_{31}}{\varepsilon_{33}^T s_{13}^E}\right), \\
d &= s_{13}^E\left(1 - \frac{d_{33}d_{31}}{\varepsilon_{33}^T s_{13}^E}\right)
\end{aligned}
\tag{21.10}
$$

$$T_2 = \frac{2h^3 n_{22}}{3} k_1 R_1 \frac{1}{A_2^3} \frac{\partial^3 w}{\partial \alpha_2^3}, \qquad S = -\frac{2h^3}{3} n_{22} \frac{1}{A_1} \frac{\partial}{\partial \alpha_1} \frac{R_1}{A_2^3} \frac{\partial^3 w}{\partial \alpha_2^3}$$

$$N_2 = -\frac{2h^3}{3} \frac{n_{22}}{A_2^3} \frac{\partial^3 w}{\partial \alpha_2^3}, \qquad G_1 = -\frac{2h^3}{3} n_{12} \frac{1}{A_2^2} \frac{\partial^2 w}{\partial \alpha_2^2}$$

$$G_2 = -\frac{2h^3}{3} \frac{n_{22}}{A_2^2} \frac{\partial^2 w}{\partial \alpha_2^2}. \tag{21.11}$$

If the faces are electrode-covered, formulas 21.7 and 21.9 remain valid. For the edge $\alpha_2 = \alpha_{20}$ formulas 21.8 are replaced by

$$n_{22}\varepsilon_2 + n_{21}\varepsilon_1 = 0, \qquad T_1 = 2h(n_{11}\varepsilon_1 + n_{21}\varepsilon_2)$$

$$S = \frac{2h}{s_{44}^E}\omega, \qquad D_1^{(0)} = \frac{d_{15}}{2h} S,$$

$$D_2^{(0)} = \frac{1}{2h} d_{31} T_1. \tag{21.12}$$

Calculations are made by 21.11 supplemented by the formulas

$$\frac{1}{A_2^4} \frac{\partial^4 w}{\partial \alpha_2^4} + 4g^4 w = 0,$$

$$4g^4 = \frac{3}{h^2 n_{22} s_{11}^E R_1^2}$$

$$u_2 = -\frac{n_{22} s_{11}^E R_1^2}{3} \left(\frac{1}{R_2} + \frac{\nu_2}{R_2} \right) \frac{1}{A_2^3} \frac{\partial^3 w}{\partial \alpha_2^3},$$

$$\nu_2 = \frac{n_{12}}{n_{22}}$$

$$T_1 = -\frac{2h}{s_{11}^E} \frac{w}{R_1}. \tag{21.13}$$

22 BOUNDARY CONDITIONS FOR THE PRINCIPAL ELECTROELASTIC STATE AND SIMPLE EDGE EFFECT

Let us consider an example of dividing the boundary conditions of the electroelastic shells theory into the boundary conditions for the principal electroelastic state and the boundary conditions for simple edge effect. We choose an electroelastic state of a shell with tangential polarization and no electrodes on the faces. We then suppose that all the conditions for breaking the electroelastic state into the membrane electroelastic state and simple edge effects hold true.

We represent the sought-for quantities as a sum of two terms:

$$P = P^{(m)} + \eta^a P^{(e)}. \tag{22.1}$$

Here, P is any of the quantities (displacements, forces, moments, etc.) defining the electroelastic state of the shell. The subscripts (m) and (e) show that the quantity belongs to the membrane state or simple edge effect, respectively. The solution of the membrane problem is found from the system of nonhomogeneous equations, and the solution to the simple edge effect problem is found from the homogeneous equations. Therefore, we use a scale factor η^a before $P^{(e)}$, where a is the same number for all the needed quantities and is chosen depending on the boundary conditions (a defines the intensity of the edge effect).

We assume that the shell has two electrode-covered edges on which the potential difference is specified. One of the edges is rigidly clamped. On this edge ($\alpha_1 = \alpha_{10}$), the boundary conditions are

$$u_1 = 0, \qquad u_2 = 0, \qquad w = 0, \qquad \gamma_1 = 0, \qquad \psi = V\cdot \tag{22.2}$$

From the results obtained in Section 16, it follows that for a membrane electroelastic state the displacements u_i and w, and rotation angles γ_i have the following asymptotic order:

$$\left(\frac{u_1}{R}, \frac{u_2}{R}, \frac{d_{33}}{R}\psi\right) \approx O\left(\frac{d_{33}}{R}V\right), \qquad \frac{w}{R} \approx O\left(\eta^{-s}\frac{u_i}{R}\right),$$
$$\gamma_i \approx O\left(\eta^{-2s}\frac{u_i}{R}\right).$$

We write these equations in a more convenient form and replace the sought-for quantities by dimensionless quantities with asterisks so that the introduced quantities have the same asymptotic order:

$$u_{i*}^{(m)} = \frac{u_i}{R}, \qquad w_*^{(m)} = \eta^{-s}\frac{w}{R}, \qquad \gamma_{i*}^{(m)} = \eta^{-2s}\gamma_i^{(m)},$$
$$\psi_* = \frac{d_{33}}{R}\psi, \qquad V_* = \frac{d_{33}}{R}V. \tag{22.3}$$

From the equations of Section 21, it follows that the quantities

$$u_{1*}^{(e)}, u_{2*}^{(e)}, w_*^{(e)}, \gamma_{1*}^{(e)}, \psi_*^{(e)}$$

$$u_{1*}^{(e)} = \eta^{1/2}\frac{u_1^{(e)}}{R}, \qquad u_{2*}^{(e)} = \eta^{1}\frac{u_2^{(e)}}{R}, \qquad w_*^{(e)} = \frac{w^{(e)}}{R}, \qquad \gamma_{1*}^{(e)} = \eta^{-1/2}\gamma_1^{(e)}$$
$$\psi_*^{(e)} = \eta^{\frac{1}{2}}\frac{d_{33}}{R}\psi \tag{22.4}$$

have the same asymptotic order. Allowing for 22.1 and replacing the sought-for quantities by their asymptotic representations, we rewrite the boundary conditions in the form

$$\eta^{-s}w_*^{(m)} + \eta^a w_*^{(e)} = 0, \qquad u_{1*}^{(m)} + \eta^{1/2+a}u_{1*}^{(e)} = 0$$

$$u_{2*}^{(m)} + \eta^{1+a}u_{2*}^{(e)} = 0, \qquad \eta^{-2s}\gamma_{1*}^{(m)} + \eta^{-1/2+a}\gamma_{1*}^{(e)} = 0$$

$$\psi_*^{(m)} + \eta^{1/2+a}\psi_*^{(e)} = V_*.$$

This problem can be solved by taking $a = -s$. We preserve only the principal terms and find the boundary conditions for the membrane theory

$$u_1^{(m)} = 0, \qquad u_2^{(m)} = 0, \qquad \psi^{(m)} = V \tag{22.5}$$

and for the simple edge effect

$$w^{(e)} = -w^{(m)}, \qquad \gamma_1^{(e)} = 0. \tag{22.6}$$

If the edges of the shell have no electrodes, conditions for the edge effect 22.6 remain valid and one electrical condition

$$u_1^{(m)} = 0, \qquad u_2^{(m)} = 0, \qquad D_1^{(m)} = 0 \tag{22.7}$$

changes in the membrane theory.

In order to receive the equations 22.5 to 22.7 we neglected by the terms of order of $O(\eta^{1/2-s})$. It gives an error of the boundary conditions for the membrane theory of $O(\eta^{1/2-s})$.

In both cases, the order of the shell design is similiar to the theory of non-electrical shells. First, we find the solution to the membrane problem with the conditions 22.5 or 22.7, and then we compute the simple edge effect with nonhomogeneous boundary conditions 22.6.

In a similar way, we can show that for shells with electrode-covered faces having a membrane electroelastic state the boundary conditions can be divided just like in the theory of nonelectrical shells. The tangential conditions are met when calculating the membrane electroelastic state and nontangential conditions are met when calculating the simple edge effect. If necessary, we can use this technique to divide the boundary conditions for any electroelastic shell and any type of edge fixation.

Chapter 4

APPROXIMATE METHODS FOR COMPUTING FREE AND FORCED VIBRATIONS OF ELECTROELASTIC SHELLS

23 FREE VIBRATIONS OF SHELLS WITH THICKNESS POLARIZATION

The vibration theory is well developed for nonelectrical shells: different types of vibrations are distinguished, approximate theories are constructed for describing every type, and the errors are estimated [23]. By analogy with the theory of nonelectrical shells, we will construct a classification of vibration types for piezoceramic shells. We will use the asymptotic method for integrating the dynamic equations of the piezoceramic shells theory.

We will assume that the vibrations obey the law $e^{-i\omega t}$, where ω is the angular frequency of the vibrations and t is the time. Since the vibrations are harmonic, we will describe them by equations written for the amplitude values of the needed quantities.

Consider a shell whose faces are covered with short-circuited electrodes. Then the electrical condition

$$\psi\Big|_{\gamma=\pm h} = 0 \tag{23.1}$$

should be met. We write the initial equations as the

Equilibrium equations

$$\begin{gathered}\frac{1}{A_i}\frac{\partial T_i}{\partial \alpha_i} + \frac{1}{A_j}\frac{\partial S}{\partial \alpha_j} + k_j(T_i - T_j) - \frac{N_i}{R_i} + 2k_iS + 2h\rho\omega^2 u_i = 0 \\ \frac{T_1}{R_1} + \frac{T_2}{R_2} + \frac{1}{A_1}\frac{\partial N_1}{\partial \alpha_1} + \frac{1}{A_2}\frac{\partial N_2}{\partial \alpha_2} + k_2N_1 + k_1N_2 + 2h\rho\omega^2 w = 0\end{gathered} \tag{23.2}$$

$$N_i = \frac{1}{A_i}\frac{\partial G_i}{\partial \alpha_i} - \frac{1}{A_j}\frac{\partial H_{ij}}{\partial \alpha_j} + k_j(G_i - G_j) - k_i(H_{ij} + H_{ji}). \tag{23.3}$$

Strain-displacement formulas

$$\begin{gathered}\varepsilon_i = \frac{1}{A_i}\frac{\partial u_i}{\partial \alpha_i} + k_iu_j - \frac{w}{R_i} \\ \omega = \frac{1}{A_1}\frac{\partial u_2}{\partial \alpha_1} + \frac{1}{A_2}\frac{\partial u_1}{\partial \alpha_2} - k_1u_1 - k_2u_2\end{gathered} \tag{23.4}$$

$$\gamma_i = -\frac{1}{A_i}\frac{\partial w}{\partial \alpha_i} - \frac{u_i}{R_i}, \qquad \kappa_i = -\frac{1}{A_i}\frac{\partial \gamma_i}{\partial \alpha_i} - k_i \gamma_i$$

$$\tau = -\frac{1}{A_i}\frac{\partial \gamma_j}{\partial \alpha_i} + k_i\gamma_i + \frac{1}{R_i}\left(\frac{1}{A_j}\frac{\partial u_i}{\partial \alpha_j} - k_j u_j\right). \qquad (23.5)$$

Allowing for 23.1, we write the electroelasticity relations 7.23 as

$$T_i = \frac{2h}{s_{11}^E(1-\nu^2)}(\varepsilon_i + \nu\varepsilon_j), \qquad S = S_{21} = S_{12} = \frac{2h}{s_{66}^E}\omega \qquad (23.6)$$

$$H_{ij} = \frac{4h^3}{3s_{66}^E}\left(\tau - \frac{\omega}{2R_j}\right), \qquad G_i = -\frac{2h^3 B}{3}(\kappa_i + \sigma\kappa_j). \qquad (23.7)$$

Equations 23.2 to 23.7 form a complete system that differs from the system of equations by the sense of the constant factors in the elasticity relations. Therefore, we arrive at a classification of dynamic integrals similar to that of the nonelectrical shells theory. We do not give the formulas for the electrical quantities. They are like those in Section 7 and can be used for computing the electrical quantities after the mechanical problem has been solved.

We will classify the free vibrations into

1. Quasitransverse vibrations with small variability $0 \le s < 1/2$;
2. Quasitransverse vibrations with variability $s = 1/2$;
3. Quasitransverse vibrations with great variability $1/2 < s < 1$;
4. Quasitangential vibrations with $0 < s < 1$;
5. Low frequency vibrations of the Rayleigh-type with $0 \le s < 1/2$.

In the initial equations, we extend the scales along the coordinate lines usually done in the asymptotic methods:

$$\alpha_i = \eta^s R\xi_i. \qquad (23.8)$$

We choose the asymptotics

$$\frac{w}{R} = \eta^a w_*, \qquad \frac{u_i}{R} = \eta^{s+b} u_{i*},$$

$$\frac{T_i s_{11}^E}{2h} = \eta^b T_{i*}, \qquad \frac{S s_{11}^E}{2h} = \eta^b S_*$$

$$\frac{H_{ij} s_{11}^E}{2h} = \eta^{2-2s+a} H_{ij*}, \qquad \frac{G_i s_{11}^E}{2hR} = \eta^{2-2s+a} G_{i*}$$

$$\frac{N_i s_{11}^E}{2h} = \eta^{2-3s+a} N_{i*}, \qquad \lambda = \rho\omega^2 R^2 s_{11}^E = \eta^{2r}\lambda_*. \qquad (23.9)$$

The quantities with asterisks in 23.9 are dimensionless and of the same order. The numbers a, b, and c assume different values depending on the type of vibration.

For each vibration type, the asymptotic representation of the sought-for quantities is chosen so that it has physical sense and leads to a noncontradictory system in the original approximation, with the number of equations equal to the number of unknowns. In cases where the complete dynamic system of shell theory is divided into two approximate systems, the boundary conditions of both problems should be broken so that after the problems have been solved, the boundary conditions for the complete problem are approximately satisfied. We also agree, as in the theory of nonelectrical shells, to call the approximate edge problem, whose equations are used to find the natural frequencies, the principal problem and to call the second problem the auxiliary one.

QUASITRANSVERSE VIBRATIONS WITH SMALL VARIABILITY $0 \leq s < 1/2$

The asymptotics of quasitransverse vibrations with small variability can be derived from equation 23.9, if

$$a = b = r = 0. \tag{23.10}$$

We substitute equations 23.9 and 23.10 into equations 23.2 to 23.7 and change the variables by equation 23.8. In the resultant equation, the order of each term is defined by the preceding factor η having an appropriate degree:

$$\frac{1}{A_i}\frac{\partial T_{i*}}{\partial \xi_i} + \frac{1}{A_j}\frac{\partial S_*}{\partial \xi_j} + \eta^s Rk_j\left(T_{i*} - T_{j*}\right) - \eta^{2-2s}\frac{R}{R_i}N_{i*}$$
$$+2\eta^s Rk_i S_* + \eta^{2s}\lambda_* u_{i*} = 0$$

$$T_{1*}\frac{R}{R_1} + T_{2*}\frac{R}{R_2} + \eta^{2-4s}\left(\frac{1}{A_1}\frac{\partial N_{1*}}{\partial \xi_1} + \frac{1}{A_2}\frac{\partial N_{2*}}{\partial \xi_2}\right)$$
$$+\eta^{2-3s}R\left(k_2 N_{1*} + k_1 N_{2*}\right) + \lambda_* w_* = 0. \tag{23.11}$$

$$\begin{aligned}
\varepsilon_{i*} &= \frac{1}{A_i}\frac{\partial u_{i*}}{\partial \xi_i} + \eta^s Rk_i u_{j*} - \frac{R}{R_i}w_* \\
\omega_* &= \frac{1}{A_1}\frac{\partial u_{2*}}{\partial \xi_1} + \frac{1}{A_2}\frac{\partial u_{1*}}{\partial \xi_2} - \eta^s R(k_1 u_{1*} + k_2 u_{2*})
\end{aligned} \tag{23.12}$$

$$\begin{aligned}
\gamma_{i*} &= -\frac{1}{A_i}\frac{\partial w_*}{\partial \xi_i} - \eta^{2s}\frac{R}{R_i}u_{i*}, \qquad \kappa_{i*} = -\frac{1}{A_i}\frac{\partial \gamma_{i*}}{\partial \xi_i} - \eta^s Rk_i\gamma_{j*} \\
\tau_* &= -\frac{1}{A_i}\frac{\partial \gamma_{j*}}{\partial \xi_i} + \eta^s Rk_i\gamma_{i*}
\end{aligned} \tag{23.13}$$

$$T_{i*} = \frac{1}{1-\nu^2}\left(\varepsilon_{i*} + \nu\varepsilon_{j*}\right), \qquad S_* = \frac{s_{11}^E}{s_{66}^E}\omega_*$$
$$H_{ij*} = \frac{4s_{11}^E}{3s_{66}^E}\left(\tau_* - \eta^{2s}\frac{R}{2R_j}\omega_*\right), \qquad G_{i*} = -\frac{2Bs_{11}^E}{3}(\kappa_{i*} + \sigma\kappa_{j*}). \quad (23.14)$$

In these equations, we only keep the greatest terms and obtain the approximate equations of the principal problem by writing them without asterisks:

$$\frac{1}{A_i}\frac{\partial T_i}{\partial \alpha_i} + \frac{1}{A_j}\frac{\partial S}{\partial \alpha_j} + k_j(T_i - T_j) + 2k_iS + 2h\rho\omega^2 u_i = 0$$
$$\frac{T_1}{R_1} + \frac{T_2}{R_2} + 2h\rho\omega^2 w = 0$$

$$\varepsilon_i = \frac{1}{A_i}\frac{\partial u_i}{\partial \alpha_i} + k_i u_j - \frac{w}{R_i}$$
$$\omega = \frac{1}{A_1}\frac{\partial u_2}{\partial \alpha_1} + \frac{1}{A_2}\frac{\partial u_1}{\partial \alpha_2} - k_1u_1 - k_2u_2$$
$$T_i = \frac{2h}{s_{11}^E(1-\nu^2)}(\varepsilon_i + \nu\varepsilon_j)$$
$$S = \frac{2h}{s_{66}^E}\omega. \quad (23.15)$$

When passing from equations 23.11 to 23.14 to the principal problem equations 23.15, we neglected the transverse forces in the equilibrium equations. The factor η^{2-4s} before the transverse forces shows that this introduces an error of the order $O(\eta^{2-4s})$. Also, when constructing the shell theory equations, we neglected the terms of order $O(\eta^1)$, which gives a complete error of the principal problem equations of the order $O(\eta^{2-4s} + \eta^1)$. The error of the edge problem consists of the error in the equations and the error in the boundary conditions. We showed in Section 22 that the principal problem boundary conditions error is $O(\eta^{1/2-s})$; and, since the total error is equal to the greatest of all the errors, the principal problem error ε is defined by

$$\varepsilon = O(\eta^{1/2-s}).$$

System 23.15 consists of dynamic membrane equations. In the membrane system, only two boundary conditions are met at the edge of the shell. The errors appearing due to the two unsatisfied conditions can be eliminated without changing the natural frequencies found when solving the principal problem. With this aim in view, we construct an additional stressed-strained state described by the auxiliary problem. We imply a stressed-strained state quickly varying along the direction normal to the edge. In statics and quasistatics it coincides with simple edge effect. In dynamics, it can either damp down or oscillate. The auxiliary problem equations are constructed in the same way as in statics (see Section 20), and they lead to equations 20.8 to 20.13 where we should take

$$g^4 = \frac{3(1 - \rho\omega^2 R_2^2 s_{11}^E)}{4h^2 B s_{11}^E R_2^2}.$$

The behavior of the auxiliary problem solution is greatly dependent on the value of g^4. If $g^4 > 0$, the principal problem solution decreases with the distance from the edge just like a simple edge effect in statics. If $g^4 < 0$, the auxiliary problem solution oscillates and extends through the entire shell. Generally speaking, the auxiliary problem equations can be unsuitable for the inner region. This is because, as the distance from the edge grows, the assumption of the fast variability of the sought-for quantities along the coordinate line normal to the edge may turn wrong.

As in the theory of nonelectrical shells, the boundary conditions are divided. When solving the principal problem, the tangential conditions are met and the errors in the nontangential conditions are eliminated by solving the auxiliary problem. The boundary conditions are then satisfied within the quantities of the order of $O(\eta^{1/2-s})$.

QUASITRANSVERSE VIBRATIONS WITH GREAT VARIABILITY $1 > s > 1/2$

For brevity, we will write only the asymptotics of the stressed-strained state and the asymptotic estimates of the accuracy of the approximate theories for the remaining types of vibration. The reader can check the results by substituting the asymptotic representations of the needed quantities into system equations 23.2 to 23.7.

In order to obtain the principal problem asymptotics, we set $a = b = 0$ and $2r = 2 - 4s$ in 23.9. The principal problem error is $O(\eta^{4s-2} + \eta^s + \eta^{2-2s})$.

The principal problem equations for quasitransverse vibrations with great variability have the form

$$\frac{1}{A_1}\frac{\partial N_1}{\partial \alpha_1} + \frac{1}{A_2}\frac{\partial N_2}{\partial \alpha_2} + 2h\rho\omega^2 w = 0$$

$$N_i = \frac{1}{A_i}\frac{\partial G_i}{\partial \alpha_i} - \frac{1}{A_j}\frac{\partial H}{\partial \alpha_j}$$

$$\kappa_i = \frac{1}{A_i^2}\frac{\partial^2 w}{\partial \alpha_i^2},$$

$$
\begin{aligned}
\tau &= \frac{1}{A_i A_j}\frac{\partial^2 w}{\partial\alpha_i \partial\alpha_j},\\
\gamma_i &= -\frac{1}{A_i}\frac{\partial w}{\partial\alpha_i}\\
G_i &= -\frac{2h^3 B}{3}(\kappa_i + \sigma\kappa_j),\\
H = H_{12} = H_{21} &= \frac{4h^3}{3s_{66}^E}\tau.
\end{aligned}
\tag{23.16}
$$

The principal problem equations formally coincide with the equations for bending vibrations of a plate. They should be integrated, taking into account the nontangential boundary conditions.

The auxiliary problem equations are nonhomogeneous equations of the static plane problem of elasticity theory, where w should be treated as a known quantity found from the principal problem solution:

$$
\begin{aligned}
\frac{1}{A_i}\frac{\partial T_i}{\partial\alpha_i} + \frac{1}{A_j}\frac{\partial S}{\partial\alpha_j} &= 0,\\
\frac{T_1}{R_1} + \frac{T_2}{R_2} &= 0\\
\varepsilon_i &= \frac{1}{A_i}\frac{\partial u_i}{\partial\alpha_i} - \frac{w}{R_i},\\
\omega &= \frac{1}{A_1}\frac{\partial u_2}{\partial\alpha_1} + \frac{1}{A_2}\frac{\partial u_1}{\partial\alpha_2}\\
T_i &= \frac{2h}{s_{11}^E(1-\nu^2)}(\varepsilon_i + \nu\varepsilon_j),\\
S &= \frac{2h}{s_{66}^E}\omega.
\end{aligned}
\tag{23.17}
$$

When integrating the auxiliary problem equations, the tangential boundary conditions should be met. In the approximate equations of the principal and auxiliary problems, we can neglect any function as compared to its derivative within quantities of order ε; and we can treat the quantities that describe the shell geometry as constants.

QUASITRANSVERSE VIBRATIONS WITH VARIABILITY $s = 1/2$

These vibrations are in between the membrane and the bending vibrations. The asymptotics for the quasitransverse vibrations with variability $s = 1/2$ is obtained from equation 23.9 for $a = b = r = 0$ and $s = 1/2$. The error of the approximate theory is $\varepsilon = O(\eta^{1/2})$.

In the initial approximation, the quasitransverse vibrations with variability $s = 1/2$ can be investigated using the equations that are dynamic analogues of static equations with great variability. We do not divide the problem into principal and auxiliary ones and integrate it meeting all the conditions at the shell's edge. The system of equations 23.2 to 23.9 becomes much simpler because we can neglect the functions as compared to their derivatives and treat the quantities describing the shell's geometry as constants within our approximation. We write the system for the quasitransverse vibrations with variability $s = 1/2$ in the form

$$\frac{T_1}{R_1} + \frac{T_2}{R_2} + \frac{1}{A_1}\frac{\partial N_1}{\partial \alpha_1} + \frac{1}{A_2}\frac{\partial N_2}{\partial \alpha_2} + 2h\rho\omega^2 w = 0$$

$$N_i = \frac{1}{A_i}\frac{\partial G_i}{\partial \alpha_i} - \frac{1}{A_j}\frac{\partial H}{\partial \alpha_j},$$

$$\frac{1}{A_i}\frac{\partial T_i}{\partial \alpha_i} + \frac{1}{A_j}\frac{\partial S}{\partial \alpha_j} = 0$$

$$\kappa_i = \frac{1}{A_i^2}\frac{\partial^2 w}{\partial \alpha_i^2},$$

$$\tau = \frac{1}{A_i A_j}\frac{\partial^2 w}{\partial \alpha_i \partial \alpha_j},$$

$$\gamma_i = -\frac{1}{A_\iota}\frac{\partial w}{\partial \alpha_i}$$

$$\varepsilon_i = \frac{1}{A_i}\frac{\partial u_i}{\partial \alpha_i} - \frac{w}{R_i},$$

$$\omega = \frac{1}{A_1}\frac{\partial u_2}{\partial \alpha_1} + \frac{1}{A_2}\frac{\partial u_1}{\partial \alpha_2}$$

$$G_i = -\frac{2h^3 B}{3}(\kappa_i + \sigma\kappa_j),$$

$$H = \frac{4h^3}{3s_{66}^E}\tau$$

$$T_i = \frac{2h}{s_{11}^E(1-\nu^2)}(\varepsilon_i + \nu\varepsilon_j),$$

$$S = \frac{2h}{s_{66}^E}\omega. \tag{23.18}$$

QUASITANGENTIAL VIBRATIONS

These vibrations are characterized by the relations

$$u_i \gg w, \qquad 0 < s < 1.$$

The principal problem asymptotics for quasitangential vibrations can be obtained by putting $a = 2s$, $b = 0$, and $r = -s$ in formulas 23.9. The principal error is a quantity of order ε, where

$$\varepsilon = O(\eta^{2s}).$$

The equations of quasitangential vibrations coincide with those of the plane problem and have the form

$$\begin{aligned}
&\frac{1}{A_i}\frac{\partial T_i}{\partial \alpha_i} + \frac{1}{A_j}\frac{\partial S}{\partial \alpha_j} + k_j(T_i - T_j) + 2h\rho\omega^2 u_i = 0 \\
&\varepsilon_i = \frac{1}{A_i}\frac{\partial u_i}{\partial \alpha_i} + k_i u_j, \\
&\omega = \frac{1}{A_1}\frac{\partial u_2}{\partial \alpha_1} + \frac{1}{A_2}\frac{\partial u_1}{\partial \alpha_2} - k_1 u_1 - k_2 u_2 \\
&T_i = \frac{2h}{s_{11}^E(1-\nu^2)}(\varepsilon_i + \nu\varepsilon_j), \qquad S = \frac{2h}{S_{66}^E}\omega.
\end{aligned} \tag{23.19}$$

The principal problem system 23.19 is integrated, taking into account the tangential conditions. The auxiliary problem equation consists of integrating 20.8 to 20.13 where we put

$$g^4 = -\frac{3\rho\omega^2}{4h^2 B} < 0.$$

The stressed state, described by the auxiliary problem, is oscillating. Its variability in the direction normal to the edge is $(1/2 + s/2)$ and exceeds the variability of the simple edge effect, which is 1/2.

RAYLEIGH-TYPE VIBRATIONS WITH $0 < s < 1/2$

These vibrations are realized in a shell whose edges are fixed in such a way that its middle surface can bend, say, when all the edges of the shell are free. The principal problem system of equations is an analogue of the dynamic equations of the pure moment state described in Section 11. This system looks like

$$\begin{aligned}
&\frac{1}{A_i}\frac{\partial T_i}{\partial \alpha_i} + \frac{1}{A_j}\frac{\partial S}{\partial \alpha_j} + k_j(T_i - T_j) - \frac{N_i}{R_i} + 2h\rho\omega^2 u_i = 0 \\
&\frac{T_1}{R_1} + \frac{T_2}{R_2} + \frac{1}{A_1}\frac{\partial N_1}{\partial \alpha_1} + \frac{1}{A_2}\frac{\partial N_2}{\partial \alpha_2} + k_2 N_1 + k_1 N_2 + 2h\rho\omega^2 w = 0
\end{aligned}$$

$$N_i = \frac{1}{A_i}\frac{\partial G_i}{\partial \alpha_i} - \frac{1}{A_j}\frac{\partial H_{ij}}{\partial \alpha_j} + k_j(G_i - G_j) - k_i(H_{ij} + H_{ji})$$

$$
\begin{aligned}
\varepsilon_i &= \frac{1}{A_i}\frac{\partial u_i}{\partial \alpha_i} + k_i u_j - \frac{w}{R_i},\\
\omega &= \frac{1}{A_1}\frac{\partial u_2}{\partial \alpha_1} + \frac{1}{A_2}\frac{\partial u_1}{\partial \alpha_2} - k_1 u_1 - k_2 u_2\\
\kappa_i &= -\frac{1}{A_i}\frac{\partial \gamma_i}{\partial \alpha_i} - k_i \gamma_j,\\
\gamma_i &= -\frac{1}{A_i}\frac{\partial w}{\partial \alpha_i} - \frac{u_i}{R_i}\\
\tau &= -\frac{1}{A_i}\frac{\partial \gamma_j}{\partial \alpha_i} + k_i\gamma_i + \frac{1}{R_i}\left(\frac{1}{A_j}\frac{\partial u_i}{\partial \alpha_j} - k_j u_j\right)\\
\varepsilon_i &= 0,\\
\omega &= 0\\
G_i &= -\frac{2h^3 B}{3}(\kappa_i + \sigma\kappa_j),\\
H_{ij} &= \frac{4h^3}{3s^E_{66}}\left(\tau - \frac{\omega}{2R_j}\right).
\end{aligned}
\tag{23.20}
$$

The auxiliary problem equations coincide with those for simple edge effect 20.8 to 20.13. The partition of the boundary conditions at the free edge leads to two boundary conditions for the principal problem, which are a combination of the edge forces and moments, and to nonhomogeneous nontangential conditions for the auxiliary problem as in the theory of nonelectric shells [15].

The principal problem asymptotics is described by 23.9 for $a = 0$, and $r = b = 2 - 4s$. Its error is a quantity of the order $\varepsilon = O(\eta^{1/2-s})$.

For all the vibration types, the order of the frequency parameter λ depends on its asymptotics defined by 23.9. Let us write the asymptotics for each vibration type:

1. For quasitransverse vibrations with small variability $0 \le s < 1/2$

$$\lambda \simeq O(\eta^0)$$

2. For quasitransverse vibrations with variability $s = 1/2$

$$\lambda \simeq O(\eta^0)$$

3. For quasitransverse vibrations with variability $1/2 < s < 1$

$$\lambda \simeq O(\eta^{2-4s})$$

4. For quasitangential vibrations with variability $0 < s < 1$

$$\lambda \simeq O(\eta^{-2s})$$

5. For low frequency Rayleigh-type vibrations $0 \leq s < 1/2$

$$\lambda \simeq O(\eta^{2-4s})$$

The transition from λ to ω is performed by

$$\lambda = \rho\omega^2 R^2 s_{11}^E.$$

Consider the free vibrations of a shell with disconnected electrodes. For mechanical quantities, the asymptotics 23.9 remain valid; and we supplement it with the asymptotics for the unknown potential difference on the electrodes:

$$d_{31}\frac{V}{h} = \eta^{2s+b} k_{31}^2 V_*, \qquad k_{31}^2 = \frac{d_{31}^2}{\varepsilon_{33}^T s_{11}^E}.$$

After simple transformations, the condition on the disconnected electrodes (see Section 7) and the electroelasticity relations for the forces will be written as

$$\int_\Omega (T_{1*} + T_{2*})\, d\Omega = V_*\Omega$$

$$T_{i*} = \frac{1}{1-\nu^2}(\varepsilon_{i*} + \nu\varepsilon_{j*}) + \eta^{2s+b}\frac{k_{31}^2}{1-\nu}V_*. \qquad (23.21)$$

The last term in 23.21 gives the greatest contribution $O(\eta^{2s})$ for quasitransverse vibrations with small variability: it is of the order of $O(\eta^0)$ for $s = 0$. For quasitransverse vibrations with great variability $s \geq 1/2$, it is of the order of $O(\eta^{2s})$. For quasitangential vibrations, it is of the order of $O(\eta^{2s})$, and for Rayleigh-type vibrations, it is $O(\eta^{2-2s})$.

We can see from these estimates that for our approximation of quasitransverse vibrations with small variability, the principal problem equations 23.21 should be supplemented by the formula

$$\int_\Omega (T_1 + T_2)\, d\Omega = \frac{2\varepsilon_{33}^T}{d_{31}} V$$

and the electroelasticity relation for the forces T_i should be written as

$$T_i = \frac{2h}{s_{11}^E(1-\nu^2)}(\varepsilon_i + \nu\varepsilon_j) + \frac{2d_{31}}{s_{11}^E(1-\nu)}V.$$

In other approximate theories, the quantity V can be neglected. The resultant problem completely coincides with that for a shell with short-circuited electrodes. For shells without electrodes on the faces, the given classification of vibration types remains true, but the elasticity relations 23.6 should be replaced using

$$T_i = 2hB(\varepsilon_i + \sigma\varepsilon_j)$$

by the formulas of Section 8.

For the auxiliary problem, quasitransverse vibrations with small variability, the quasitangential vibrations and the Rayleigh-type vibrations equations 20.19 and 20.20 are valid where g^4 is calculated depending on the vibration type as

$$g^4 = \frac{3(1-\sigma^2)}{4h^2R_2^2}\left(1 - \frac{\rho\omega^2R_2^2}{B(1-\sigma^2)}\right)$$

for quasitransverse vibrations with small variability,

$$g^4 = \frac{3\rho\omega^2}{4h^2B}$$

for quasitangential vibrations, and

$$g^4 = \frac{3(1-\sigma^2)}{h^2R_2^2}$$

for Rayleigh-type vibrations.

24 FREE VIBRATIONS OF SHELLS WITH TANGENTIAL POLARIZATION

Consider a shell pre-polarized along the α_2-lines and having short-circuited electrodes on the faces. Since $E_3 = 0$, the electroelasticity relations 10.6 assume the form

$$T_i = 2h(n_{ii}\varepsilon_i + n_{ij}\varepsilon_{ij}), \qquad S = \frac{2h}{s_{44}^E}\omega$$
$$G_i = \frac{2h^3}{3}(n_{ii}\kappa_i + n_{ij}\kappa_j), \qquad H_{ij} = \frac{4h^3}{3s_{44}^E}\tau. \qquad (24.1)$$

The electroelasticity relations 24.1 do not contain electrical quantities and differ from the elasticity relations 23.6 and 23.7 in constant coefficients. Therefore, the classification of vibration types given in the previous section for shells with short-circuited electrodes on the faces remains valid.

If the electrodes covering the faces are disconnected, it can be shown that, within the quantities of order $O(\eta^1)$, the problem can be reduced to a mechanical one. The potential difference on the disconnected electrodes can be computed after the mechanical problem has been solved by using the formula

$$V = \frac{3d_{15}}{4\varepsilon_{11}^T \Omega} \int_\Omega N_2 d\Omega$$

where the integration is carried out over the middle surface Ω of the shell.

Consider the vibrations of a shell that has no electrodes on its faces [84]. We write the electroelasticity relations 9.7 and electrostatics equations 9.8 (for brevity we will omit indexes (0) of the electrical quantities)

$$T_i = 2hn_{ii}(\varepsilon_i + \nu_i \varepsilon_j) - 2hc_i E_2, \qquad S = S_{ij} = \frac{2h}{s_{44}^E}(\omega - d_{15}E_1) \quad (24.2)$$

$$G_i = -\frac{2h^3 n_{ii}}{3}(\kappa_i + \nu_i \kappa_j), \qquad H = H_{ij} = \frac{2h^3}{3s_{44}^E}\tau \quad (24.3)$$

$$D_1 = \varepsilon_{11}^T E_1 + \frac{d_{15}}{2h}S, \qquad D_2 = \varepsilon_{33}^T E_2 + \frac{d_{31}}{2h}T_1 + \frac{d_{33}}{2h}T_2 \quad (24.4)$$

$$\varepsilon_{11}^T \frac{\partial}{\partial \alpha_1}\frac{A_2}{A_1}\frac{\partial \psi}{\partial \alpha_1} + \varepsilon_{33}^T \frac{\partial}{\partial \alpha_2}\frac{A_1}{A_2}\frac{\partial \psi}{\partial \alpha_2} = \frac{d_{15}}{2h}\frac{\partial}{\partial \alpha_1}A_2 S + \frac{d_{31}}{2h}\frac{\partial}{\partial \alpha_2}A_1 T_1 + \frac{d_{33}}{2h}\frac{\partial}{\partial \alpha_2}A_1 T_2 \quad (24.5)$$

$$E_i = -\frac{1}{A_i}\frac{\partial \psi}{\partial \alpha_i}, \qquad \nu_i = \frac{n_{12}}{n_{ii}}. \quad (24.6)$$

In order to obtain a complete system, we add the equilibrium equations 23.2 to 23.3 and strain-displacement formulas 23.4 to 23.5 to equations 24.2 to 24.6.

We saw in Section 9 that for shells with pre-polarization of the considered kind and no electrodes on the faces the complete two-dimensional problem cannot, generally speaking, be partitioned into electrical and mechanical parts. The system of differential equations of the electro-elastic shells theory has the tenth order. Therefore, five boundary conditions, four mechanical (like in the theory of nonelectrical shells) and one electrical should be met at each edge.

We restrict ourselves to the shells with two types of electrical conditions at the edges: (1) at the edge without electrodes in vacuum or air, the component of the electric induction vector that is normal to the edge surface should be equal to zero, (2) at the electrode-covered edge with short-circuited electrodes, the electrical potential is equal to zero.

The free vibrations of an arbitrary piezoceramic shell can be subdivided into quasitransverse, quasitangential, and low frequency Rayleigh-type vibrations. Though the principal problem and auxiliary problem equations include electrical quantities, it is expedient to stick to the terminology of the theory of nonelectrical shells.

In the initial equations, we extend the scale by 23.8 as is commonly done in asymptotic methods and replace the needed quantities by those with asterisks using the formulas

$$\begin{aligned}
\frac{u_i}{R} &= \eta^{s+b} u_{i*} \\
\frac{w}{R} &= \eta^0 w_* \\
\left(\frac{T_i}{2hn_{11}}, \frac{S}{2hn_{11}}\right) &= \eta^{c+b} (T_{i*}, S_*) \\
\left(\frac{G_i}{2hRn_{11}}, \frac{H}{2hRn_{11}}\right) &= \eta^{2-2s} (G_{i*}, H_*) \\
\frac{N_i}{2hn_{11}} &= \eta^{2-3s} N_{i*} \\
\frac{\varepsilon_{11}^T}{d_{15}n_{11}R}\psi &= \eta^{s+b+c}\psi_* \\
\frac{\rho\omega^2 R^2}{n_{11}} &= \eta^r \lambda_*.
\end{aligned} \tag{24.7}$$

The quantities with asterisks in 24.7 are dimensionless and have the same order. The numbers r, b, and c assume different values depending on the type of vibration:

1. For quasitransverse vibrations with small variability $0 \leq s < 1/2$

$$r = b = c = 0$$

2. For quasitransverse vibrations with variability $s = 1/2$

$$r = b = c = 0$$

3. For quasitransverse vibrations with great variability $1/2 < s < 1$

$$r = 2 - 4s, \qquad b = c = 0$$

4. For quasitangential vibrations with variability $0 < s < 1$

$$r = -2s, \qquad b = -2s, c = 0$$

5. For low frequency Rayleigh-type vibrations $(0 \leq s < 1/2)$

$$r = c = 2 - 4s, \qquad b = 0.$$

We substitute the asymptotics into equations 23.2 to 23.5; equations 24.2 to 24.6 retain only the principal terms; and, as in Section 23, we get the approximate equations for each vibration type. Omitting the simple but cumbersome transformations, we write the final equations.

QUASITRANSVERSE VIBRATIONS WITH SMALL VARIABILITY

The principal problem equations are

$$\frac{1}{A_i}\frac{\partial T_i}{\partial \alpha_i} + \frac{1}{A_j}\frac{\partial S}{\partial \alpha_j} + k_j(T_i - T_j) + 2k_i S + 2h\rho\omega^2 u_i = 0$$

$$\frac{T_1}{R_1} + \frac{T_2}{R_2} + 2h\rho\omega^2 w = 0$$

$$\varepsilon_i = \frac{1}{A_i}\frac{\partial u_i}{\partial \alpha_i} + k_i u_j - \frac{w}{R_i}$$

$$\omega = \frac{1}{A_1}\frac{\partial u_2}{\partial \alpha_1} + \frac{1}{A_2}\frac{\partial u_1}{\partial \alpha_2} - k_1 u_1 - k_2 u_2$$

$$T_i = 2hn_{ii}(\varepsilon_i + \nu_i \varepsilon_j) - 2hc_i E_2$$

$$S = \frac{2h}{s_{44}^E}(\omega - d_{15}E_1)$$

$$D_1 = \varepsilon_{11}^T E_1 + \frac{d_{15}}{2h} S$$

$$D_2 = \varepsilon_{33}^T E_2 + \frac{d_{31}}{2h} T_1 + \frac{d_{33}}{2h} T_2$$

$$\varepsilon_{11}^T \frac{\partial}{\partial \alpha_1}\frac{A_2}{A_1}\frac{\partial \psi}{\partial \alpha_1} + \varepsilon_{33}^T \frac{\partial}{\partial \alpha_2}\frac{A_1}{A_2}\frac{\partial \psi}{\partial \alpha_2} = \frac{d_{15}}{2h}\frac{\partial}{\partial \alpha_1} A_2 S + \frac{d_{31}}{2h}\frac{\partial}{\partial \alpha_2} A_1 T_1 + \frac{d_{33}}{2h}\frac{\partial}{\partial \alpha_2} A_1 T_2$$

$$E_i = -\frac{1}{A_i}\frac{\partial \psi}{\partial \alpha_i}. \tag{24.8}$$

This system is a dynamic analogue of the membrane theory of the electroelastic static state given in Section 17. The membrane electroelastic state near the edges should be supplemented with an electroelastic state quickly varying in the direction orthogonal to the edge, which is described by the auxiliary problem equations.

For the auxiliary problem, we use equations 21.1 to 21.4 where we take

$$g^4 = \frac{3}{4h^2 n_{11}}\left(\frac{n_{22}(1 - \nu_1\nu_2)}{R_2^2} - \rho\omega^2\right) \tag{24.9}$$

for the edge $\alpha_1 = \alpha_{10}$, and

$$g^4 = \frac{3}{4h^2 n_{22}} \left[\frac{n_{11}(1 - \nu_1 \nu_2)}{R_1^2} \frac{\varepsilon_{33}^T}{\varepsilon_{33}^T + d_{31}(\nu_2 c_2 - c_1)} - \rho\omega^2 \right] \tag{24.10}$$

for the edge $\alpha_2 = \alpha_{20}$.

For the equations of the principal boundary problem, three boundary conditions are met at each edge. The errors due to the two neglected boundary conditions can be eliminated using the integration constants of the auxiliary electroelastic state.

The principal problem and auxiliary problem boundary conditions are obtained using the scheme given in Section 22. By partitioning the complete boundary conditions, we get two tangential mechanical conditions for the principal edge problem (which coincide with the conditions of the membrane theory for nonelectrical shells) and one electrical condition. By solving the auxiliary problem, we diminish the errors in the two remaining mechanical nontangential conditions. Specifically, for a rigidly fixed edge $\alpha_1 = \alpha_{10}$ without electrodes, the boundary conditions are partitioned as

$$u_1^{(p)} = 0, \qquad u_2^{(p)} = 0, \qquad D_1^{(p)} = 0 \tag{24.11}$$

$$w^{(a)} = -w^{(p)}, \qquad \gamma_1^{(a)} = 0. \tag{24.12}$$

The superscripts (p) and (a) show that the quantity belongs to the principal or auxiliary problem, respectively. The conditions in equation 24.11 are met when solving the principal problem. The conditions in equation 24.12 are met when solving the auxiliary problem.

Analysis of the errors made in deriving the equations and boundary conditions shows that the principal problem and auxiliary problem have been constructed with an error of the order $O(\eta^{1/2-s})$.

QUASITRANSVERSE VIBRATIONS WITH A VARIABILITY OF 1/2

The approximate system of equations has the form

$$\frac{T_1}{R_1} + \frac{T_2}{R_2} + \frac{1}{A_1}\frac{\partial N_1}{\partial \alpha_1} + \frac{1}{A_2}\frac{\partial N_2}{\partial \alpha_2} + 2h\rho\omega^2 w = 0$$

$$N_i = \frac{1}{A_i}\frac{\partial G_i}{\partial \alpha_i} - \frac{1}{A_j}\frac{\partial H}{\partial \alpha_j}, \qquad \frac{1}{A_i}\frac{\partial T_i}{\partial \alpha_i} + \frac{1}{A_j}\frac{\partial S}{\partial \alpha_j} = 0$$

$$\kappa_i = \frac{1}{A_i^2}\frac{\partial^2 w}{\partial \alpha_i^2}, \qquad \tau = \frac{1}{A_i A_j}\frac{\partial^2 w}{\partial \alpha_i \partial \alpha_j},$$

$$\gamma = -\frac{1}{A_i}\frac{\partial w}{\partial \alpha_i} \qquad \varepsilon_i = \frac{1}{A_i}\frac{\partial u_i}{\partial \alpha_i} - \frac{w}{R_i},$$

$$\omega = \frac{1}{A_1}\frac{\partial u_2}{\partial \alpha_1} + \frac{1}{A_2}\frac{\partial u_1}{\partial \alpha_2} \qquad G_i = -\frac{2h^3 n_{ii}}{3}(\kappa_i + \nu_i \kappa_j),$$

$$H_{ij} = \frac{4h^3}{3s_{44}^E}\tau \qquad T_i = 2hn_{ii}(\varepsilon_i + \nu_i \varepsilon_j) - 2hc_1 E_2,$$

$$S = \frac{2h}{s_{44}^E}(\omega - d_{15}E_1) \qquad D_1 = \varepsilon_{11}^E E_1 + \tfrac{d_{15}}{2h} S,$$

$$D_2 = \varepsilon_{22}^T E_2 + \frac{d_{31}}{2h}T_1 + \frac{d_{33}}{2h}T_2$$

$$\frac{\varepsilon_{11}^T}{A_1^2}\frac{\partial^2 \psi}{\partial \alpha_1^2} + \frac{\varepsilon_{33}^T}{A_2^2}\frac{\partial^2 \psi}{\partial \alpha_2^2} = \frac{d_{15}}{2h}\frac{1}{A_1}\frac{\partial S}{\partial \alpha_1} + \frac{d_{31}}{2h}\frac{1}{A_2}\frac{\partial T_1}{\partial \alpha_2} + \frac{d_{33}}{2h}\frac{1}{A_2}\frac{\partial T_2}{\partial \alpha_2}$$

$$E_i = -\frac{1}{A_i}\frac{\partial \psi}{\partial \alpha_i}. \tag{24.13}$$

The system can be simplified if we neglect the needed function compared to its derivative and treat the quantities that characterize the shell geometry within the required approximation as constants.

The approximate system cannot be subdivided into a principal problem and an auxiliary problem, and it has the tenth order. When integrating it, we should satisfy five boundary conditions. The error of the theory is of the order $o(\eta^{1/2})$.

QUASITRANSVERSE VIBRATIONS WITH GREAT VARIABILITY $1 > s > 1/2$

The principal problem equations have the form

$$\frac{1}{A_1}\frac{\partial N_1}{\partial \alpha_1} + \frac{1}{A_2}\frac{\partial N_2}{\partial \alpha_2} + 2h\rho\omega^2 w = 0$$

$$N_i = \frac{1}{A_i}\frac{\partial G_i}{\partial \alpha_i} - \frac{1}{A_j}\frac{\partial H}{\partial \alpha_j}, \qquad H = \frac{4h^3}{3s_{44}^E}\tau$$

$$G_i = -\frac{2h^3 n_{ii}}{3}(\kappa_i + \nu_i \kappa_j) \qquad \kappa_i = -\frac{1}{A_i}\frac{\partial \gamma_i}{\partial \alpha_i},$$

$$\tau = -\frac{1}{A_j}\frac{\partial \gamma_i}{\partial \alpha_j}, \qquad \gamma_i = -\frac{1}{A_i}\frac{\partial w}{\partial \alpha_i}. \tag{24.14}$$

The principal problem system formally coincides with the equations for bending vibrations of elastic plates. The auxiliary problem equations have the form

$$\frac{1}{A_i}\frac{\partial T_i}{\partial \alpha_i} + \frac{1}{A_j}\frac{\partial S}{\partial \alpha_j} = 0$$

$$\begin{aligned}
\varepsilon_i &= \frac{1}{A_i}\frac{\partial u_i}{\partial \alpha_i} - \frac{w}{R_i},\\
\omega &= \frac{1}{A_1}\frac{\partial u_2}{\partial \alpha_1} + \frac{1}{A_2}\frac{\partial u_1}{\partial \alpha_2}\\
T_i &= 2hn_{ii}(\varepsilon_i + \nu_i\varepsilon_j) - 2hc_iE_2\\
S &= \frac{2h}{s_{44}^E}(\omega - d_{15}E_1)
\end{aligned}$$

$$\frac{\varepsilon_{11}^T}{A_1^2}\frac{\partial^2\psi}{\partial \alpha_1^2} + \frac{\varepsilon_{33}^T}{A_2^2}\frac{\partial^2\psi}{\partial \alpha_2^2} = \frac{d_{15}}{2h}\frac{1}{A_1}\frac{\partial S}{\partial \alpha_1} + \frac{1}{2h}\frac{1}{A_2}\frac{\partial}{\partial \alpha_2}(d_{31}T_1 + d_{33}T_2)$$

$$\begin{aligned}
D_1 &= \varepsilon_{11}^T E_1 + \frac{d_{15}}{2h}S\\
D_2 &= \varepsilon_{33}^T E_2 + \frac{d_{31}}{2h}T_1 + \frac{d_{33}}{2h}T_2,\\
E_i &= -\frac{1}{A_i}\frac{\partial \psi}{\partial \alpha_i}.
\end{aligned} \tag{24.15}$$

The auxiliary problem equations formally coincide with the equations of the nonhomogeneous electroelasticity plane problem.

We first integrate the principal problem system and then solve the auxiliary problem, treating w as a known function. The principal problem system has the fourth order, and the auxiliary problem system has the sixth order. When integrating the principal problem two nontangential mechanical boundary conditions should be met at each edge. When integrating the auxiliary problem, two tangential mechanical conditions and one electrical condition should be met. Specifically, for a rigidly fixed edge $\alpha_1 = \alpha_{10}$ without electrodes, the boundary conditions are partitioned as

$$u^{(p)} = 0, \qquad \gamma_1^{(p)} = 0, \qquad u_1^{(a)} = u_2^{(a)} = D_1^{(a)} = 0.$$

The resultant principal problem and auxiliary problem solutions are accurate to the quantities of the order $O(\eta^s)$, where $1/2 < s < 1$.

QUASITANGENTIAL VIBRATIONS ($0 < s < 1$)

The principal problem equations are

$$\frac{1}{A_i}\frac{\partial T_i}{\partial \alpha_i} + \frac{1}{A_j}\frac{\partial S}{\partial \alpha_j} + k_j(T_i - T_j) + 2h\rho\omega^2 u_i = 0$$

$$\varepsilon_i = \frac{1}{A_i}\frac{\partial u_i}{\partial \alpha_i} + k_i u_j$$
$$\omega = \frac{1}{A_1}\frac{\partial u_2}{\partial \alpha_1} + \frac{1}{A_2}\frac{\partial u_1}{\partial \alpha_2} - k_1 u_1 - k_2 u_2$$
$$T_i = 2hn_{ii}(\varepsilon_i + \nu_i \varepsilon_j) - 2hc_1 E_2$$
$$S = \frac{2h}{s_{44}^E}(\omega - d_{15}E_1)$$
$$D_1 = \varepsilon_{11}^T E_1 + \frac{d_{15}}{2h}S$$
$$D_2 = \varepsilon_{22}^T E_2 + \frac{d_{31}}{2h}T_1 + \frac{d_{33}}{2h}T_1$$
$$\varepsilon_{11}^T \frac{\partial}{\partial \alpha_1}\frac{A_2}{A_1}\frac{\partial \psi}{\partial \alpha_1} + \varepsilon_{33}^T \frac{\partial}{\partial \alpha_2}\frac{A_1}{A_2}\frac{\partial \psi}{\partial \alpha_2} = \frac{d_{15}}{2h}\frac{\partial}{\partial \alpha_1}A_2 S + \frac{d_{31}}{2h}\frac{\partial}{\partial \alpha_2}A_1 T_1 + \frac{d_{33}}{2h}\frac{\partial}{\partial \alpha_2}A_1 T_2$$
$$E_i = -\frac{1}{A_i}\frac{\partial \psi}{\partial \alpha_i}. \tag{24.16}$$

The principal problem system formally coincides with the dynamic equations of the plane problem of electroelasticity theory. The auxiliary problem equations describe a quickly varying electroelastic state where formulas 21.1 to 21.4 remain valid with

$$g^4 = -\frac{3\rho\omega^2}{4h^2 n_{ii}} \qquad \text{(for the edge } \alpha_1 = \alpha_{10}\text{)}. \tag{24.17}$$

When integrating the principal problem equations, three homogeneous conditions should be met at each edge. There are two tangential conditions for the mechanical quantities and one electrical condition. The errors in the nontangential conditions are allowed for when integrating the auxiliary problem.

RAYLEIGH-TYPE VIBRATIONS

These vibrations may appear in a shell whose edges are all free.

The principal problem equations include equations 23.2 to 23.5, 24.3; and

$$\varepsilon_i = 0, \qquad \omega = 0 \tag{24.18}$$

are used instead of the electroelasticity relations 24.2. The principal problem equations are a dynamic analogue of the pure moment electroelastic state.

Note that the electrical quantities are only contained in equations 24.4 to 24.6. The remaining equations in the principal problem constitute a complete system with respect to the mechanical quantities, which coincides up to the constant coefficients with the system for free nonelectrical shells. Therefore, it is expedient to integrate the principal problem equations in two stages: first, find the solution to the mechanical problem and second, integrate equation 24.5 with respect to the electrical potential ψ, treating the forces as known quantities. E_i and D_i are directly found by equations 24.6 and 24.4, knowing ψ and the forces. The auxiliary problem equations coincide with equations 21.1 to 21.4 which describe simple edge effect.

Let us partition the boundary conditions into those for the principal edge problem and those for simple edge effect. All the sought-for quantities in the boundary conditions should be represented as sums of the principal problem quantities and simple edge effect quantities in asymptotic representations. Specifically, for the edge $\alpha_1 = \alpha_{10}$ without electrodes the boundary conditions will be written as

$$\begin{aligned}
\eta^{2-4s} T_{1*}^{(p)} + \eta^{a+1/2} T_{1*}^{(a)} &= 0 \\
\eta^{2-4s} S_{*}^{(p)} + \eta^{a+1/2-s} S_{*}^{(a)} &= 0 \\
\eta^{2-4s} D_{1*}^{(p)} + \eta^{a+1/2-s} D_{1*}^{(a)} &= 0 \\
\eta^{2-2s} G_{1*}^{(p)} + \eta^{a+1} G_{1*}^{(a)} &= 0 \\
\eta^{2-3s} N_1^{\prime(p)} + \eta^{a+1/2} N_{1*}^{(a)} &= 0.
\end{aligned}$$

Here, N_1' is a reduced edge transverse force

$$N_1' = N_1 - \frac{1}{A_2} \frac{\partial H_{21}}{\partial \alpha_2}.$$

The simple edge effect quantities are found from the homogeneous equations. It explains the scale factor η^a before them, where a is chosen so that the nonhomogeneous boundary conditions should be valid for the simple edge effect. In our case, $a = 1 - 2s$. Then the boundary conditions for the simple edge effect assume the form

$$G_1^{(a)} = -G_1^{(p)}, \qquad N_1^{(a)} = -\eta^{1/2-s} N_1'^{(p)}.$$

As in [15], we solve the simple edge effect problem and express $T_1^{(a)}, S^{(a)}$, and $D_1^{(a)}$ in terms of $G_1^{(p)}$ and $N_1'^{(p)}$. For the principal boundary problem, we get two mechanical boundary conditions 24.19 coinciding with the respective conditions of the theory of nonelectrical shells and one electrical condition 24.20 or 24.21 for the edge $\alpha_i = \alpha_{i0}$:

$$T_i^{(p)} + \frac{1}{A_j} \frac{\partial}{\partial \alpha_j} \frac{1}{A_j} \frac{\partial}{\partial \alpha_j} R_j G_i^{(p)} + k_j R_j N_i^{(p)} = 0$$

$$S^{(p)} + k_j \frac{1}{A_j} \frac{\partial}{\partial \alpha_j} \left(R_j G_i^{(p)} \right) - \frac{1}{A_j} \frac{\partial}{\partial \alpha_j} \left(R_j N_i^{(p)} \right) = 0 \qquad (24.19)$$

$$\psi^{(p)} = \frac{d_{15} - d_{33}}{2h\varepsilon_{11}^T} \frac{A_1}{A_2} \frac{\partial}{\partial \alpha_2} \left(\frac{R_2}{A_1} G_1^{(p)} \right) \qquad (\alpha_1 = \alpha_{10})$$

$$\psi^{(p)} = -\frac{R_1}{2h\varepsilon_{33}^T} \left(d_{31} N_2'^{(p)} + d_{33} k_1 G_2^{(p)} \right) \qquad (\alpha_2 = \alpha_{20}) \tag{24.20}$$

$$D_1^{(p)} = \frac{1}{2h} \frac{1}{A_2} \frac{\partial}{\partial \alpha_2} \left(d_{31} k_2 R_2 G_1^{(p)} + d_{33} R_2 N_1'^{(p)} \right) \quad (\alpha_1 = \alpha_{10})$$

$$D_2^{(p)} = -\frac{1}{2h} \left(\frac{\varepsilon_{11}^T d_{31}}{\varepsilon_{33}^T} + d_{15} \right) \frac{1}{A_1} \frac{\partial}{\partial \alpha_1} \frac{A_2}{A_1} \frac{\partial}{\partial \alpha_1} \frac{R_1}{A_2} G_2^{(p)} (\alpha_2 = \alpha_{20}). \tag{24.21}$$

The conditions 24.20 hold for electrode-covered edges, while conditions 24.21 hold for edges without electrodes.

Our asymptotic analysis shows that the free vibrations of piezoceramic shells with preliminary polarization along the α_2-lines and faces without electrodes may be subdivided into

1. Quasitransverse vibrations with small variability $0 \le s < 1/2$
2. Quasitransverse vibrations with great variability $1/2 < s < 1$
3. Quasitransverse vibrations with variability $s = 1/2$
4. Quasitangential vibrations with $0 < s < 1$
5. Superlow frequency vibrations of the Rayleigh type with $0 \le s < 1/2$.

Every vibration type is described by an appropriate system of equations. This classification is physically clear and simplifies the computation of natural frequencies and other quantities.

Note that, though the classification of free vibrations is like that in the theory of nonelectrical shells, the systems of the principal and auxiliary problems qualitatively differ in the greater order of the equations and greater number of the initial quantities and boundary conditions. Therefore, this classification of the free vibrations of piezoceramic shells should be considered as a generalization of the classification of the free vibrations of nonelectrical shell theory.

25 THE REFINED MEMBRANE DYNAMIC THEORY

When designing actual piezoceramic elements, the first natural frequencies attract the most attention. The principal problem equations for quasitransverse vibrations with small variability have been obtained up to the quantities $O(\eta^{2-4s} + \eta^1)$; the error of the principal problem boundary conditions is $O(\eta^{1/2-s})$; and the total error, which is equal to the largest of all the errors, has the order of $O(\eta^{1/2-s})$.

For the problem under consideration, the accuracy of the boundary conditions can be improved by introducing the correction $O(\eta^{1/2-s})$ into the principal problem boundary conditions as a result of discarding the auxiliary problem terms in the tangential boundary conditions. At the same time, since the auxiliary problem equations are integrated in quadratures and all the required auxiliary problem quantities are expressed through the edge values of the principal problem tangential quantities, there is no necessity to discard the auxiliary problem quantities in the tangential conditions. By keeping them in the principal problem boundary conditions, we will construct the principal problem quantities, including the natural frequencies, up to the quantities $O(\eta^{1-2s})$. The principal problem solution can be obtained by an iterative process as is done in [88].

Let us consider a shell with thickness polarization and electrode-covered faces. We assume that the rigidly fixed edges of the shell coincide with the lines $\alpha_1 = \alpha_{10}$ and $\alpha_1 = \alpha_{11}$. We choose the vibration frequency of the shell ω so that in the auxiliary problem resultant equation $4g^4 > 0$.

Within the quantities $O(\eta^{1-2s})$, the boundary conditions at the edge $\alpha_1 = \alpha_{10}$ are written as

$$u_1^{(p)} + u_1^{(a)} = 0, \qquad u_2^{(p)} = 0. \tag{25.1}$$

Within the quantities $O(\eta^{1/2-s})$, the auxiliary problem boundary conditions are written as

$$w^{(a)} = -w^{(p)}, \qquad \gamma_1^{(a)} = 0. \tag{25.2}$$

The solution 20.15 near the edge $\alpha_1 = \alpha_{10}$ has the form ($\alpha_1 \leq \alpha_{10}$):

$$w^{(a)} = [F_1 \cos A_1 g(\alpha_1 - \alpha_{10}) + F_2 \sin A_1 g(\alpha_1 - \alpha_{10})]e^{A_1 g(\alpha_1 - \alpha_{10})}.$$

Meeting the conditions 25.2, we find

$$F_1 = -F_2 = -w^{(p)}\Big|_{\alpha_1=\alpha_{10}} \qquad u_1^{(a)}\Big|_{\alpha_1=\alpha_{10}} = -\left(\frac{1}{R_1} + \frac{1}{R_2}\right)\frac{w^{(p)}}{2g}\Big|_{\alpha_1=\alpha_{10}}$$

By substituting the obtained value of $u_1^{(a)}$ into 25.1, we get the refined boundary conditions for the principal problem:

$$u_1^{(p)} - \left(\frac{1}{R_1} + \frac{1}{R_2}\right)\frac{1}{2g}w^{(p)} = 0, \qquad u_2^{(p)} = 0 \qquad (\alpha_1 = \alpha_{10}).$$

Thus, by adding the refined boundary conditions to the membrane equations we diminish the error of the principal electroelastic state to the quantities $O(\eta^{1-2s})$.

Other boundary conditions can be refined in a similar way. For example, the refined conditions at a rigidly fixed edge $\alpha_i = \alpha_{i0}$ of a shell polarized along the α_2-lines and having no electrodes on the faces have the form

$$u_i^{(p)} - \left(\frac{1}{R_i} + \frac{\nu_i + b_i}{R_j}\right)\frac{1}{g_i}w^{(p)} = 0, \qquad u_j^{(p)} = 0 \tag{25.3}$$

$$\psi^{(p)}\Big|_{\alpha_i=\alpha_{i0}} = V \tag{25.4}$$

$$D_1^{(p)} - \frac{2h^2}{3} n_{11} d_{33}^T \frac{1}{A_2} \frac{\partial}{\partial \alpha_2} [R_2 g_1^3 w^{(p)}] = 0, \qquad (\alpha_1 = \alpha_{10}) \tag{25.5}$$

$$D_2^{(p)} + \frac{2h^2}{3} n_{22} d_{33} k_1 R_1 g_2^3 w^{(p)} = 0 \qquad (\alpha_2 = \alpha_{20}). \tag{25.6}$$

Here

$$4g_i^4 = \frac{3}{h^2 n_{ii}} \left(\frac{n_{jj} a_i}{R_j^2} - \rho\omega^2 \right) \qquad n_{11} = s_{33}^E/\delta,$$
$$n_{22} = s_{11}^E/\delta, \qquad n_{ij} = s_{13}^E/\delta,$$
$$\delta = s_{11}^E s_{33}^E - (s_{13}^E)^2 \qquad \nu_i = n_{12}/n_{ii},$$
$$b_1 = 0, \qquad b_2 = c_2 d_{31} n_{11} a_2/(n_{22}\varepsilon_{33}^T)$$
$$a_1 = 1 - \nu_1\nu_2, \qquad a_2 = a_1\varepsilon_{33}^T/[\varepsilon_{33}^T + d_{31}(\nu_2 c_2 - c_1)]$$
$$c_1 = (d_{31}s_{33}^E - d_{33}s_{13}^E)/\delta, \qquad c_2 = (d_{33}s_{11}^E - d_{31}s_{13}^E)/\delta.$$

The condition in 25.4 should hold at the electrode-covered edge. Those in 25.5 and 25.6 should hold at the edges without electrodes, $\alpha_1 = \alpha_{10}$ and $\alpha_2 = \alpha_{20}$, respectively.

26 FORCED VIBRATIONS OF SHELLS WITH THICKNESS POLARIZATION

Consider the vibrations of a piezoceramic shell with electrode-covered faces to which we apply a potential difference $2Ve^{-i\omega t}$ harmonically varying with time t.

The electroelasticity relations for the shell are given in Section 7 and have the form 7.23 and 7.24. The normal component E_3 of the electric field strength is

$$E_3 = -\frac{V}{h}$$

where E_3 and V do not depend on the coordinates.

We assume that the geometry and fixation of the shell are such that a membrane electroelastic state appears at a distance from the edges if all the other conditions for the applicability of the membrane theory for a static load are fulfilled. For definiteness, we assume that the shell has two edges passing along the α_2-lines. We do not consider the frequencies of the forcing load that coincide with the natural frequencies of the vibrations of the piezoceramic shell.

We extend the scale along the coordinate lines:

$$\alpha_1 = \eta^s R\xi_1, \qquad \alpha_2 = \eta^\theta R\xi_2 \tag{26.1}$$

and introduce a dimensionless frequency parameter

$$\lambda = \rho\omega^2 R^2 s_{11}^E = \eta^{2r}\lambda_*. \tag{26.2}$$

Let us investigate the forced vibrations depending on the parameter r. Let $r \geq 0$. In this case, the internal stressed state is composed of the slowly varying stressed state (principal problem) and the quickly varying stressed state (auxiliary problem). Since the electrical load is independent of the coordinates, we choose $s = 0$ and $\theta = 0$ in the principal problem and change the variables:

$$d_{31}E_3 = E_{3*}, \qquad \frac{w}{R} = w_*, \qquad \frac{u_i}{R} = u_{i*}$$
$$\left(T_i, S_{ij}\right)\frac{s_{11}^E}{2h} = \left(T_{i*}, S_{ij*}\right),$$
$$\left(G_i, H_{ij}, N_i R\right)\frac{s_{11}^E}{2hR} = \eta^2\left(G_{i*}, H_{ij*}, N_{i*}\right). \tag{26.3}$$

We substitute the asymptotics 26.3 and the formulas for extending of the scale with respect to the coordinates 26.1 and time 26.2 in the system of the piezoceramic shell theory with thickness polarization. By discarding the small terms in the resultant system, we arrive at the dynamic membrane equations. If $r > 0$, we may disregard the inertial terms within the quantities of the order $O(\eta^2 + \eta^{2r})$ and obtain the static membrane equations.

The auxiliary problem equations are given in Sections 20 and 23. The complete boundary conditions are subdivided into the boundary conditions for the principal problem and for the auxiliary problem (like in the case of free vibrations).

Consider the high frequency vibrations with $r < 0$. We represent the needed quantities (displacements, forces, and moments) as sums of three terms:

$$P = P^{(r)} + \eta^a P^{(tr)} + \eta^b P^{(ta)}. \tag{26.4}$$

The superscript (r) shows that the quantity is found as a particular solution of the nonhomogeneous equations. The superscripts (tr) and (ta) indicate that the quantity is found from the homogeneous equations of quasitransverse vibrations with great variability and quasitangential vibrations, respectively. The factors η^a and η^b are chosen depending on the boundary conditions. The first factor defines the intensity of the quasitransverse vibrations, and the second factor defines the intensity of the quasitangential vibrations.

When constructing the homogeneous equations, we will take $\theta = 0$ in 26.1. For the quasitransverse vibrations we have the asymptotics

$$\frac{T_1 s_{11}^E}{2h} = \left(\eta^{s-2c} + \eta^{2-2s}\right) T_{1*}, \qquad \frac{S_{ij} s_{11}^E}{2h} = \eta^{s-2c} S_{ij*},$$
$$\frac{T_2 s_{11}^E}{2h} = \eta^{-c} T_{2*} \qquad \frac{w}{R} = \eta^{c} w_*,$$
$$\frac{u_1}{R} = \eta^{s-c} u_{1*}, \qquad \frac{u_2}{R} = \eta^{2s-2c} u_{2*}$$
$$R\kappa_1 = \eta^{-2s+c}\kappa_{1*}, \qquad R\kappa_2 = \eta^{-s+c}\kappa_{2*},$$
$$R\tau = \eta^{-s}\tau_* \qquad \frac{G_i s_{11}^E}{2hR} = \eta^{2-2s+c} G_{i*},$$
$$\frac{H_{ij} s_{11}^E}{2hR} = \eta^{2-s} H_{ij*}, \qquad \frac{N_1 s_{11}^E}{2h} = \eta^{2-3s+c} N_{1*}$$
$$\frac{N_2 s_{11}^E}{2h} = \eta^{2-2s+c} N_{2*}. \tag{26.5}$$

The number c is assumed to be zero. We will need it in what follows. Using the condition for the system in the initial approximation to be noncontradictory, we get the relation between s and r: $2s = 1 - r$.

Allowing for the asymptotics 26.5, we get the final equation for the quasitransverse vibrations with great variability:

$$\frac{1}{A_1^4}\frac{\partial^4 w}{\partial \alpha_1^4} - 4g^4 w = 0, \qquad 4g^4 = \frac{3\rho\omega^2}{h^2 B}. \tag{26.6}$$

Its solution has the form

$$w = c_1 e^{\lambda_0 A_1 \alpha_1} + c_2 e^{-\lambda_0 A_1 \alpha_1} + c_3 \cos\lambda_0 A_1 \alpha_1 + c_4 \sin\lambda_0 A_1 \alpha_1 \tag{26.7}$$

where

$$\lambda_0^4 = 4g^4.$$

For the rest of the quantities the formulas 20.12 to 20.13 remain valid with the exception of the relation for T_1, which assumes the form

$$\frac{1}{A_1}\frac{\partial T_1}{\partial \alpha_1} = k_2 T_2 + \frac{1}{R_1} N_1 - 2h\rho\omega^2 u_1.$$

For the quasitangential vibrations, we choose the asymptotics 26.5, change the variables by the formulas 26.1, and put $c = s$, $s = r$, and $\theta = 0$.

By the scheme employed in the asymptotic methods, we get the approximate equations

$$\frac{1}{A_1^2}\frac{\partial^2 u_1}{\partial \alpha_1^2} + \lambda_1^2 u_1 = 0, \qquad \lambda_1^2 = (1-\nu^2)\rho\omega^2 s_{11}^E$$
$$\frac{1}{A_2^2}\frac{\partial^2 u_2}{\partial \alpha_2^2} + \lambda_2^2 u_2 = 0, \qquad \lambda_2^2 = \rho\omega^2 s_{66}^E,$$

whose solution is in the form

$$u_1 = c_5 \cos \lambda_1 A_1 \alpha_1 + c_6 \sin \lambda_1 A_1 \alpha_1$$
$$u_2 = c_7 \cos \lambda_2 A_1 \alpha_1 + c_8 \sin \lambda_2 A_1 \alpha_1.$$

The greatest forces are found as

$$T_1 = \frac{2h}{s_{11}^E(1-\nu^2)}\frac{1}{A_1}\frac{\partial u_1}{\partial \alpha_1}, \qquad T_2 = \nu T_1, \qquad S_{ij} = \frac{2h}{s_{66}^E}\frac{1}{A_1}\frac{\partial u_2}{\partial \alpha_1}.$$

We find a particular integral. The asymptotics of the particular solution is ($s = \theta = 0$):

$$\frac{w}{r} = \eta^{-2r} w_*, \qquad \frac{u_i}{R} = \eta^{-4r} u_{i*}, \qquad d_{31}E_3 = E_{3*}$$
$$\frac{T_i s_{11}^E}{2h} = T_{i*}, \left(\frac{G_i}{R}, \frac{H_{ij}}{R}, N_i\right)\frac{s_{11}^E}{2h} = \eta^{2-2r}\left(G_{i*}, H_{ij*}, N_{i*}\right). \qquad (26.8)$$

The greatest quantities in equations 26.8 are the normal forces and the deflection. In the roughest approximation, they are defined as

$$T_i = -\frac{2hd_{31}}{s_{11}^E(1-\nu)}E_3, \qquad w = \frac{d_{31}}{\omega^2\rho s_{11}^E(1-\nu)}\left(\frac{1}{R_1} + \frac{1}{R_2}\right)E_3.$$

We break the boundary conditions into those for the quasitransverse vibrations and those for the quasitangential vibrations. As in the theory of nonelectrical shells, we get for the rigidly fixed edge $\alpha_1 = \alpha_{10}$

$$(a = -2r, \quad b = -2r + s)$$
$$w^{(tr)} = -w^{(r)}, \qquad \gamma_1^{(tr)} = 0, \qquad u_1^{(ta)} = -u_1^{(tr)}, \qquad u_2^{(ta)} = 0.$$

For the free edge

$$(a = \min(s, 2-2s), \quad b = a + 2s)$$
$$G_1^{(tr)} = 0, \qquad N_1^{(tr)} = 0, \qquad T_1^{(ta)} = -T_1^{(r)} - T_1^{(tr)}, \qquad S_{ij}^{(ta)} = -S_{ij}^{(tr)}.$$

For the hinge-supported edge

$$(a = -2r, \quad b = -r)$$
$$w^{(tr)} = -w^{(r)}, \qquad G_1^{(tr)} = 0, \qquad T_1^{(ta)} = -T_1^{(r)}, \qquad u_2^{(ta)} = 0.$$

Thus, we have reduced the solutions to the quasistatic and dynamic problems to finding the membrane integrals meeting the tangential boundary conditions. The errors arising in the nontangential conditions are eliminated using the arbitrary integration functions of the auxiliary problem. The auxiliary problem in the quasistatic case coincides with the simple edge effect. The solution is complete for very high frequency vibrations.

Chapter 5

SOME DYNAMIC PROBLEMS IN THE THEORY OF PIEZOCERAMIC PLATES AND SHELLS

27 FORCED VIBRATIONS OF A CIRCULAR CYLINDRICAL SHELL WITH LONGITUDINAL POLARIZATION (AXISYMMETRIC PROBLEM)

We choose a system of orthogonal curvilinear dimensionless coordinates ξ, and φ. The ξ-line coincides with the generatrix, and the φ-line coincides with the directrix of the cylinder [87].

The equations for the axially-symmetric problem are the

Equilibrium equations

$$\frac{dT_{1*}}{d\xi} + \lambda u_* = 0, \qquad T_{2*} + \varepsilon^2 N_{1*} + \lambda w_* = 0$$

$$N_{1*} = \frac{\partial G_{1*}}{\partial \xi} \tag{27.1}$$

Electroelasticity relations (see Section 9)

$$\begin{aligned} T_{1*} &= \varepsilon_{1*} + \nu_2 \varepsilon_{2*} + \frac{d\psi_*}{d\xi}, \\ T_{2*} &= \alpha(\varepsilon_{2*} + \nu_1 \varepsilon_{1*}) + c_{12}\frac{d\psi_*}{d\xi}, \\ G_{1*} &= -\varepsilon^2 \kappa_{1*} \\ D_{1*} &= \frac{\varepsilon_{33}^T}{c_2 d_{31}} E_{1*} + T_{2*} + \frac{d_{33}}{d_{31}} T_{1*} \\ E_{1*} &= -\frac{d\psi_*}{d\xi} \end{aligned} \tag{27.2}$$

Electrostatics equation

$$\frac{dD_{1*}}{d\xi} = 0 \tag{27.3}$$

Strain-displacement formulas

$$\varepsilon_{1*} = \frac{du_*}{d\xi}, \qquad \varepsilon_{2*} = -qw_*, \qquad \kappa_{1*} = \frac{d^2 w_*}{d\xi^2} \tag{27.4}$$

To make the computation more convenient, we have used the dimensionless quantities with asterisks:

$$u_* = \frac{u}{r}, \qquad w_* = \frac{w}{r},$$
$$\frac{T_1}{2hn_{22}} = T_{1*}, \qquad \frac{T_2}{2hn_{22}} = T_{2*}$$
$$\psi_* = \frac{c_2}{n_{22}}\psi, \qquad E_{1*} = \frac{c_2}{n_{22}}E_1 = -\frac{d\psi_*}{d\xi}$$
$$D_{1*} = \frac{D_1}{d_{31}n_{22}},$$
$$\frac{N_1}{2hn_{22}} = \varepsilon^2 N_{1*}, \qquad \frac{G_1}{2hRn_{22}} = \varepsilon^2 G_{1*}$$
$$\lambda = \frac{\rho\omega^2 r^2}{n_{22}}, \qquad \varepsilon^2 = \frac{h^2}{3r^2}$$
$$\nu_1 = \frac{n_{12}}{n_{11}}, \qquad \nu_2 = \frac{n_{12}}{n_{22}},$$
$$\alpha = \frac{n_{11}}{n_{22}}, \qquad c_{12} = \frac{c_1}{c_2}. \tag{27.5}$$

The radius of the cylinder is r, and the number q is equal to one or to zero depending on the value of the angular frequency ω of the electrical load.

We will assume that the shell edges are covered with electrodes, and their electrical potentials are given. (Here, and below, we take into account that the needed quantities vary as $e^{-i\omega t}$ and write all the equations for the amplitude values)

$$\psi_*\big|_{\xi=\pm l/r} = \pm 1 \tag{27.6}$$

where $2l$ is the length of the shell.

System 27.1 to 27.4 can be transformed into three equations with respect to the unknowns

$$\left[\frac{d^2}{d\xi^2} + \lambda\right] u_* - q\nu_2\frac{dw_*}{d\xi} + \frac{d^2\psi_*}{d\xi^2} = 0$$
$$\alpha\nu_1\frac{du_*}{d\xi} - \left[\varepsilon^2\frac{d^4}{d\xi^4} - (\lambda - q\alpha)\right] w_* + c_{12}\frac{d\psi_*}{d\xi} = 0$$
$$a_1\frac{d^2u_*}{d\xi^2} - a_2 w_* + a_3\frac{d^2\psi_*}{d\xi^2} = 0 \tag{27.7}$$

where

$$a_1 = \alpha\nu + \frac{d_{33}}{d_{31}}$$

$$a_2 = \left[\alpha + \nu_2 \frac{d_{33}}{d_{31}}\right]$$

$$a_3 = -\frac{\varepsilon_{33}^T}{d_{31}c_2} + c_{12} + \frac{d_{33}}{d_{31}}.$$

We introduce a new unknown function Φ defined by

$$D\Phi = 0,$$

$$D = \begin{vmatrix} \frac{d^2}{d\xi^2} + \lambda & -q\nu_2 \frac{d}{d\xi} & \frac{d^2}{d\xi^2} \\ \alpha\nu_1 \frac{d}{d\xi} & -\left[\varepsilon^2 \frac{d^4}{d\xi^4} - (\lambda - \alpha q)\right] & c_{12}\frac{d}{d\xi} \\ a_1 \frac{d^2}{d\xi^2} & -a_2 q \frac{d}{d\xi} & a_3 \frac{d^2}{d\xi^2} \end{vmatrix}. \qquad (27.8)$$

The quantities u_*, w_*, and ψ_* can be expressed in terms of Φ:

$$u_* = D_1\Phi, \qquad w_* = D_2\Phi, \qquad \psi_* = D_3\Phi$$

$$D_1 = \begin{vmatrix} -q\nu_2 & 1 \\ -\left[\varepsilon^2 \frac{d^4}{d\xi^4} - (\lambda - \alpha q)\right] & c_{12} \end{vmatrix} \frac{d^2}{d\xi^2}$$

$$D_2 = \begin{vmatrix} \frac{d^2}{d\xi^2} + \lambda & \frac{d^2}{d\xi^2} \\ \alpha\nu_1 \frac{d}{d\xi} & c_{12}\frac{d}{d\xi} \end{vmatrix}$$

$$D_3 = \begin{vmatrix} \frac{d^2}{d\xi^2} + \lambda & -q\nu_2 \frac{d}{d\xi} \\ \alpha\nu_1 \frac{d}{d\xi} & -\left[\varepsilon^2 \frac{d^4}{d\xi^4} - (\lambda - q\alpha)\right] \end{vmatrix}. \qquad (27.9)$$

We assume that the shell is subjected either to an edge mechanical axisymmetrical load acting in the longitudinal direction or to an electrical load 27.6 varying by the law $e^{-i\omega t}$.

The electroelastic state of the shell will be studied as a function of the frequency parameter λ. An asymptotic analysis of the equations for the forced vibrations allows us to essentially simplify equations 27.8 to 27.9. We break the range of the frequency parameter λ into four intervals: (1) $0 \leq \lambda \ll \lambda_0$, (2) $|\lambda - \lambda_0| \leq \varepsilon$, (3) $\lambda_0 \ll \lambda < \lambda_1$, and (4) $\lambda > \lambda_1$. The values of λ_0 and λ_1 will be found later.

In the first interval, the complete problem is divided into the membrane principal problem and simple edge effect. To get the principal problem equations, we put $q = 1$ and $\varepsilon = 0$ in equations 27.8 and 27.9. By equating ε to zero, we discard the terms with moments.

In the second interval, the complete problem cannot be broken into simpler problems, because for $|\lambda - \lambda_0| \leq \varepsilon$ the principal problem solution stops to be quickly varying in the direction of the generatrix. The number λ_0 is found from the condition $g^4 = 0$ (24.10) in the auxiliary problem equation given in

Section 21,

$$\lambda_0 = \alpha \frac{(1 - \nu_1 \nu_2)\varepsilon_{33}^T}{\varepsilon_{33}^T + d_{31}(\nu_2 c_2 - c_1)}.$$

In this case, we should use the shell theory (equations 27.8 and 27.9) for $q = 1$.

In the third interval, the complete problem is divided into the membrane principal problem and the auxiliary problem. The auxiliary problem solution will oscillate without damping down because for $\lambda > \lambda_0$, the quantity g^4 (formula 24.17) is negative. In order to get the membrane principal problem equations, we put $q = 1$ and $\varepsilon = 0$ in equations 27.8 and 27.9.

In the fourth interval, the equations are divided into the principal problem describing the tangential vibrations and the auxiliary problem. The auxiliary problem solution oscillates with variability index $(1 + s)/2$, where s is the variability index to the principal problem solution whose equations are obtained from equations 27.8 and 27.9 for $q = 0$ and $\varepsilon = 0$. By equating these quantities to zero, we discard w compared to u.

We will perform the computation using the approximate equations constructed as described above. First, this partition of the electroelastic state gives a qualitative picture of the shell vibrations; and second, it essentially decreases the computation time.

When the complete problem can be broken into the principal problem and auxiliary problem, we will restrict ourselves to solving the principal problem because the greatest quantities (stresses, displacements, electrical potential) of the electroelastic state described near the edge by the auxiliary problem equations are r/h times less than the respective greatest values of the principal problem.

A very important characteristic of the performance of piezoceramic elements is the electromechanical coupling coefficient (EMCC). It gives the ratio of the electrical (mechanical) energy stored in the volume of a piezoceramic body and capable of conversion to the total mechanical (electrical) energy supplied to the body [26].

There are a few methods for calculating the EMCC. For example, Berlincourt et al. [8] suggested defining the EMCC as a ratio of the mutual elastic and electrical energy U_m to the geometric mean of the elastic U_e and electrical U_d energy densities:

$$k = \frac{U_m}{\sqrt{U_e U_d}}. \tag{27.10}$$

For our problem, the U_m, U_e, and U_d are computed as

$$U_m = \int_s (d_{31} T_2 E_1 + d_{33} T_1 E_1 + d_{15} S_{12} E_2) ds$$

$$U_e = \int_s (s_{11}^E T_2^2 + 2s_{12}^E T_1 T_2 + s_{33}^E T_1^2 + s_{44}^E S_{12}^2) ds$$

$$U_d = \int_s (\varepsilon_{11}^T E_2^2 + \varepsilon_{33}^T E_1^2) ds. \tag{27.11}$$

We integrate over the middle surface of the shell. Formula 27.10 in [8] is three-dimensional. We obtain 27.11 by only integrating with respect to the thickness coordinate, discarding the small quantities of shell theory, and passing to the notation of shell theory. Since we are dealing with an axisymmetrical problem, let us put $S_{12} = 0$ and $E_2 = 0$. We will need these quantities to solve the nonaxially-symmetric problem.

The dynamic EMCC is introduced as

$$k_d = \frac{\sqrt{\omega_a^2 - \omega_r^2}}{\omega_a} \tag{27.12}$$

where ω_r is the resonance frequency and ω_a is the antiresonance frequency. Another energy formula for the EMCC was suggested by W.J. Toulis [107], but R.S. Wollett [113, 114] showed that it was erroneous. Using some reasonable approaches of W.J Toulis [107], V.T. Grinchenko et al. [26] gave their own version of the formula for computing the EMCC:

$$k_e^2 = \frac{U^{(d)} - U^{(sh)}}{U^{(d)}} \tag{27.13}$$

where $U^{(d)}$ is the internal energy of the body when the electrodes are disconnected and $U^{(sh)}$ is the internal energy for short-circuited electrodes. In order to find $U^{(d)}$ and $U^{(sh)}$, we should solve equation 27.3, assuming that the strains have been found from the solution of the initial problem. The arbitrary integration functions for $U^{(d)}$ are found from the conditions

$$\int_{S_e} D_1^{(d)} dS_e = 0 \tag{27.14}$$

where S_e is the surface of one of the electrodes. For $U^{(sh)}$ we use the conditions

$$\psi^{(sh)}|_{\xi=\pm l/r} = 0. \tag{27.15}$$

Suppose that the initial problem with conditions 27.6 has been solved. In order to determine $U^{(d)}$ and $U^{(sh)}$, we should solve two more problems.

PROBLEM 1

Let us find $U^{(d)}$. The electrodes at the edge of the cylinder are disconnected and condition 27.14 holds for each electrode. Since the problem is axisymmetric, 27.14 is equivalent to

$$D_{1*}^{(d)}|_{\xi=\pm 1} = 0. \tag{27.16}$$

We integrate equation 27.3, taking into account 27.2 and get

$$D_{1*}^{(d)} = \frac{\varepsilon_{33}^T}{c_2 d_{31}} E_{1*}^{(d)} + T_{2*}^{(d)} + \frac{d_{33}}{d_{31}} T_{1*}^{(d)} = 0 \qquad (27.17)$$

$$T_{1*}^{(d)} = \alpha(\varepsilon_{2*} + \nu_1 \varepsilon_{1*}) - c_{12} E_{1*}^{(d)}$$

$$T_{2*}^{(d)} = \varepsilon_{1*} + \nu_2 \varepsilon_{2*} - E_{1*}^{(d)}. \qquad (27.18)$$

The quantities ε_{1*} and ε_{2*} in 27.18 are known as the solutions of the initial problem. We substitute 27.18 into equation 27.17 and solve the latter with respect to $E_1^{(d)}$ to obtain

$$E_{1*}^{(d)} = \frac{1}{a_3}(a_1 \varepsilon_{1*} + a_2 \varepsilon_{2*}).$$

We compute $U_*^{(d)}$ by the formula

$$U_*^{(d)} = \int_{S_e} (\varepsilon_{1*} T_{1*}^{(d)} + \varepsilon_{2*} T_{2*}^{(d)}) dS_e. \qquad (27.19)$$

PROBLEM 2

We find $U^{(sh)}$. Condition 27.15 holds on the electrodes. Allowing for the formulas

$$T_{1*}^{(sh)} = \alpha(\varepsilon_{2*} + \nu_1 \varepsilon_{1*}) - c_{12} E_{1*}^{(sh)}$$

$$T_{2*}^{(sh)} = \varepsilon_{1*} + \nu_2 \varepsilon_{2*} - E_{1*}^{(sh)}, \qquad E_{1*}^{(sh)} = -\frac{\partial \psi_*^{(sh)}}{\partial \xi} \qquad (27.20)$$

we write equation 27.3 in the form

$$\frac{d^2 \psi_*^{(sh)}}{d\xi^2} = -\frac{1}{a_3}(a_1 \varepsilon_{1*} + a_2 \varepsilon_{2*}).$$

We solve this equation meeting the conditions 27.15 and find $E_{1*}^{(sh)}, D_{1*}^{(sh)}, T_{1*}^{(sh)}$, and $T_{2*}^{(sh)}$. Then we calculate $U_*^{(sh)}$ as

$$U_*^{(sh)} = \int_{Se} \left(\varepsilon_{1*} T_{1*}^{(sh)} + \varepsilon_{2*} T_{2*}^{(sh)} + \frac{n_{22} d_{31}}{c_2} E_{1*}^{(sh)} D_{1*}^{(sh)} \right) dS.$$

The computation was performed for a shell made of PZT-4 of length $2l = 2R, h/r = 0.025$. Figure 5 gives the amplitude values of forces and displacements near (a) the first resonance for a shell and (b) the second resonance as functions of the longitudinal coordinate in the interval $[-l, +l]$ (the edges of the shell are not clamped). Each dimensionless quantity in Figure 5 is divided by

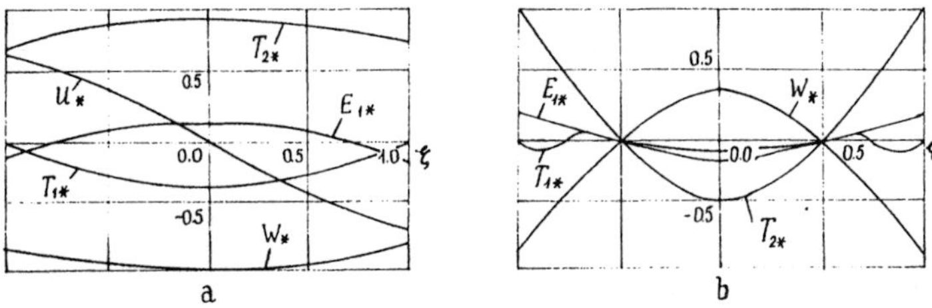

FIGURE 5. Distribution of dimensionless forces, displacements, electric strength components along the generatrix of a cylindrical shell near (a) the first and (b) the second resonances

the same number that corresponds to the greatest of all the values they assume. Figure 6 shows different electro-mechanical coupling coefficients as functions of the frequency parameter λ. The dashed line is for the values of k computed by 27.10; the solid line gives k_e computed by 27.13; and the circles are used for k_d computed by 27.12. Since k_d gives the EMCC values in the interval $[\lambda^{(r)}, \lambda^{(a)}]$, the circle in Figure 6 is situated in the middle of this interval. We see from the figure that the k_d-values lie on the curve k_e. The computation using formula 27.10 gives qualitatively different results. We know that k_d is a reliable and well-checked characteristic widely used in applications; therefore, the obtained coincidence confirms 27.13 for computing k_e. Figure 7 gives the current versus $I_*(I_* = I/(\varepsilon_{11}^T Vr\omega))$ the frequency parameter. We use the following formula for amplitude value of current [8]:

$$I = -i\omega \int_{Se} D_1 dS. \qquad (27.21)$$

Note that the number k_d is computed from the current values at resonance and antiresonance frequencies corresponding to the infinite and zero current values, respectively.

The computational results for the first three natural frequencies depending on the shell length and the respective EMCC values found by different formulas are given in Table 2. The superscript (r) or (a) shows that the value was obtained for the resonance or antiresonance frequency, respectively. We see from the table that k_d is the averaged EMCC for the whole range of the frequency parameter λ from the resonance to the antiresonance. Formula 27.13 allows us to determine the EMCC's for static and dynamic cases for each value of λ.

Thus, for computing the forced vibrations, we have broken the range of the frequency parameter λ into four intervals. For λ lying in the first and third intervals, the computation is done using the membrane theory (the shell performs quasitransverse vibrations with small variability). For λ lying in the second interval, we use the moment theory (the shell performs quasitransverse vibrations with greater variability). For λ from the fourth interval, we use the

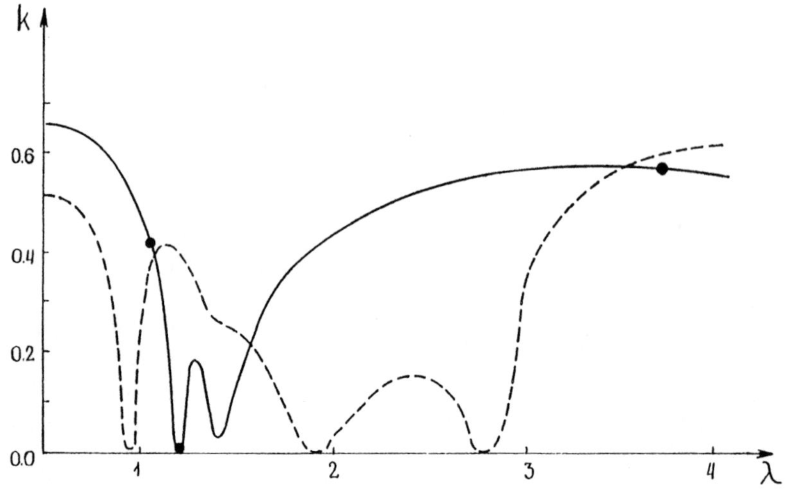

FIGURE 6. Electromechanical coupling coefficients as functions of the frequency parameter λ

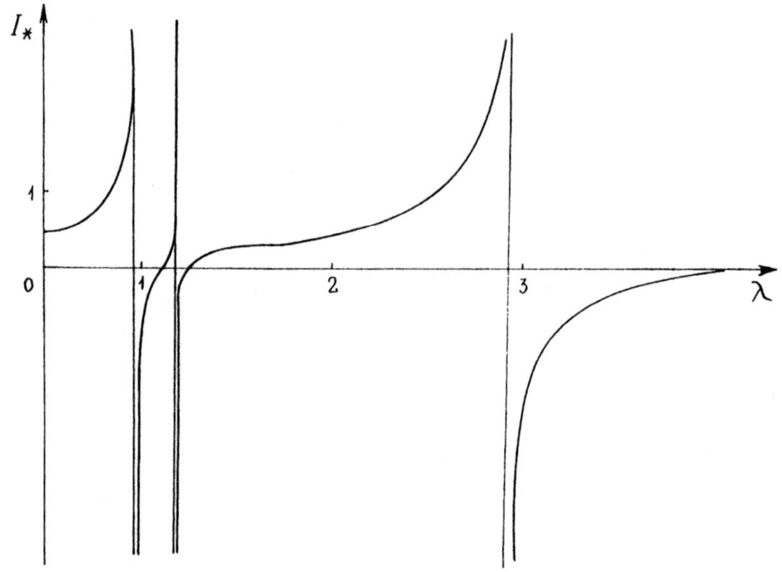

FIGURE 7. Current as a function of the frequency parameter λ

TABLE 2.
Resonance and antiresonance frequency parameters and the corresponding EMCC's as functions of the shell length (axisymmetric problem)

l/r	$\lambda^{\langle r\rangle}$	$\lambda^{\langle a\rangle}$	k_d	$k_e^{\langle r\rangle}$	$k_e^{\langle a\rangle}$	$k^{\langle r\rangle}$	$k^{\langle a\rangle}$
	1.06	1.20	0.345	0.436	0.261	0.000	0.338
0.5	1.25	1.25	0.013	0.007	0.019	0.323	0.318
	10.68	16.72	0.599	0.615	0.578	0.285	0.661
	0.98	1.18	0.413	0.522	0.279	0.100	0.392
1.0	1.20	1.21	0.041	0.021	0.081	0.337	0.331
	2.94	4.28	0.559	0.564	0.558	0.263	0.646
	0.76	1.10	0.556	0.598	0.332	0.112	0.413
1.5	1.19	1.20	0.068	0.045	0.110	0.334	0.336
	1.68	2.03	0.418	0.398	0.432	0.186	0.583

equations describing the quasitangential vibrations. Our computation shows that the EMCC assumes the greatest values for the quasitangential vibrations, decreasing as the shell length grows. The EMCC values for λ from the second interval are close to zero.

28 FORCED VIBRATIONS OF A CIRCULAR CYNDRICAL SHELL WITH LONGITUDINAL POLARIZATION (NONAXISYMMETRIC PROBLEM)

Consider a shell with longitudinal polarization, which performs forced vibrations caused by an electrical load acting on the shell's edges [87]. The edges are covered by dissected electrodes and the electrical potential on each is given. We expand the electrical load and all the sought-for quantities in the Fourier series in the circular coordinate φ:

$$\psi_*|_{\xi=\pm l/r} = \pm\sum_{n=1}^{\infty} t_n \sin n\varphi, \qquad P_1 = \sum_{n=1}^{\infty} P_{1n} \sin n\varphi, \qquad P_2 = \sum_{n=1}^{\infty} P_{2n} \cos n\varphi. \tag{28.1}$$

Here, P_1 is any of the quantities $\psi_*, T_{i*}, u_*, w_*, E_{1*}$, and D_{1*}; and P_2 is any of the quantities S_*, v_*, E_{2*}, and D_{2*}; t_n are constants. Both P_{1n} and P_{2n} are functions of the longitudinal coordinate ξ.

We substitute expansions 28.1 into the equations of shell theory, equating the coefficients at the same trigonometric functions to zero. We find for every n the system

$$\frac{dT_{1n*}}{d\xi} - nS_{n*} + \lambda u_{n*} = 0$$
$$\frac{dS_{n*}}{d\xi} + nT_{2n*} + \lambda v_{n*} = 0$$
$$T_{2n*} + \lambda w_{n*} = 0 \tag{28.2}$$

$$T_{1n*} = \varepsilon_{1n*} + \nu_2 \varepsilon_{2n*} + \frac{d\psi_{n*}}{d\xi}$$
$$T_{2n*} = \alpha(\varepsilon_{2n*} + \nu_1 \varepsilon_{1n*}) + c_{12}\frac{d\psi_{n*}}{d\xi}$$
$$S_{n*} = \frac{1}{s_{44}^E n_{22}}\left[w_{n*} + n\frac{d_{15}n_{22}}{c_2}\psi_{n*}\right] \tag{28.3}$$

$$D_{1n*} = -\frac{\varepsilon_{33}^T}{c_2 d_{31}}\frac{d\psi_{n*}}{d\xi} + T_{2n*} + \frac{d_{33}}{d_{31}}T_{1n*}$$
$$D_{2n*} = -\frac{\varepsilon_{11}^T}{c_2 d_{31}} n\psi_{n*} + \frac{d_{15}}{d_{31}}S_{n*} \tag{28.4}$$

$$\frac{dD_{1n*}}{d\xi} - nD_{2n*} = 0 \tag{28.5}$$

$$\varepsilon_{1n*} = \frac{du_{n*}}{d\xi}, \qquad \varepsilon_{2n*} = -nv_n - qw_{n*}, \qquad \omega_{n*} = \frac{dv_{n*}}{d\xi} + nu_{n*}. \tag{28.6}$$

The numbers $\alpha, \lambda, c_{12}, \nu_1$, and ν_2 are taken from Section 27.

The dimensionless quantities with asterisks are introduced by formulas 27.5, to which we should add

$$E_{2*} = \frac{c_2}{n_{22}}E_2, \qquad D_{2*} = \frac{D_2}{d_{31}n_{22}}.$$

The initial equations are membrane. This means that we do not consider the vibrations described using the moment equations of the theory of piezoceramic shells. We can do this because, as it was shown in Section 27, the frequency parameters λ, for which the problem cannot be broken into the principal problem, and auxiliary problem correspond to the EMCC values close to zero.

The effect of the electrical load at the edges of the shell is like that of a longitudinal load, applied to the edges. For low frequencies of the load, the shell will perform quasitransverse vibrations, while for $\lambda >> 1$ the vibrations are quasitangential. The quasitransverse vibrations with small variability are described by the membrane system 28.1 to 28.6 for $q = 1$. The equations for the quasitangential vibrations are obtained from the same system for $q = 0$.

We solve system 28.1 to 28.6 with respect to $u_{n*}, v_{n*}, \psi_{n*}$:

$$\left[b_1 \frac{d^2}{d\xi^2} + b_2\right] u_{n*} + b_3 \frac{dv_{n*}}{d\xi} + \left[b_4 \frac{d^2}{d\xi^2} + b_5\right] \psi_{n*} = 0$$

$$e_1 \frac{du_{n*}}{d\xi} + \left[e_2 \frac{d^2}{d\xi^2} + e_3\right] v_{n*} + e_4 \frac{d\psi_{n*}}{d\xi} = 0$$

$$\left[d_1 \frac{d^2}{d\xi^2} + d_2\right] u_{n*} + d_3 \frac{dv_{n*}}{d\xi} + \left[d_4 \frac{d^2}{d\xi^2} + d_5\right] \psi_{n*} = 0 \tag{28.7}$$

$$b_1 = 1 - \frac{q}{q\alpha - \lambda}\alpha\nu_1\nu_2, \qquad b_2 = \lambda - \frac{n^2}{s_{44}^E n_{22}}$$

$$b_3 = n\left[\frac{q}{q\alpha - \lambda}\alpha\nu_2 - \nu_2 - \frac{1}{s_{44}^E n_{22}}\right]$$

$$b_4 = 1 - \frac{q\nu_2 c_{12}}{\alpha q - \lambda}, \qquad b_5 = -\frac{n^2 d_{15}}{s_{44}^E c_2}$$

$$a = 1 - \frac{q\alpha}{q\alpha - \lambda},$$

$$e_1 = n\left[\frac{1}{s_{44}^E n_{22}} + \alpha\nu_1 a\right], \qquad e_2 = \frac{1}{s_{44}^E n_{22}}$$

$$e_3 = -n^2\alpha a + \lambda, \qquad e_4 = n\left[\frac{d_{15}}{s_{44}^E c_2} + c_{12} a\right]$$

$$d_1 = \alpha\nu_1 \frac{d_{31}}{\varepsilon_{33}^T} a + \frac{d_{33}}{\varepsilon_{33}^T} b_1, \qquad d_2 = -n^2 \frac{d_{15}}{\varepsilon_{33}^T s_{44}^E n_{22}}$$

$$d_3 = -n\left[\left(\alpha\frac{d_{31}}{\varepsilon_{33}^T} + \nu_2 \frac{d_{33}}{\varepsilon_{33}^T}\right) a + \frac{d_{15}}{s_{44}^E \varepsilon_{33}^T n_{22}}\right]$$

$$d_4 = \left[-\frac{1}{c_2} + \frac{d_{31} c_{12}}{\varepsilon_{33}^T} a + \frac{d_{33}}{\varepsilon_{33}^T} b_4\right], \qquad d_5 = \frac{n^2}{c_2 \varepsilon_{33}^T}\left(\varepsilon_{11}^T - \frac{d_{15}^2}{s_{44}^E}\right). \tag{28.8}$$

Let us introduce an unknown function Φ, defined as

$$D\Phi = 0, \qquad D = \begin{vmatrix} b_1 \frac{d^2}{d\xi^2} + b_2 & b_3 \frac{d}{d\xi} & b_4 \frac{d^2}{d\xi^2} + b_5 \\ e_1 \frac{d}{d\xi} & e_2 \frac{d^2}{d\xi^2} + e_3 & e_4 \frac{d}{d\xi} \\ d_1 \frac{d^2}{d\xi^2} + d_2 & d_3 \frac{d}{d\xi} & d_4 \frac{d^2}{d\xi^2} + d_{15} \end{vmatrix}. \tag{28.9}$$

The quantities $u_{n*}, v_{n*}, \psi_{n*}$ are expressed in terms of Φ as

$$u_{n*} = D_1\Phi, \qquad v_{n*} = D_2\Phi, \qquad \Psi_{n*} = D_3\Phi$$

$$D_1 = \begin{vmatrix} b_3\frac{d}{d\xi} & b_4\frac{d^2}{d\xi^2} + b_5 \\ e_2\frac{d^2}{d\xi^2} + e_3 & e_4\frac{d}{d\xi} \end{vmatrix}$$

$$D_2 = \begin{vmatrix} b_1\frac{d^2}{d\xi^2} + b_2 & b_4\frac{d^2}{d\xi^2} + b_5 \\ e_1\frac{d}{d\xi} & e_4\frac{d}{d\xi} \end{vmatrix}$$

$$D_3 = \begin{vmatrix} b_1\frac{d^2}{d\xi^2} + b_2 & b_3\frac{d}{d\xi} \\ e_1\frac{d}{d\xi} & e_2\frac{d^2}{d\xi^2} + e_3 \end{vmatrix}. \qquad (28.10)$$

For the nth expansion of ψ_*, the following electrical conditions should hold at the edges:

$$\psi_{n*}|_{\xi=\pm l/r} = \pm t_n.$$

For definiteness, we consider the case when there are two electrodes at each edge:

$$\psi_*|_{\xi=\pm l/r} = \begin{cases} \pm 1, & 0 < \varphi < \pi \\ \mp 1, & \pi < \varphi < 2\pi \end{cases}$$

or

$$\psi_{n*}|_{\xi=\pm l/r} = \pm t_n, \qquad t_n = \frac{2}{\pi n}\left(1 - (-1)^n\right). \qquad (28.11)$$

We will also assume that the shell edges are not clamped:

$$T_{n*}|_{\xi=\pm l/r} = 0, \qquad S_{n*}|_{\xi=\pm l/r} = 0. \qquad (28.12)$$

As in Section 27, we compute the EMCC using different formulas 27.10 to 27.13. The values of k and k_d are found directly, after the solution of the initial problem has been found. In order to find k_e, we should solve two additional problems.

PROBLEM 1

Suppose that the electrodes are disconnected. Since every electrode is an exponential surface, its electrical potential is an unknown constant. In our problem with two electrodes at every edge and the same mechanical conditions at the edges $\xi = \pm l/r$, it is sufficient to introduce one unknown constant c:

$$\psi_*^{(d)}|_{\xi=\pm l/r} = \begin{cases} \pm c, & 0 < \varphi < \pi \\ \mp c, & \pi < \varphi < 2\pi \end{cases}. \qquad (28.13)$$

In order to find $\psi_*^{(d)}$, we integrate equation 28.5 under the assumption that the strains are known. It is more convenient to determine $\psi_{n*}^{(d)}$ by writing equation 28.5 in the form

$$\left[d_4\frac{d^2}{d\xi^2}+d_5\right]\psi_{n*}^{(d)}=-d_3\frac{dv_{n*}}{d\xi}-\left[d_1\frac{d^2}{d\xi^2}+d_2\right]u_{n*}. \tag{28.14}$$

We know functions v_{n*} and u_{n*} from the solution of the initial problem. The arbitrary integration constants in equation 28.14 are found from the conditions 28.13.

Having found all the terms of the trigonometric series for $\psi_*^{(d)}$, we substitute the resultant series into the integral condition 27.14. By integrating the coordinate φ, we find c.

The energy $U_*^{(d)}$ is computed by the formula

$$U_*^{(a)}=\int_s\left[\varepsilon_{1*}T_{1*}^{(a)}+\varepsilon_{2*}T_{2*}^{(a)}+\omega_*S_*^{(a)}+\frac{n_{22}d_{31}}{c_2}\left(E_{1*}^{(a)}D_{1*}^{(a)}+E_{2*}^{(a)}D_{2*}^{(a)}\right)\right]ds \tag{28.15}$$

where

$$\begin{aligned}
T_{1*}^{(a)}&=\varepsilon_{1*}+\nu_2\varepsilon_{2*}-E_{1*}^{(a)}\\
T_{2*}^{(a)}&=\alpha(\varepsilon_{2*}+\nu_1\varepsilon_{1*})-c_{12}E_{1*}^{(a)}\\
S_*^{(a)}&=\frac{1}{s_{44}^E n_{22}}\left[\omega_*-\frac{d_{15}n_{22}}{c_2}E_{2*}^{(a)}\right]\\
D_{1*}^{(a)}&=\frac{\varepsilon_{33}^T}{c_2d_{31}}E_{1*}^{(a)}+T_{2*}^{(a)}+\frac{d_{33}}{d_{31}}T_{1*}^{(a)}\\
D_{2*}^{(a)}&=\frac{\varepsilon_{11}^T}{c_2d_{31}}E_{2*}^{(a)}+\frac{d_{15}}{d_{31}}S_*^{(a)}.
\end{aligned} \tag{28.16}$$

Here, we put $a=d$.

PROBLEM 2

The shell electrodes are short-circuited. In order to find $\psi_*^{(sh)}$, we must find the solution to equation 28.14; change the superscript (d) for (sh); and satisfy conditions 27.15 on the electrodes. The energy $U^{(sh)}$ is found from 28.15 and 28.16, where we put $a=sh$.

We calculated the first natural frequencies for different shell lengths and placed them in Table 3. The first column gives the shell's length divided by its radius. The second column gives the resonance frequency. The third gives the antiresonance frequency. The remaining columns give the EMCC values computed using different formulas. The superscripts (r) and (a) show that the values were computed for the resonance and antiresonance frequencies respectively.

TABLE 3.
Resonance and antiresonance frequency parameters and the corresponding EMCC's as functions of the shell length (nonaxisymmetric problem)

l/r	$\lambda^{\langle r\rangle}$	$\lambda^{\langle a\rangle}$	k_d	$k_e^{\langle r\rangle}$	$k_e^{\langle a\rangle}$	$k^{\langle r\rangle}$	$k^{\langle a\rangle}$
	2.07	2.24	0.282	0.370	0.206	0.060	0.275
0.50	7.00	7.04	0.071	0.118	0.072	0.267	0.282
	9.86	10.80	0.310	0.401	0.200	0.268	0.359
	12.34	12.44	0.086	0.106	0.069	0.422	0.424
	1.91	2.10	0.300	0.394	0.218	0.110	0.305
0.75	4.63	4.76	0.166	0.238	0.108	0.278	0.319
	5.87	5.27	0.135	0.163	0.106	0.420	0.431
	7.87	8.88	0.336	0.402	0.270	0.154	0.399
	1.64	1.82	0.312	0.410	0.232	0.265	0.388
1.00	3.13	3.19	0.042	0.078	0.056	0.264	0.271
	4.48	4.53	0.110	0.168	0.060	0.471	0.475
	5.33	5.93	0.317	0.377	0.263	0.128	0.382
	1.37	1.52	0.325	0.398	0.258	0.283	0.426
1.25	2.68	2.70	0.086	0.089	0.074	0.229	0.230
	4.07	4.20	0.171	0.341	0.050	0.111	0.428
	4.26	4.34	0.250	0.165	0.235	0.502	0.453

As in Section 27, the k_e-values are in good agreement with k_d averaged over the interval $[\lambda^{(r)}, \lambda^{(a)}]$; k assumes quite different values in the same interval. Figures 8 to 10 show the needed quantities as functions of coordinates for a shell with length $2r$. Figure 8 gives the displacements, forces, and electrical quantities for the first resonance; Figure 9 gives the same quantities for the second resonance; and Figure 10 gives the quantities for the third resonance. Each dimensionless quantity in Figures 8–10 is divided by the same number that corresponds to the greatest of all the values they assume. The solid line in Figure 11 is for k_e versus λ; the dashed line is for k versus. λ; and the circles denote k_d. We see that all the circles lie on the k_e curve. Figure 12 presents the current $I_*(I_* = I/(\varepsilon_{11}^T Vr\omega))$ as a function of the frequency parameter λ.

For the nonaxially-symmetric problem, the EMCC dependence on λ is more complicated than in the axially-symmetric case. But there is the same tendency for the EMCC to assume the greatest values for the quasitangential vibrations.

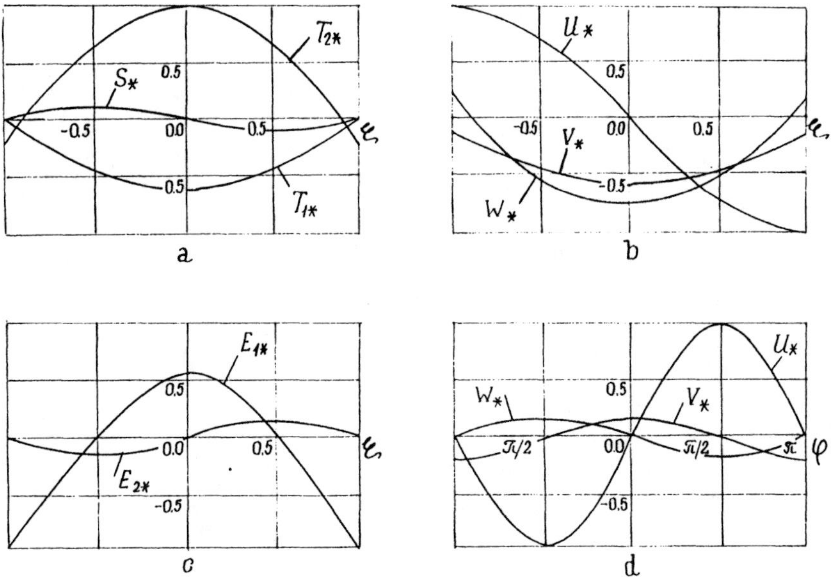

FIGURE 8. Distribution of dimensionless (a) forces, (b) displacements, and (c) electric strength components along the generatrix of a cylindrical shell near the first resonance ($\varphi = \pi/4$); (*d*) displacements along the edge line $\xi = 1$ as functions of the φ-coordinate (the first resonance)

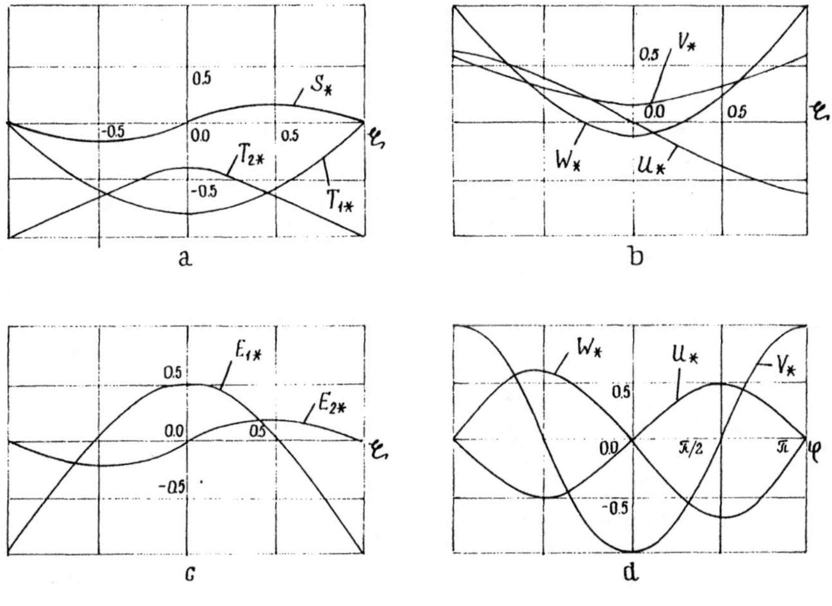

FIGURE 9. Distribution of dimensionless (a) forces, (b) displacements, and (c) electric strength components along the generatrix of a cylindrical shell near the first resonance ($\varphi = \pi/4$); (*d*) displacements along the edge line $\xi = 1$ as functions of the φ-coordinate (the second resonance)

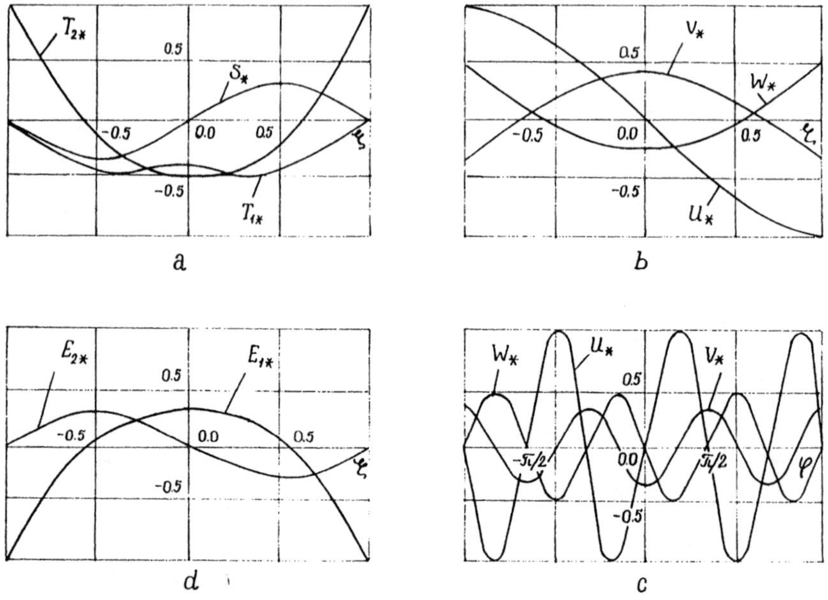

FIGURE 10. Distribution of dimensionless (a) forces, (b) displacements, and (c) electric strength components along the generatrix of a cylindrical shell near the first resonance ($\varphi = \pi/4$); (*d*) displacements along the edge line $\xi = 1$ as functions of the φ-coordinate (the third resonance)

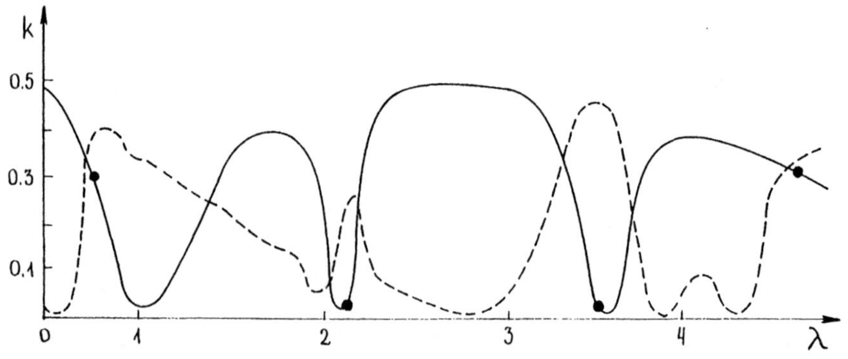

FIGURE 11. Electromechanical coupling coefficient as a function of the frequency parameter for a shell with cutted electrodes

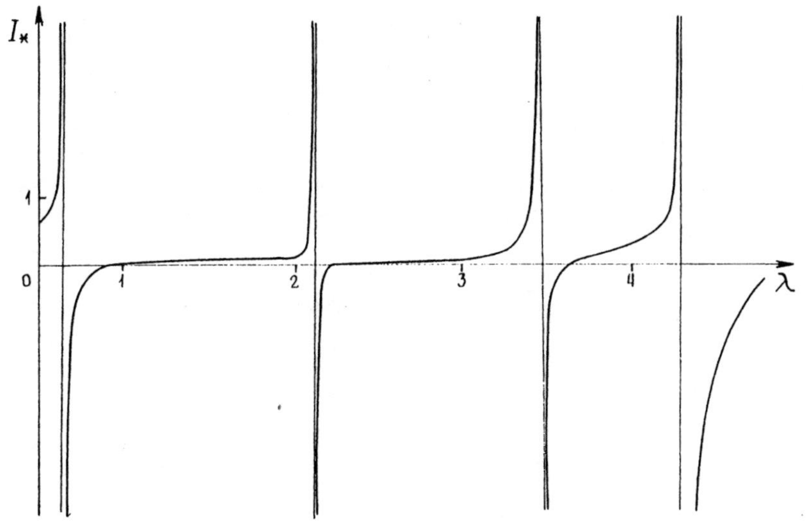

FIGURE 12. Current as a function of the frequency parameter for a shell with cutted electrodes

29 TWO NONCLASSICAL PROBLEMS FOR SHELLS WITH TANGENTIAL POLARIZATION

PROBLEM 1

Consider a spherical shell. We place its middle surface in a geographic coordinate system where the position of a point is specified by the polar distance θ and longitude φ. In parametric form, the equations of the spherical surface will be

$$x = r\sin\theta\cos\varphi, \qquad y = r\sin\theta\sin\varphi, \qquad z = r\cos\theta.$$

We change the independent variables

$$\alpha = \ln\tan\frac{\theta}{2}, \qquad \varphi = \beta.$$

Then the coefficients of the first quadratic form have the form [21]:

$$A = B = \frac{r}{\cosh\alpha}$$

$$\cos\chi = 0,$$

$$k_1 = \frac{1}{AB}\frac{\partial A}{\partial\beta} = 0$$

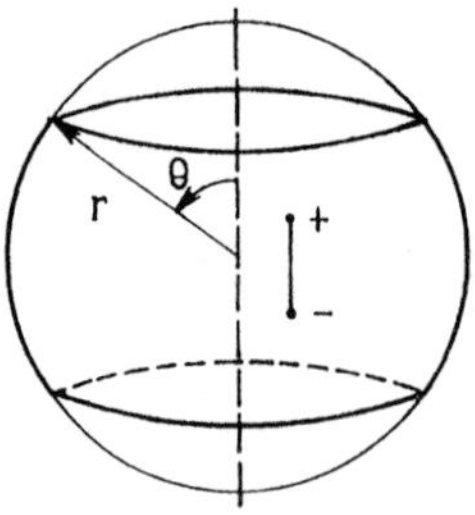

FIGURE 13. Part of a spherical shell polarized in meridional direction

$$k_2 = \frac{1}{AB}\frac{\partial B}{\partial \alpha} = -\frac{\sinh \alpha}{r}. \tag{29.1}$$

Let the shell be pre-polarized along the θ-lines, and its surfaces be covered with electrodes on which we are given the potential

$$\psi|_{\gamma=\pm h} = \pm V e^{-i\omega t}. \tag{29.2}$$

Our shell is part of a sphere with two rigidly fixed edges: $\theta = \theta_1$ and $\theta = \theta_2, \theta_1 \leq \theta \leq \theta_2$ (see Figure 13). We restrict ourselves to calculating the electroelastic state in the roughest approximation.

We saw in Chapter 4 that the internal electroelastic state can be represented as a sum of a slowly varying electroelastic state and a quickly varying auxiliary electroelastic state. The greatest stresses of the principal electroelastic state are computed from the moments that in the first approximation can be determined without solving the problem by using the formulas

$$\begin{aligned} G_{\alpha *} &= -\frac{n_{21}}{n_{22}} r k_2 = \frac{n_{21}}{n_{22}} \sinh \alpha \\ G_{\beta *} &= -\frac{n_{11}}{n_{22}} r k_2 = \frac{n_{11}}{n_{22}} \sinh \alpha \\ G_{i*} &= \frac{3G_i}{2h^3 n_{22} d_{15} E_3} \\ E_3 &= -\frac{V}{h} \qquad (i = \alpha, \beta). \end{aligned} \tag{29.3}$$

The resultant equation for the auxiliary electroelastic state was obtained in Section 24 and has the form

$$\frac{1}{A^4}\frac{\partial^4 w_*}{\partial \alpha^4} + 4g^4 w_* = 0 \qquad \left(w_* = \frac{w}{r}\right) \tag{29.4}$$

where

$$4g^4 = \frac{3n_{22}(1-\nu_1\nu_2)}{h^2r^2n_{11}}\left[1-\frac{\rho\omega^2r^2}{n_{22}(1-\nu_1\nu_2)}\right]$$
$$\nu_1 = \frac{n_{21}}{n_{11}}, \qquad \nu_2 = \frac{n_{21}}{n_{22}}. \tag{29.5}$$

To be concrete, we will only count the values of the frequency ω such that

$$1-\rho\frac{\omega^2r^2}{n_{22}(1-\nu_1\nu_2)} = 1-d < 0$$

and the solution to equation 29.4 will be written as

$$w_* = c_1e^{k\alpha}+c_2e^{-k\alpha}+c_3\cos k\alpha+c_4\sin k\alpha \tag{29.6}$$

where

$$k = \sqrt{-2g^2A}.$$

The conditions at the rigidly clamped edges $\theta = \theta_1$ and $\theta = \theta_2$ for the auxiliary problem will be written as (see Section 10)

$$w_*|_{\theta=\theta_1,\theta_2} = 0, \qquad \gamma_{1*}|_{\theta=\theta_1,\theta_2} = 1.$$

The computation was made for $h/R = 0.01, \theta_1 = \pi/4, \theta_2 = 3\pi/4$, and $d = -1.5$ for a shell made of PZT-4.

Figure 14 depicts the bending moments $G_{\theta *}$. The dashed line gives the moment computed from the principal problem, and the solid line gives the moment computed as a sum of the principal problem and auxiliary problem. We see in Figure 14 that the bending moments of the auxiliary electroelastic state are $\sqrt{R/h}$ times greater than the respective moments of the principal electroelastic state. The transverse forces of the auxiliary problem are R/h times greater than the transverse forces of the principal electroelastic state.

Thus, the greatest stresses generated by the smooth surface electrical load are found from the equations of the auxiliary problem. This situation is impossible in the theory of nonelectrical shells because the greatest stresses of the auxiliary stressed-strained state do not exceed the greatest stresses of the principal stressed-strained state for any smooth surface load.

PROBLEM 2

Consider a shell that is a part of an ellipsoid whose equations have the form

$$x = a\sin\theta\cos\varphi, \qquad y = b\sin\theta\sin\varphi, \qquad z = c\cos\theta.$$

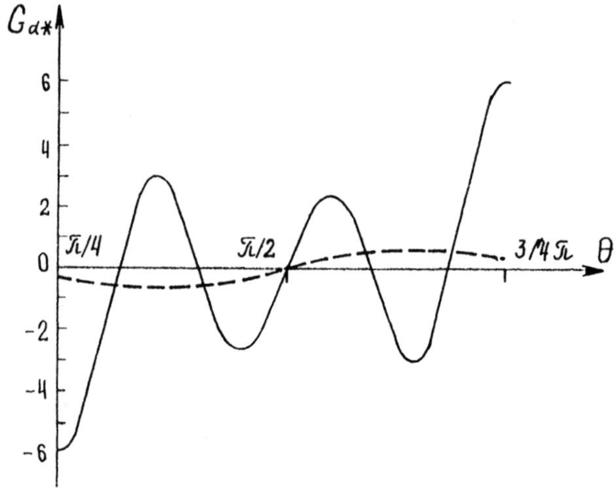

FIGURE 14. Distribution of the bending moment along the meridian of a spherical shell (the dashed line presents the principal problem solution, the solid line combines solutions of the principal and auxiliary problems)

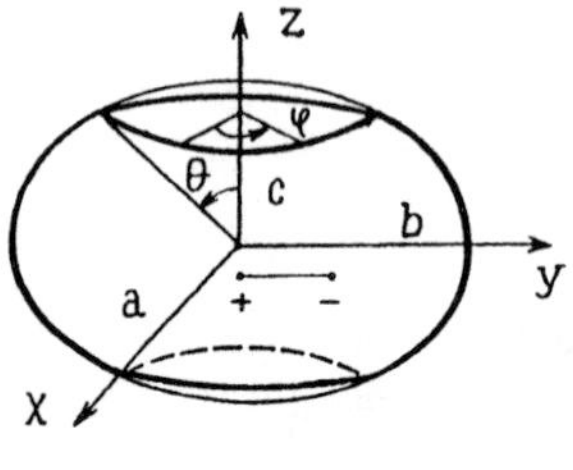

FIGURE 15. Element of an ellipsoidal shell polarized along parallels

Both edges $\theta = \theta_1$ and $\theta = \theta_2$ of the shell are rigidly clamped (Figure 15).

The coefficients A and B of the first quadratic form and the geodesic curvatures of the θ- and φ-lines are defined by the formulas

$$A^2 = a^2(\cos^2\theta\cos^2\varphi + \sin^2\theta) + b^2\cos^2\theta\sin^2\varphi$$
$$B^2 = a^2\sin^2\theta\sin^2\varphi + b^2\sin^2\theta\cos^2\varphi$$
$$k_1 = \frac{1}{AB}\frac{\partial A}{\partial\varphi}, \qquad k_2 = \frac{1}{AB}\frac{\partial B}{\partial\theta}.$$

We will assume that the shell is pre-polarized along the φ-lines, and its faces are covered with electrodes fot which we are given an electrical potential 29.2.

We saw in Section 10 that the internal electroelastic state of such a shell can be partitioned into the principal and auxiliary electroelastic states. As in the first problem, we will perform the computation in the first approximation.

In order to find the greatest stresses, we need the bending and transverse forces. For the principal electroelastic state, we have

$$G_{\theta *} = -k_1, \qquad N_{\theta *} = -\frac{1}{A}\frac{\partial k_1}{\partial \theta},$$
$$G_{\theta *} = \frac{3G_\theta}{2h^3 n_{11} d_{15} E_3}, \qquad N_{\theta *} = \frac{1}{A}\frac{\partial G_{\theta *}}{\partial \theta}. \tag{29.7}$$

For the auxiliary electroelastic state we keep equation 29.4 where

$$4g^4 = \frac{3(1-\nu_1\nu_2)n_{11}}{h^2 r_\varphi^2 n_{22}}\left(1 - \frac{r_\varphi^2}{a^2}d\right), \qquad d = \frac{\rho\omega^2 a^2}{n_{11}(1-\nu_1\nu_2)}$$

and r_φ is the curvature radius for the φ-lines.

Nonclassical boundary conditions for the considered case were discussed in Section 10 and were obtained in Section 45 using the asymptotic method. For the auxiliary problem, the boundary conditions for the rigidly clamped edges $\theta = \theta_1$ and $\theta = \theta_2$ have the form

$$w_* = 0, \qquad \gamma_{1*} = -h\rho_1 k_1.$$

The constant ρ_1 was calculated for PZT-4 (see Table 1) using the formulas given in Section 50. It turned out to be equal to -2.03 for

$$c = a, \qquad b = 2a, \qquad \theta_1 = \pi/4, \qquad \theta_2 = 3\pi/4$$
$$h/r_\varphi(\theta_2) = h/r_\varphi(\theta_2) = 0.01, \qquad d = -1.5.$$

Figure 16 shows $G_{\theta *}$ and $N_{\theta *}$ as functions of the variable θ for $\varphi = \pi/4$. The dashed line gives the computation results obtained by using 29.7 (the principal electroelastic state) without allowing for the boundary layer corrections. The solid line gives the transverse forces and moments computed allowing for the boundary layer corrections (the transverse forces and moments are found as sums of the principal and auxiliary electroelastic states). Comparing the solutions where the boundary layer corrections are and are not allowed for, we see that the latter variant gives absolutely the wrong results.

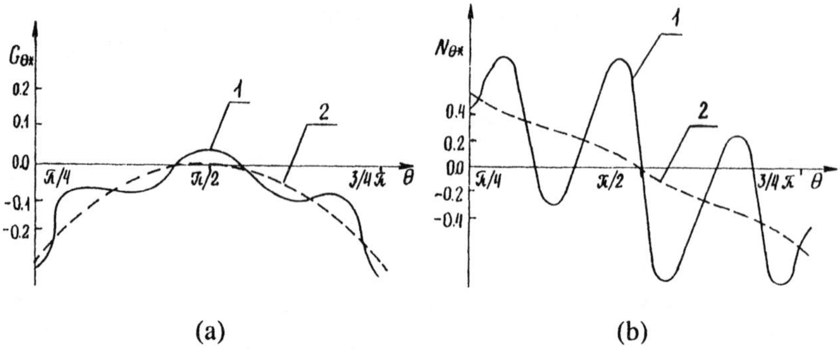

FIGURE 16. Distribution of (a) the bending moment, (b) transverse force along the meridian of a ellipsoidal shell with allowing for the boundary layer corrections (solid line) and without the corrections (dashed line)

30 AXISYMMETRIC TANGENTIAL VIBRATIONS OF CIRCULAR PLATES WITH THICKNESS POLARIZATION

Some devices employ elements whose surfaces are only partially covered with electrodes that may either be solid or have cuts.

Let us investigate the effect produced by different electrical conditions given for the plate faces on the dynamic electroelastic state. We will consider the free and forced vibrations of plates whose faces are (a) completely covered with electrodes, (b) have no electrodes, and (c) are partially covered with electrodes. We will solve the problems using the constitutive relations of the theory of piezoceramic plates given in Section 13.

Let us consider a circular plate with thickness polarization. Its radius is R and its thickness is $2h$. The numerical computation will be carried out for the PZT-5 piezoceramic. Let us write the equilibrium equations, strain-displacement formulas, and electrostatics equations in the polar coordinates, which have the same form for all our problems.

The equilibrium equations are

$$\frac{dT_r}{dr} + \frac{T_r - T_\theta}{r} + 2h\rho\omega^2 u = 0. \tag{30.1}$$

The strain-displacement formulas are

$$\varepsilon_r = \frac{du}{dr}, \qquad \varepsilon_\theta = \frac{u}{r}. \tag{30.2}$$

The electroelasticity relations can have different forms depending on the electrical conditions on the faces. We will write them separately for each case.

PROBLEM 1 A Plate with Electrodes.

Consider the free and forced vibrations of a plate with electrode-covered faces due to an electrical load. The edge of the plate is not fixed and is free from the edge forces:

$$T_r|_{r=R} = 0. \tag{30.3}$$

The electroelasticity relations are

$$\begin{aligned} T_r &= \frac{2h}{s_{11}^E(1-\nu^2)}(\varepsilon_r + \nu\varepsilon_\theta) + \frac{2d_{31}}{s_{11}^E(1-\nu)}V \\ T_\theta &= \frac{2h}{s_{11}^E(1-\nu^2)}(\varepsilon_\theta + \nu\varepsilon_r) + \frac{2d_{31}}{s_{11}^E(1-\nu)}V \\ D_3 &= -\varepsilon_{33}^T\frac{V}{h} + \frac{d_{31}}{2h}(T_r + T_\theta). \end{aligned} \tag{30.4}$$

Here, we have taken into account that the electrical potential on the electrodes is given as

$$\psi|_{\gamma=\pm h} = \pm V. \tag{30.5}$$

We express the stresses in terms of the displacements and substitute them into the equilibrium equations to get the equation

$$r^2\frac{d^2u}{dr^2} + r\frac{du}{dr} + \left[(\lambda r)^2 - 1\right]u = 0, \qquad \lambda^2 = \omega^2\rho s_{11}^E(1-\nu^2) \tag{30.6}$$

whose solution is a first-kind Bessel function

$$u = C_1 J_1(\lambda r). \tag{30.7}$$

We find the arbitrary integration constant from the condition in 30.3:

$$C_1 = -\frac{(1+\nu)Rd_{31}}{\lambda R J_0(\lambda R) - (1-\nu)J_1(\lambda R)}\frac{V}{h}. \tag{30.8}$$

The forces, strains, and electrical quantities are found from the formulas

$$\begin{aligned} T_r &= \frac{2hC_1}{s_{11}^E(1-\nu^2)}\left[\lambda J_0(\lambda r) - \frac{1-\nu}{r}J_1(\lambda r)\right] + \frac{2d_{31}}{s_{11}^E(1-\nu)}V \\ T_\theta &= \frac{2hC_1}{s_{11}^E(1-\nu^2)}\left[\nu\lambda J_0(\lambda r) + \frac{1-\nu}{r}J_1(\lambda r)\right] + \frac{2d_{31}}{s_{11}^E(1-\nu)}V \\ D_3 &= -\varepsilon_{33}^T(1-k_p^2)\frac{V}{h} + \frac{d_{31}}{s_{11}^E(1-\nu)}C_1\lambda J_0(\lambda r), \\ k_p^2 &= \frac{2d_{31}^2}{s_{11}^E\varepsilon_{33}^T(1-\nu)}. \end{aligned} \tag{30.9}$$

In order to determine the amplitude of the current, we integrate the component D_3 of the electric induction vector over the surface Ω of the electrode on one of the plate faces:

$$I = -i\omega \int_{\Omega} D_3 d\Omega = i\omega\pi R^2 \varepsilon_{33}^T \left[\left(1 - k_p^2\right) \frac{V}{h} - \frac{k_p^2}{Rd_{31}} C_1 J_1(\lambda R) \right]. \tag{30.10}$$

Here we have used the formula

$$\int_0^R (\varepsilon_r + \varepsilon_\theta) r dr = \int_0^R d(ru) = Ru\Big|_{r=R}.$$

Now we can use the solution obtained for a numerical computation.

The resonance frequencies are found from

$$\lambda R J_0(\lambda_*) - (1 - \nu) J_1(\lambda_*) = 0. \qquad \lambda_* = \lambda R \tag{30.11}$$

The first four roots of equation 30.11 for a plate made of PZT-5 are 2.08, 5.40, 8.58, 11.34.

The antiresonance frequencies for which the current through the piezoceramic plate vanishes are important characteristics. Equation 30.10 implies that the antiresonances are defined by the roots of the equation

$$\lambda R J_0(\lambda_*) - \left[(1 - \nu) - (1 + \nu) \frac{k_p^2}{1 - k_p^2} \right] J_1(\lambda_*) = 0 \qquad \lambda_* = \lambda R. \tag{30.12}$$

The first four roots of this equation for a plate made of PZT-5 are 2.46, 5.54, 8.67, 11.80.

The obtained solution is used in Figure 17, showing the dependence of the quantities

$$T_{r*} = T_r / \left| T_{r\max} \right|, \qquad u_* = u / \left| u_{\max} \right|$$

on the radial coordinate. Each curve was constructed for a fixed frequency, and the dimensionless quantities $T_{r*}(u_*)$ were found by dividing the force $T_r\,(u)$ by, the plate force (displacement) with the greatest absolute value for the given frequency. We carried out the computations for frequencies close to the first three resonances. The curves marked by 1, 2, and 3 give the quantities T_{r*}, u_* as functions of $x = r/R$ near the first, second, and third resonance frequencies, respectively.

The EMCC is of special interest in designing electroelastic elements. We will compute it using the Mason formula 27.12 and the energy formula 27.13.

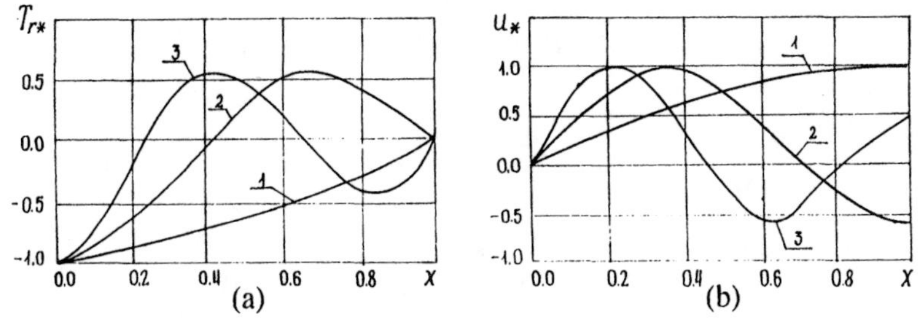

FIGURE 17. Distribution of dimensionless (a) force T_{r*} and (b) displacement u_* along the plate radius near the first three resonances

The Mason formula has the form

$$k_d^2 = \frac{\omega_a^2 - \omega_r^2}{\omega_a^2} \tag{30.13}$$

and gives the following values of k_d^2 near the first four resonances: 0.282, 0.051, 0.021, 0.011. We compute k_e^2 by the formula

$$k_e^2 = \frac{U^{(d)} - U^{(sh)}}{U^{(d)}} \tag{30.14}$$

where

$$U^{(d)} = \int_\Omega \left(\varepsilon_r T_r^{(d)} + \varepsilon_\theta T_\theta^{(d)} + 2hE_3^{(d)} D_3^{(d)}\right) d\Omega$$

$$U^{(sh)} = \int_\Omega \left(\varepsilon_r T_r^{(sh)} + \varepsilon_\theta T_\theta^{(sh)}\right) d\Omega. \tag{30.15}$$

Here, $U^{(d)}$ is the internal energy of the plate with disconnected electrodes, and $U^{(sh)}$ is the internal energy of the plate with short-circuited electrodes. In both cases of computing the internal energy in equation 30.15, we use the strains ε_r and ε_θ that were found from the solution of the initial problem.

Let us find the internal energy of the plate with disconnected electrodes. The potential difference V on the disconnected electrodes is found from

$$\int_\Omega D_3^{(d)} d\Omega = 0. \tag{30.16}$$

We substitute the formulas

$$D_3^{(d)} = \varepsilon_{33}^T E_3^{(d)} + \frac{d_{31}}{2h}\left(T_r^{(d)} + T_\theta^{(d)}\right),$$

$$T_r^{(d)} = \frac{2h}{s_{11}^E(1-\nu^2)}(\varepsilon_r + \nu\varepsilon_\theta) + \frac{2d_{31}V^{(d)}}{s_{11}^E(1-\nu)},$$
$$E_3^{(d)} = -\frac{V^{(d)}}{h} \tag{30.17}$$

into equation 30.16 and find $V^{(d)}$:

$$\frac{V^{(d)}}{h} = C_1 \frac{k_p^2}{1-k_p^2} \frac{J_1(\lambda_*)}{Rd_{31}}. \tag{30.18}$$

We find $U^{(d)}$ from equation 30.15 by integrating over the surface of one of the electrodes, taking into account equations 30.17 and 30.18.

For a plate with short-circuited electrodes, we should put $V = 0$ in 30.9:

$$T_r^{(sh)} = \frac{2h}{s_{11}^E(1-\nu^2)}(\varepsilon_r + \nu\varepsilon_\theta)$$
$$T_\theta^{(sh)} = \frac{2h}{s_{11}^E(1-\nu^2)}(\varepsilon_\theta + \nu\varepsilon_r). \tag{30.19}$$

We substitute equations for $T_r^{(sh)}$ and $T_\theta^{(sh)}$ into the second formula 30.15 and integrate over the electrode surface to obtain the internal energy $U^{(sh)}$.

The difference between the energy of the plate with disconnected electrodes and the short-circuited electrodes under invariable strains is the electrical energy that can be converted. The energy possessed by the plate with disconnected electrodes is equal to the sum of its mechanical and electrical energies. By dividing $U^{(d)} - U^{(sh)}$ by $U^{(d)}$, we get k_e^2.

In Figure 18, we present the EMCC values computed for different frequency parameters λ_* for a PZT-5 piezoceramic plate. The solid line gives the k_e^2 values and the asterisks give the k_d^2 values. Since every value of k_d^2 characterizes the EMCC within some variation range of the frequency parameter, i.e., from resonance to antiresonance, we will place the asterisks in the middle of the interval. We see that all the k_d^2 values lie on the k_e^2 curve.

The problem on the tangential vibrations of a circular piezoceramic plate without computing k_e by the Kirchhoff theory was considered in [26,103].

PROBLEM 2 A Plate Without Electrodes.

This problem is well-studied in elasticity theory, and we give its solution here in order to compare its behavior with that of plates having electrodes. We will also use the solution when considering plates with faces partially covered by electrodes.

The plate vibrates due to an edge load directed along the radius

$$T_r\big|_{r=R} = T_0. \tag{30.20}$$

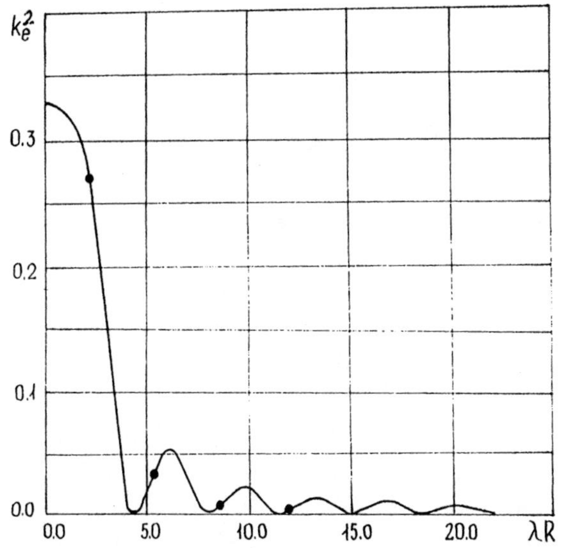

FIGURE 18. Electromechanical coupling coefficient k_e^2 as function of the frequency parameter for a plate with solid electrodes (the k_d^2-values are marked by circles)

The constitutive relations have the form

$$T_r = 2hB(\varepsilon_r + \sigma\varepsilon_\theta)$$
$$T_\theta = 2hB(\varepsilon_\theta + \sigma\varepsilon_r) \tag{30.21}$$

where B and σ are defined by 7.27.

We substitute equations 30.21 and 30.2 into equation 30.1 and obtain the equation

$$r^2\frac{d^2u}{dr^2} + r\frac{du}{dr} + [(\eta r)^2 - 1]u = 0, \qquad \eta^2 = \frac{\rho\omega^2}{B}$$

whose solution has the form

$$u = C_1 J_1(\eta r). \tag{30.22}$$

The formulas for the forces and strains are as follows:

$$T_r = 2hBC_1\left[\eta J_0(\eta r) - \frac{1-\sigma}{r}J_1(\eta r)\right]$$
$$T_\theta = 2hBC_1\left[\sigma\eta J_0(\eta r) + \frac{1-\sigma}{r}J_1(\eta r)\right]$$

$$\varepsilon_r = C_1\eta\left[J_0(\eta r) - \frac{1}{\eta r}J_1(\eta r)\right], \qquad \varepsilon_\theta = C_1\frac{J_1(\eta r)}{r}. \qquad (30.23)$$

We find the arbitrary constant from equation 30.20:

$$C_1 = \frac{T_0}{2hB}\frac{R}{R\eta J_0(\eta_*) - (1-\sigma)J_1(\eta_*)}, \qquad \eta_* = \eta R.$$

The resonance frequencies are found by equating the denominator to zero:

$$\eta R J_0(\eta_*) - (1-\sigma)J_1(\eta_*) = 0.$$

The first four resonance frequencies for a PZT-5 plate are 2.57, 6.40, 10.13, 13.84.

PROBLEM 3 Vibrations of a plate with electrode-covered faces.

The electrodes are disconnected. Suppose that a dynamic edge load 30.20 is acting on the plate. The electroelasticity relations have form 30.4, where V is an unknown constant determined from the following integral condition on the surface Ω of one of the electrodes:

$$-i\omega\int_\Omega D_3 d\Omega = 0.$$

After simple transformations, we get

$$V = \frac{h}{R}\frac{k_p^2}{d_{31}(1-k_p^2)}u\Big|_{r=R} = \frac{h}{R}\frac{k_p^2}{d_{31}(1-k_p^2)}C_1 J_1(\lambda r).$$

We find C_1 from the edge condition 30.20:

$$C_1 = \frac{T_0}{2h}\frac{Rs_{11}^E(1-\nu^2)}{\lambda R J_0(\lambda_*) - (1-\nu-a)J_1(\lambda_*)}$$

$$a = \frac{(1+\nu)k_p^2}{1-k_p^2}.$$

By equating the denominator to zero, we get the equation for determining the resonance frequencies. The first four resonance frequencies are 2.46, 5.54, 8.67, 11.80.

PROBLEM 4 A round plate with circular electrodes.

Suppose that the electrodes cover the area $R_0 \le r \le R$ of the plate faces, and we are given the electrical potential (equation 30.5) on them. The remaining parts of the faces have no electrodes, and there is no mechanical load.

For the part with the electrodes, equations 30.2 and 30.4 and 30.6 describe the problem. It has a solution

$$u^{(e)} = C_2 J_1(\lambda r) + C_3 Y_1(\lambda r). \tag{30.24}$$

The superscripts (e) and (n) will refer to the plate parts covered and not covered by the electrodes respectively.

For the part without electrodes, we will have equations 30.2 and 30.23 and solution 30.22. The arbitrary constants for these solutions are found using the conditions at the interface $r = R_0$ between the electrode-covered part of the plate and the part without electrodes

$$\begin{aligned} u^{(n)}\Big|_{r=R_0} &= u^{(e)}\Big|_{r=R_0} \\ T_r^{(n)}\Big|_{r=R_0} &= T_r^{(e)}\Big|_{r=R_0} \end{aligned} \tag{30.25}$$

and the conditions at the plate edge $r = R$

$$T_r\Big|_{r=R} = 0. \tag{30.26}$$

We get the system

$$\sum_{i=1}^{3} C_i a_{ij} = b_j, \qquad j = 1, 2, 3$$

$$\begin{aligned}
a_{11} &= J_1(\eta R_0), \\
a_{12} &= -J_1(\lambda R_0), \\
a_{13} &= -Y_1(\lambda R_0) \\
a_{21} &= B s_{11}^E \left[\eta J_0(\eta R_0) - \frac{1-\sigma}{R_0} J_1(\eta R_0)\right] \\
a_{22} &= -\frac{1}{1-\nu^2}\left[\lambda J_0(\lambda R_0) - \frac{1-\nu}{R_0} J_0(\lambda R_0)\right] \\
a_{23} &= -\frac{1}{1-\nu^2}\left[\lambda Y_0(\lambda R_0) - \frac{1-\nu}{R_0} Y_1(\lambda R_0)\right] \\
a_{31} &= 0, \\
a_{32} &= \frac{1}{1-\nu^2}\left[\lambda J_0(\lambda_*) - \frac{1-\nu}{R} J_1(\lambda_*)\right] \\
a_{33} &= \frac{1}{1-\nu^2}\left[\lambda Y_0(\lambda_*) - \frac{1-\nu}{R} Y_1(\lambda_*)\right], \\
b_1 &= 0 \\
b_2 &= \frac{d_{31}}{1-\nu}\frac{V}{h} \\
b_3 &= -\frac{d_{31}}{1-\nu}\frac{V}{h}
\end{aligned}$$

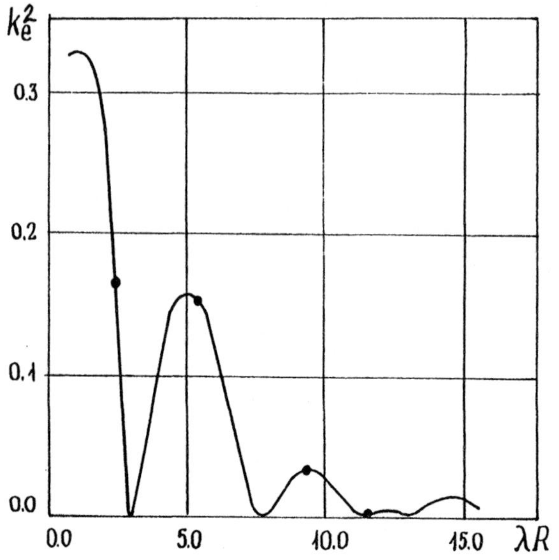

FIGURE 19. Electromechanical coupling coefficient k_e^2 as function of the frequency parameter for a plate partially covered electrodes for $R_0/R = 0.5$(the k_d^2-values are marked by circles)

for determining C_1, C_2, and C_3.

All these formulas were used to design a PZT-5 piezoceramic plate with $R_0 = R/2$. The resultant first four resonance frequencies are 2.31, 5.82, 9.28, 12.69.

The EMCC values found from equations 30.13 and 30.14 are given in Figures 19 to 21. Note that as in Problem 1, the respective values of k_d^2 and k_e^2 are in good agreement, which verifies the correctness of the energy formula for EMCC.

For a plate with faces completely covered with electrodes, the greatest EMCC is at the first resonance frequency. For a plate with circular electrodes, it is at the second resonance frequency.

Figure 22 shows the dependence of the EMCC on the ratio R_0/R for the first and second resonance frequencies.

31 ACTIVE SUPPRESSION OF VIBRATIONS IN ELECTROELASTIC BARS AND ROUND PLATES BY MEANS OF PIEZOEFFECT

In this section, we demonstrate, by solving some simple problems, that vibrations of spatial constructions can be suppressed using the electrically forced vibrations of piezoceramic and piezocomposite elements [94].

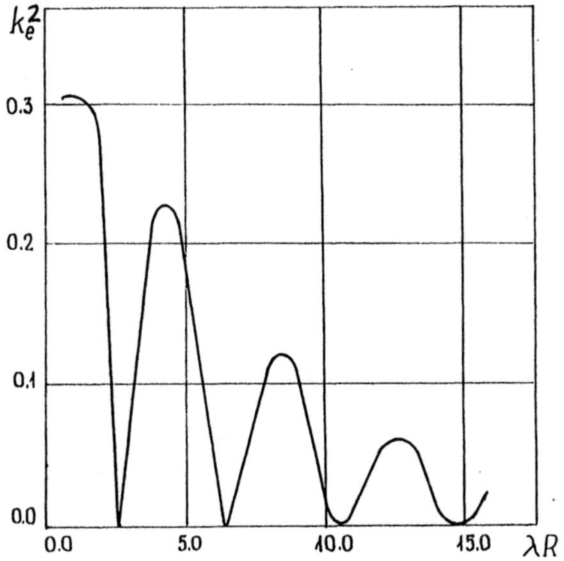

FIGURE 20. Electromechanical coupling coefficient k_e^2 as function of the frequency parameter for a plate partially covered electrodes for $R_0/R = 0.75$

PROBLEM 1

Suppose that a piezoceramic bar with longitudinal polarization performs longitudinal vibrations due to a longitudinal load applied to its edges and varying by a harmonic law $e^{-i\omega t}$, where t is the time and ω is the angular vibration frequency. As usual, we will write all equations for the amplitude values. The initial system consists of the

Equilibrium equation

$$\frac{d\sigma_*}{d\xi} + \lambda u_* = 0 \tag{31.1}$$

Constitutive relation

$$\sigma_* = \frac{du_*}{d\xi} - E_* \tag{31.2}$$

Piezoeffect equation

$$D_* = E_* + a\sigma_*, \qquad a = \frac{d_{33}^2}{s_{33}^E \varepsilon_{33}^T} \tag{31.3}$$

Electrostatics equations

$$\frac{dD_*}{d\xi} = 0, \qquad E_* = -\frac{d\psi_*}{d\xi} \tag{31.4}$$

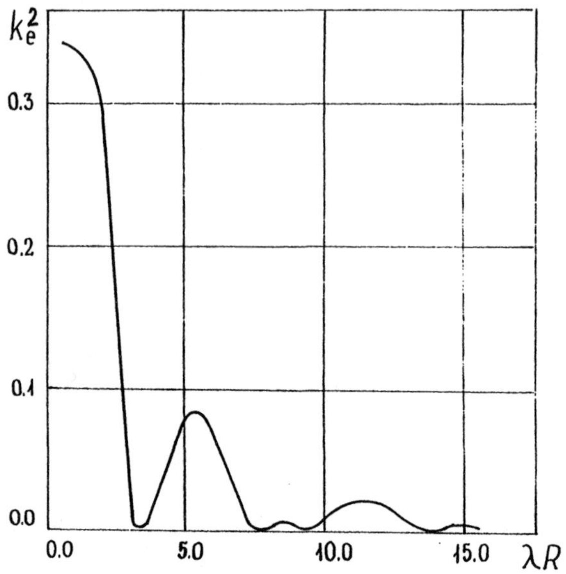

FIGURE 21. Electromechanical coupling coefficient k_e^2 as function of the frequency parameter for a plate partially covered electrodes for $R_0/R = 0.25$

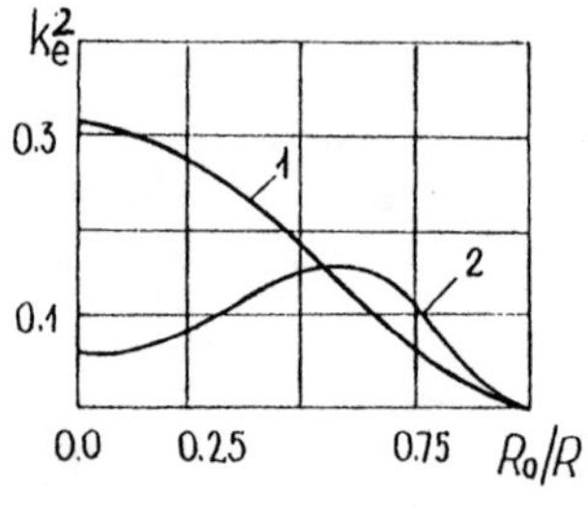

FIGURE 22. The EMCC of a plate as function of ratio R_0/R for (1) the first and (2) the second resonance frequencies

For stresses, the boundary conditions at the bar ends have the form

$$\sigma_* \Big|_{\xi=\pm 1} = 1.$$

On the electrodes covering the bar ends Ω, the boundary conditions for the electrical quantities have the form for

Disconnected electrodes

$$\int_{\Omega} D_* d\Omega = 0 \tag{31.5}$$

Short-circuited electrodes

$$\psi_* \Big|_{\Omega} = 0 \tag{31.6}$$

The potential difference on the electrodes is given as

$$\psi_* \Big|_{\Omega} = \pm V_*. \tag{31.7}$$

In equations 31.1 to 31.4, we use dimensionless quantities, dimensionless variable ξ, and dimensionless parameter λ:

$$\sigma_* = s_{33}^E \sigma, \qquad u_* = \frac{u}{l}, \qquad E_* = d_{33} E \qquad D_* = \frac{d_{33}}{\varepsilon_{33}^T} D,$$

$$\psi_* = \frac{d_{33}}{l} \psi, \qquad V_* = \frac{d_{33}}{l} V \qquad \lambda = \rho \omega^2 l^2 s_{33}^E, \qquad \xi = \frac{x}{l}$$

where $2l$ is the bar length.

Let us find the solution to the system of equations 31.1 to 31.4 for the different boundary conditions 31.5 to 31.7. When working with the conditions 31.7, we assume that the potential difference V_* on the electrodes is an unknown quantity. We will determine it under the condition that the bar ends are not displaced.

The solution of the problem has the form

$$u_* = C_2 \sin kx, \qquad \sigma_* = \frac{1}{1+a}(kC_2 \cos kx + C_1), \qquad k^2 = \frac{\lambda}{1+a}$$

where C_1 and C_2 are determined by the following formulas

$$C_1 = 0, \qquad C_2 = \frac{1+a}{k \cos k}$$

for the conditions 31.5;

$$C_1 = \frac{a(1+a)\sin k}{k \cos k + a \sin k}, \qquad C_2 = \frac{1+a}{k \cos k + a \sin k}$$

for the conditions 31.6;

$$C_1 = 1 + a, \qquad C_2 = 0$$

for the conditions 31.7.

The computational results for a bar made of PZT-4 are given in Figure 23. The solution for the boundary condition 31.5 is marked by 1, that for condition 31.6 by 2, and that for condition 31.7 by 3. We see that by properly choosing the potential difference $V_* = 1 + a$, we can eliminate any displacement. In this case, the stresses will be identically equal to one.

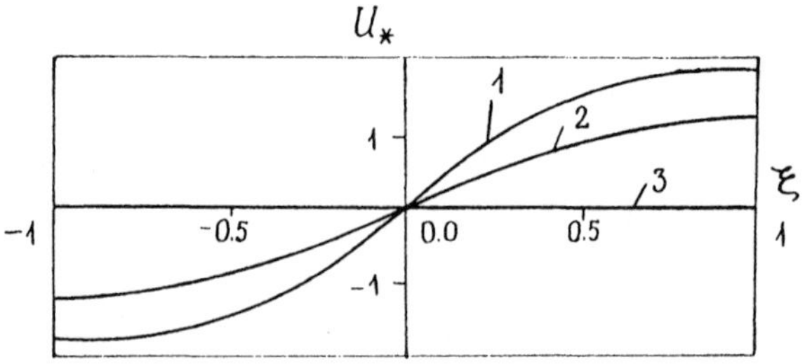

FIGURE 23. Distribution of displacements along the bar length for different boundary conditions: (1),(2) with only mechanical load and (3) with an additional specially chosen electrical load

PROBLEM 2

Consider the bending vibrations of a two-layer beam with bimorphic properties. One end of the beam is clamped and the other is loaded by a periodically varying bending moment with transverse forces. The beam is subject to a normal load uniformly distributed over its length. The faces of the beam are covered by $2n$ electrodes with cuts (see Figure 24) that are symmetric relative to the middle surface. The dimensionless length of every beam part covered with a solid electrode is $\Delta\xi = 1/n$. The beam's dimensionless length is equal to one.

For brevity, we will use only dimensionless quantities. The starting system of the differential equations has the form

$$\frac{dN_*}{d\xi} + \lambda^4 w_* = Z_*, \qquad N_* = \frac{dG_*}{d\xi}, \qquad G_* = -\kappa_* + V_*$$

$$\kappa_* = \frac{d^2 w_*}{d\xi^2}, \qquad \gamma_* = -\frac{dw_*}{d\xi} \qquad (0 \leq \xi \leq 1). \tag{31.8}$$

Here, the dimensionless quantities with asterisks were introduced by the formulas

$$w_* = \frac{w}{l}, \qquad G_* = \frac{3l}{2B_{11}h^3}G, \qquad N_* = \frac{3l^2}{2B_{11}h^3}N$$

$$V_* = \frac{3ld_{31}}{B_{11}h^2 s_{11}^E}V, \qquad \lambda^4 = \frac{3\rho\omega^2 l^4}{B_{11}h^2}.$$

Here, B_{11}, ν_* are given by 12.7 in Section 12, and l is the beam length.

We will solve two problems. In the first problem, all the electrodes are treated as short-circuited. In the second problem, a potential difference is applied to each pair of the electrodes symmetric with respect to the middle surface so that the deflections at the ends of every segment are zero. We will consider each beam segment separately.

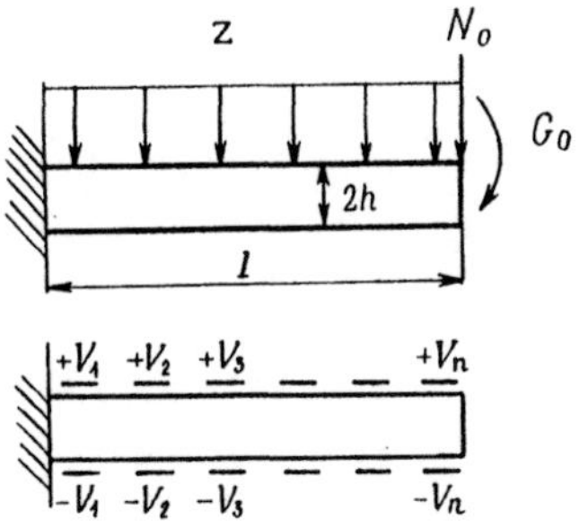

FIGURE 24. The geometry of a bimorphic bar with cutted electrodes and the mechanical load acting on it

For the ith segment covered with solid electrodes, the solution has the form

$$w_i = C_{1i}e^{-\xi\lambda} + C_{2i}e^{\xi\lambda} + C_{3i}\sin\lambda\xi + C_{4i}\cos\lambda\xi + W_i \qquad (31.9)$$

where W_i is a particular solution. The constants will be found from the boundary conditions at the beam's ends

$$w_1\Big|_{\xi=0} = 0, \qquad \gamma_1\Big|_{\xi=0} = 0, \qquad G_n\Big|_{\xi=1} = G_0, \qquad N\Big|_{\xi=1} = N_0. \qquad (31.10)$$

The conditions at the interface and the condition that the deflections at the junction points of all the segments $\xi_i = i\Delta\xi, i = 1, 2, \ldots, n-1$ are zero:

$$w_i = w_{i+1} = 0, \qquad \gamma_i = \gamma_{i+1}, \qquad G_i = G_{i+1}, \qquad N_i = N_{i+1}. \qquad (31.11)$$

The computational results for a bar of PZT-4 are plotted in Figure 25. Curve (a) in Figure 25 shows the deflection w_* as a function of ξ when the bar is subjected to only mechanical load ($N_0 = 1, G_0 = 1, Z_* = 1$). Curve (b) in Figure 25 depicts the deflection due to the electrical and mechanical loads. It turns out that, for $n = 3$, $\lambda = 1$, the deflection due to the electrical load decreases 10^4 times.

PROBLEM 3

Suppose that a circular two-layer plate (bimorph) with an opening in the center performs forced bending axisymmetrical vibrations under the action of bending moments and transverse forces distributed over the outer edge. The inner edge is rigidly clamped. The faces are completely covered by circular electrodes with cuts. As for the bar, we solve two problems: the one for short-circuited electrodes and the other where on every pair of electrodes symmetric to the

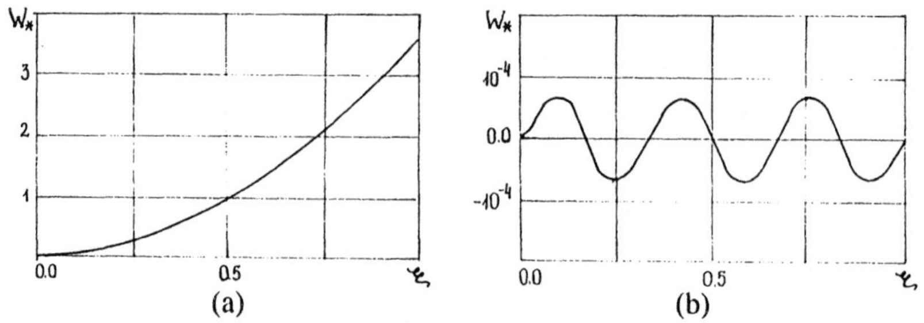

FIGURE 25. Distribution of displacements along the length of a bimorphic bar (a) under the effect of a pure mechanical load and (b) after suppressing the vibrations by an electrical load

middle surface the potential difference is chosen so that the deflections at the electrode cuts are zero.

The initial system to the dimensionless quantities consists of the *Equilibrium equations*

$$N_{1*} = \frac{1}{\xi}\frac{d}{d\xi}(\xi G_{1*}) - \frac{1}{\xi}G_{2*}, \qquad \frac{1}{\xi}\frac{d}{d\xi}(\xi N_1 *) + Z_* + \lambda^4 w_* = 0$$

Electroelasticity relations

$$G_{1*} = -(\kappa_{1*} + \nu_* \kappa_{2*}) - V_*$$
$$G_{2*} = -(\kappa_{2*} + \nu_* \kappa_{1*}) - V_*$$

Strain-displacement formulas

$$\gamma_{1*} = -\frac{dw_*}{d\xi}, \qquad \kappa_{1*} = -\frac{d\gamma_{1*}}{d\xi}, \qquad \kappa_{1*} = -\frac{1}{\xi}\gamma_{1*}$$

The dimensionless quantities with asterisks were introduced by

$$w_* = \frac{w}{R}, \qquad G_{i*} = \frac{3R}{2B_{11}h^3}G_i, \qquad N_{1*} = \frac{3R^2}{2B_{11}h^3}N_1$$
$$V_* = \frac{3Rd_{31}}{B_{11}h^2 s_{11}^E(1-\nu)}V, \qquad \lambda^4 = \frac{3\rho\omega^2 R^4}{B_{11}h^2}$$

where R is the radius and $2h$ is the thickness of the plate. The numbers B_{11} and ν_* are given by formulas 12.7. The resultant equation

$$\nabla^4 w_{i*} - \lambda^4 w_{i*} = Z_*$$
$$\nabla^2 = \frac{d^2}{d\xi^2} + \frac{1}{\xi}\frac{d}{d\xi} \qquad (31.12)$$

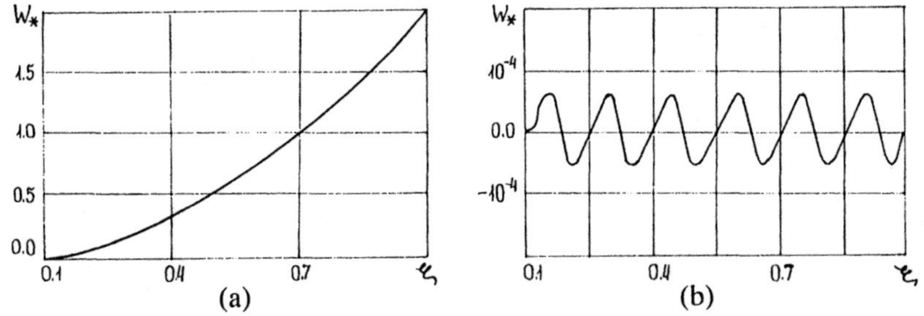

FIGURE 26. Distribution of displacements along the radius of a bimorphic plate (a) under the effect of a pure mechanical load and (b) after suppressing the vibrations by an electrical load

has a solution

$$w_{i*} = C_{1i}J_0(\lambda\xi) + C_{2i}Y_0(\lambda\xi) + C_{3i}I_0(\lambda\xi) + C_{4i}K_0(\lambda\xi) + W_i.$$

Here, W_i is particular solution of the equation 31.12.

We satisfy conditions 31.10 and 31.11 and find the integration constants and the unknown potential differences. The computational results are given in Figure 26. Curve (a) in Figure 26 shows the deflection w_* as a function of the radial coordinate when the plate is subject to only mechanical load ($N_0 = 1, G_0 = 1, Z_* = 1$). Curve (b) in Figure 26 depicts the plate's deflection after suppressing the vibrations by a specially chosen electrical load ($n = 6$). All the solutions show that piezoeffect can be used to diminish the vibration amplitude.

32 RESONANCE METHOD FOR MEASURING FLUID VISCOSITY USING A PIEZOELEMENT

It was shown in [86] that the problem of the vibration of an elastic body immersed in a viscous fluid can be essentially simplified. A simultaneous asymptotic analysis of the equations describing the elastic body and the Navier-Stoke's equations describing the viscous fluid show that the Navier-Stoke's system can be approximately replaced by simpler systems. One describes the fluid boundary layer quickly decreasing with the distance from the elastic body, and the other describes the penetrating perturbation in the fluid. The latter can be reduced to an equation similar to the Helmholtz equation.

The system for the fluid boundary layer can easily be integrated. We use the solution to express the tangential stresses at the interface between the body and the fluid in terms of the tangential displacements of points on the surface of the body.

It is shown in [92] that for shells and plates performing tangential vibrations in a fluid, the effect of the viscous fluid is allowed for by introducing a load X_i into the equations for the motion of the surface load, so that

$$\frac{1}{A_i}\frac{\partial T_i}{\partial \alpha_i} + \frac{1}{A_j}\frac{\partial S}{\partial \alpha_j} + k_j\left(T_i - T_j\right) + 2k_i S - \frac{N_i}{R_i} + 2h\rho\omega^2 u_i + X_i = 0.$$

Here,

$$X_i = 2(1+i)\omega^2\sqrt{\frac{\rho_f \mu}{2\omega}}u_i$$

ρ_s is the density of the material of the shell, ρ_f is the density of the fluid, and μ is the viscosity of the fluid.

We use ω_v to denote the resonance frequency of the tangential vibrations of the plate (shell) in a vacuum or air and ω_f to denote its resonance frequency in the viscous fluid. The system describing the vibrations in a vacuum differs from that for fluid by an inertial term: $2h\rho_s\omega_v^2 u_i$ for a vacuum and

$$2h\rho_s\omega_f^2\left(1 + \frac{1+i}{h\rho_s}\sqrt{\frac{\rho_f\mu}{2\omega_f}}\right)u_i$$

for the viscous fluid. We can now obtain the formula for the viscosity of the fluid:

$$2h\rho_s\omega_v^2 = 2h\rho_s\omega_f^2\left(1 + \frac{1+i}{h\rho_s}\sqrt{\frac{\rho_f\mu}{2\omega_f}}\right). \tag{32.1}$$

By neglecting the small quantities of the order of ε in the obtained formula, we get

$$\begin{aligned} \mu &= \delta^2(2h)^2\rho_f^2 2\omega_f/\rho_s, \\ 2\delta &= \omega_v^2/\omega_f^2 - 1. \\ \varepsilon &= \frac{1}{h\rho_s}\sqrt{\frac{\rho_f\mu}{2\omega_f}} \end{aligned} \tag{32.2}$$

The viscosity should be measured using the tangential resonance frequencies of the piezoelement. We should (1) measure the resonance frequency in the air, (2) measure the same frequency in the fluid, and (3) compute the fluid viscosity by equation 32.1 or equation 32.2.

It is difficult to excite such vibrations by a mechanical load while the electric energy in a specially chosen piezoelement can provoke vibrations in the plane of the element.

We consider the tangential vibrations of the piezoceramic plates and shear or twist vibrations of piezoceramic cylindrical shell. For example, a plate with

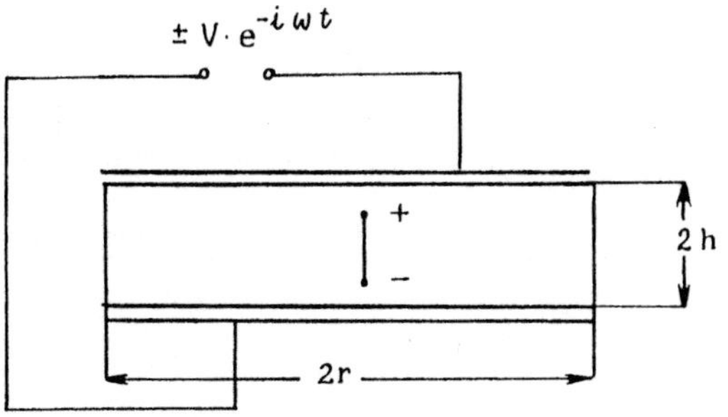

FIGURE 27. Piezoceramic circular plate with electrode-covered faces

TABLE 4.
The first resonance frequency as a function of plate sizes in vacuum and in glycerine

Hz	$r = 0.01$ m $2h = 0.0005$ m	$r = 0.01$ m $2h = 0.001$ m	$r = 0.02$ m $2h = 0.0005$ m	$r = 0.02$ m $2h = 0.001$ m
ω_v	114906.	114906.	57453.2	57453.2
ω_f	110513.	112645.	54417.5	55873.6

thickness polarization (with radius R and thickness $2h$) performs such vibrations under the effect of an electrical load applied to the electrodes on its faces (see Figure 27). The problem was solved in Section 30. In order to illustrate this theory, we compute the first resonance frequency of a plate made of PZT-4 piezoceramic in vacuum ω_v and glycerine ω_f. The results can be found in Table 4.

Let us consider one more example of a circular cylindrical shell (with radius R, length $2l$, and thickness $2h$) polarized in the circular direction (Figure 28). The twist vibrations are exited by an electrical load applied to the electrodes on the ends. The conditions at the free electrode-covered edges are written as

$$S\Big|_{x=\pm l} = 0, \qquad \psi\Big|_{x=\pm l} = \pm V. \tag{32.3}$$

The equations of the shell have the form

$$\frac{dS}{dx} + 2h\rho\omega^2 v = 0$$

$$S = \frac{2h}{s_{44}^E}\left(\frac{dv}{dx} - d_{15}E_1\right)$$

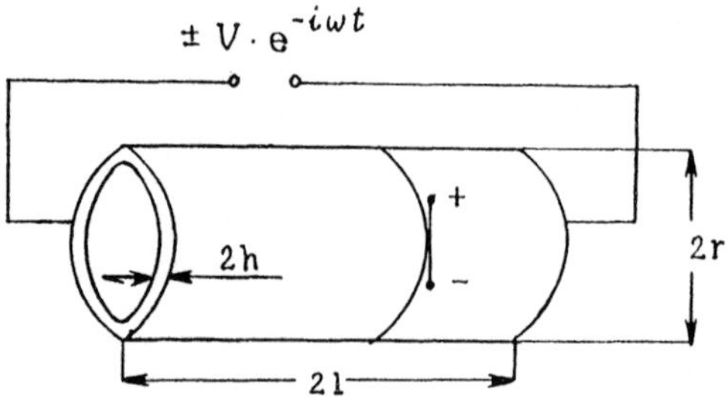

FIGURE 28. Piezoceramic cylindrical shell with electrode-covered edges

TABLE 5.
The first resonance frequency as a function of cylindrical shell sizes in vacuum and in glycerine

Hz	$l = 0.02$ m $r = 0.01$ m $2h = 0.0005$ m	$l = 0.02$ m $r = 0.01$ m $2h = 0.001$ m	$l = 0.04$ m $r = 0.02$ m $2h = 0.0005$ m	$l = 0.04$ m $r = 0.02$ m $2h = 0.001$ m
ω_v	12107.	12107.	5952.4	5952.4
ω_f	10829.	11415.	5111.1	5483.9

$$D_1 = \varepsilon_{11}^T E_1 + \frac{d_{15}}{2h} S$$

$$\frac{dD_1}{dx} = 0,$$

$$E_1 = -\frac{d\psi}{dx}. \tag{32.4}$$

Here, the x-line coincides with the generatrix of the cylindrical surface and v is the displacement in the circular direction.

We can transform system 32.4 into two equations with respect to the unknowns v, ψ:

$$\frac{d^2 v}{dx^2} + \lambda v = 0$$

$$\frac{d^2\psi}{dx^2} + \frac{d_{15}\rho\omega^2}{\varepsilon_{11}^T}\psi = 0$$

$$\lambda = (1 - k^2)\rho\omega^2 s_{44}^E$$

$$k^2 = \frac{d_{15}^2}{\varepsilon_{11}^T s_{44}^E}. \tag{32.5}$$

The solution to equation 32.4 is written as

$$v = c_1 \sin \sqrt{\lambda} x + c_2 \cos \sqrt{\lambda} x, \qquad d_{15}\psi = \frac{k^2}{1-k^2} v + c_3 x + c_4$$

where c_1, c_2, c_3, and c_4 are arbitrary constants.

By satisfying the boundary conditions 32.3 and equating the determinant of the system to zero, we obtain the equation for resonance frequency

$$\sin \sqrt{\lambda} l (\lambda \cos \sqrt{\lambda} l - k^2 \sin \sqrt{\lambda} l) = 0.$$

We calculate the first resonance frequency of a circular piezoceramic cylindrical shell made of PZT-4 in a vacuum and in glycerine for different shell sizes. The results are found in Table 5. Note that glycerine essentially reduces resonance frequencies.

The sensitivity of the piezoelement can be raised either by decreasing its thickness or by using a material with greater compliance. The advantages of this method are the simplicity of excitation and the accuracy in measuring the resonance frequencies of the piezoelement.

33 USING A PIEZOCERAMIC ELEMENT WITH THICKNESS POLARIZATION AS A STRAIN GAUGE

We suggest a method for measuring dynamic strains of a body using a piezoceramic gauge [36], which is qualitatively different from the known method for measuring static forces. In the latter method, we use the static force applied to a quartz element whose resonance frequency is changed and the resonance frequencies of the loaded and unloaded quartz element to compute the static strain. Our method is to measure the electrical signals appearing due to the piezoeffect while a harmonic wave passes through the elastic body and the piezoceramic gauge is glued to it.

Consider a $2m \times 2l \times 2h$ piezoceramic plate where $2m$ is the width, $2l$ is the length, and $2h$ is the thickness. The plate is pre-polarized to the thickness and its faces are covered by electrodes. One of the faces is glued to the investigated body with an insulating adhesive. The electrodes of the plate are connected by an external contour with known complex conductivity $Y = Y_0 + iY_1$.

We assume that (1) the plate sizes are small compared to the smallest size of the body and (2) the plate sizes are small compared to characteristic size of the body strain pattern so that within the region of the plate and body contact the plate strain can be treated as constant. The three-dimensional dynamic equations for the piezoceramic plate comprise the (see Section 2)

Equations of motion

$$\frac{\partial \sigma_{ii}}{\partial x_i} + \frac{\partial \sigma_{ij}}{\partial x_j} + \frac{\partial \sigma_{i3}}{\partial z} - \rho_0 \frac{\partial^2 v_i}{\partial t^2} = 0$$

$$\frac{\partial \sigma_{13}}{\partial x_1} + \frac{\partial \sigma_{23}}{\partial x_2} + \frac{\partial \sigma_{33}}{\partial z} - \rho_0 \frac{\partial^2 v_3}{\partial t^2} = 0 \tag{33.1}$$

Piezoeffect equations

$$\sigma_{ii} = \frac{1}{s_{11}^E(1-\nu^2)}(e_i + \nu e_j) - \frac{s_{13}^E}{(1-\nu)s_{11}^E}\sigma_{33} - \frac{d_{31}}{s_{11}^E(1-\nu)}E_3$$

$$\sigma_{ij} = \frac{1}{s_{66}^E}(m_i + m_j) \tag{33.2}$$

$$\frac{\partial v_3}{\partial z} = s_{13}^E(\sigma_{11}+\sigma_{22}) + s_{33}^E\sigma_{33} + d_{33}E_3 \tag{33.3}$$

$$\frac{\partial v_i}{\partial z} + g_i = s_{44}^E \sigma_{i3} + d_{15}E_i \tag{33.4}$$

$$D_i = \varepsilon_{11}^T E_i + d_{15}\sigma_{i3}$$

$$D_3 = \varepsilon_{33}^T E_3 + d_{31}(\sigma_{11} + \sigma_{22}) + d_{33}\sigma_{33} \tag{33.5}$$

Electrostatics equations

$$\operatorname{div} \mathbf{D} = 0, \qquad \mathbf{E} = -\operatorname{grad} \psi \tag{33.6}$$

Strain-displacement formulas

$$e_i = \frac{\partial v_i}{\partial x_i}, \qquad m_i = \frac{\partial v_i}{\partial x_j}, \qquad g_i = \frac{\partial v_3}{\partial z}$$

$$(|z| \le h, \qquad |x_1| \le m, \qquad |x_2| \le 1). \tag{33.7}$$

We assume that on the contact surface $z = -h$ and that the displacements of the plate are equal to the displacements of the body u_i, u_3,

$$v_i = u_i, \qquad v_3 = u_3, \qquad (z = -h). \tag{33.8}$$

The electrical potential on the electrodes is known (measured):

$$\psi\Big|_{z=\pm h} = \pm V. \tag{33.9}$$

On the remaining surfaces of the plate, we have the conditions

$$\sigma_{33} = \sigma_{i3} = 0 \qquad (z = +h) \tag{33.10}$$

$$D_1 = 0,$$

$$\sigma_{11} = \sigma_{13} = \sigma_{12} = 0 \qquad (x_1 = \pm m) \tag{33.11}$$

$$D_2 = 0$$

$$\sigma_{22} = \sigma_{23} = \sigma_{21} = 0 \qquad (x_2 = \pm l). \tag{33.12}$$

Since the electrodes are connected by an external contour, the integral condition

$$-i\omega\int_s D_3 ds = 2VY \tag{33.13}$$

should hold. The integral is taken over the surface of one of the electrodes. Besides, we assume that the body and the piezoplate vibrate by the law $e^{-i\omega t}$, where ω is the angular frequency of the harmonic vibrations and t is the time.

Let us design the piezoplate approximately. Suppose that $h \ll \min(m, l)$. The total electroelastic state of the plate can be represented as a sum of the internal electroelastic state and the electroelastic boundary layer. Since the plate is thin, we can discard, with some error, the boundary layer localized near the edges. We will assume that the plate strains are equal to those of the body and are constant and independent of the coordinates within the plate.

We will neglect the stresses σ_{33} compared to the principal stresses σ_{ii} in equation 33.2:

$$\sigma_{ii} = \frac{1}{s_{11}^E(1-\nu^2)}(\varepsilon_i + \nu\varepsilon_j) - \frac{d_{31}}{s_{11}^E(1-\nu^2)}E_3. \tag{33.14}$$

Here, ε_i are the extension-compression strains along the coordinates x_i.

Suppose that the electrical potential varies with thickness by a linear law

$$\psi = \psi^{(0)} + z\psi^{(1)}.$$

From conditions 33.9, we find

$$\psi^{(1)} = \frac{V}{h}, \qquad E_3 = -\frac{V}{h}. \tag{33.15}$$

We use our assumptions and substitute equations 33.14 and 33.15 into equation 33.13 to get

$$\varepsilon_1 + \varepsilon_2 = Re\left[i\frac{2VY}{\omega s}\frac{s_{11}^E(1-\nu)}{d_{31}} + d_{31}\left(\frac{1-\nu}{k^2} - 2\right)\frac{V}{h}\right],$$

$$k^2 = \frac{d_{31}^2}{\varepsilon_{33}^T s_{11}^E}. \tag{33.16}$$

The sum of strains is found using equation 33.16.

If the problem is one-dimensional or $\varepsilon_1 = \varepsilon_2$, formula 33.16 allows us to completely determine the strains, and hence the stresses, of the body below the piezoceramic plate. In order to find ε_1 and ε_2, we take a piezoceramic bar instead of the plate. The width of the bar is $2m$ and is comparable with the thickness $2h$. The length is $2l \gg 2h$.

For a piezoceramic gauge in the form of a thin plate, we can discard the boundary layer because it only spreads over a narrow zone near the edge of thickness $2h$. Its contribution into the electrical potential is of the order $h(l + m)/lm$ compared to the contribution of the inner electroelastic state. For a bar, we can compare the boundary layer contribution with that of the inner electroelastic state.

The total electroelastic state of the piezobar is represented as a sum of the inner electroelastic state and the boundary layers. We will only take into account the boundary layers appearing near the long edges of the bar $x_1 = \pm m$. The boundary layers appearing near the short edges $x_2 = \pm l$ can be neglected.

We will solve the problem in two stages. In the first stage, we determine the inner electroelastic state by equation 33.14. After solving the inner problem, we get errors in the boundary conditions 33.11 at the edges $x_1 = \pm m$ because, in this case, we have $\sigma_{11} \neq 0, \sigma_{12} \neq 0$. The errors are eliminated by introducing a boundary layer that in turn is broken into the plane and antiplane boundary layers (see Section 38). For a correct computation of the strain by equation 33.13 which allows for the boundary layer, it is sufficient to consider the plane boundary layer with conditions

$$\begin{aligned} &\sigma_{11}^{(1)} = -\sigma_{11}^{(0)}, \qquad \sigma_{13}^{(1)} = 0, \qquad D_1^{(1)} = 0 \qquad (x_1 = \pm m) \\ &\sigma_{13}^{(1)} = \sigma_{33}^{(1)} = 0 \qquad \psi^{(1)} = 0 \qquad (z_1 = +h). \end{aligned} \tag{33.17}$$

The superscripts (0) and (1) denote the quantities of the inner electroelastic state and the boundary layer, respectively.

At the surface $z = -h$, the piezoelectric gauge has contact with the elastic body. The contact conditions are

$$\begin{aligned} &\sigma_{13}^{(1)} = \tau_{13}, \qquad \sigma_{33}^{(1)} = \tau_{33}, \qquad v_1^{(1)} = u_1 \\ &v_3^{(1)} = u_3, \qquad \psi^{(1)} = 0. \end{aligned} \tag{33.18}$$

Here, $\tau_{13}, \tau_{33}, u_1$, and u_3 are the stresses and displacements of the elastic body at the contact surface.

The equations of the plane electroelastic problem for a piezoelectric gauge are given in Section 38. Let us represent D_3 as a sum

$$D_3 = D_3^{(0)} + D_3^{(1)} \tag{33.19}$$

where $D_3^{(0)}$ is computed at the first stage of the problem and $D_3^{(1)}$ is computed at the second stage as a result of the joint solution of the plane electroelastic problem and the problem for the elastic body. Since the size of the piezoelectric gauge is small compared to the size of the body, we can approximately represent the elastic body by a half-space on whose surface the gauge is placed. We rewrite the integral condition 33.13 allowing for equation 33.19:

$$\int_s D_3^{(1)} ds + \int_s D_3^{(0)} ds = i\frac{2VY}{\omega}. \tag{33.20}$$

We see from 33.17 that the second integral is proportional to $\sigma_{11}^{(0)}$. Therefore, we can introduce

$$\int_s D_3^{(1)} ds = d_{31} p \sigma_{11}^{(0)} S. \tag{33.21}$$

Here, p is a dimensionless constant computed after solving the problem with the conditions 33.17 and 33.18, where we replace the first condition 33.17 by

$$\sigma_{11}^{(1)} = -1.$$

Let us substitute equation 33.21 into equation 33.20:

$$-\varepsilon_{33}^T \frac{V^{(x_2)}}{h} + d_{31}\left(\sigma_{11}^{(0)} + \sigma_{22}^{(0)}\right) - p d_{31} \sigma_{22}^{(0)} = i\frac{2V^{(x_2)}Y}{\omega s}. \tag{33.22}$$

Here, we have taken into account that

$$D_3^{(0)} = -\varepsilon_{33}^T \frac{V}{h} + d_{31}\left(\sigma_{11}^{(0)} + \sigma_{22}^{(0)}\right).$$

The superscript (x_2) means that the gauge is placed along the x_2-axis.

We introduce another identical piezoceramic bar placed along the x_1-axis very close to the first bar gauge so that we can treat the stressed-strained state of the investigated body near the two bars as a constant quantity. Then we can write a relation of form equation 33.22 for the second gauge:

$$-\varepsilon_{33}^T \frac{V^{(x_1)}}{h} + d_{31}\left(\sigma_{11}^{(0)} + \sigma_{22}^{(0)} - p\sigma_{11}^{(0)}\right) = i\frac{2YV^{(x_1)}}{\omega s}. \tag{33.23}$$

We will omit the superscript (0) because all quantities, except p in equations 33.22 and 33.23, refer to the inner electroelastic state. We substitute equation 33.14 into equations 33.22 and 33.23 and solve the resultant system with respect to ε_1 and ε_2, to get

$$\begin{aligned}
c_1\varepsilon_1 + c_2\varepsilon_2 &= BV^{(x_2)} \\
c_2\varepsilon_1 + c_1\varepsilon_2 &= BV^{(x_1)} \\
c_1 &= \nu + 1 - p, \\
c_2 &= 1 + \nu(1 - p) \\
B &= \frac{s_{11}^E(1-\nu^2)}{d_{31}}\left[\frac{\varepsilon_{33}^T}{h}\left(1 + \frac{(p-2)k^2}{1-\nu}\right) + \frac{i2Y}{\omega s}\right].
\end{aligned} \tag{33.24}$$

We solve equations 33.24 with respect to

$$\varepsilon_1 = \left(c_1 V^{(x_2)} - c_2 V^{(x_1)}\right) \frac{B}{(c_1^2 - c_2^2)}$$
$$\varepsilon_2 = \left(c_1 V^{(x_1)} - c_2 V^{(x_2)}\right) \frac{B}{(c_1^2 - c_2^2)}. \tag{33.25}$$

The constant p needed for measuring strains can be found by solving the problem with the boundary conditions 33.17 and 33.18 or experimentally.

Our theoretical results were verified by experiments. The measurements were taken for a steel plate rigidly clamped along the edge. The plate had a 0.08 m radius, 0.0018 m thickness, and was subject to resonance vibrations. We used $19 \times 19 \times 0.4$ mm^3 square plates and $19 \times 1.2 \times 0.4$ mm^3 rectangular plates made of ЦТС-19 with thickness polarization and the following physical constants:

$$s_{11}^E = 12 \times 10^{-12}\ \mathrm{m}^2/N$$
$$\varepsilon_{33}^T = 1.63 \times 10^8\ \mathrm{F/m}$$
$$d_{31} = 125 \times 10^{-12}\ \mathrm{C/m}$$
$$\nu = 0.33.$$

The square gauge was glued in the center of a circular plate. We measured the electrical potential at three levels of resonance build-up near the first resonance ($\omega = 600$ Hz, $Y = (1 + i0.113) \times 10^{-6}\Omega^{-1}$). We then computed the respective strains by equation 33.16. We also measured the deflections in the center of the circular plate using holographic interferometry [76]. The deflections were used to compute the strains at the points where the gauges were glued by the formulas

$$\varepsilon_r = h\frac{\partial^2 w}{\partial r^2}, \qquad \varepsilon_\varphi = h\left(\frac{1}{r}\frac{\partial w}{\partial r} + \frac{1}{r^2}\frac{\partial^2 w}{\partial r^2}\right)$$

where r, φ are the polar coordinates of the plate. In this case, we have

$$\varepsilon_r = \varepsilon_\varphi$$

at the central point.

The results obtained by measuring and computing the dynamic strains in the center of the circular plate on the first axisymmetric vibration form are given in Table 6. In the first row, we give the measured electrical potential generated by the gauge at each of the three levels of the resonance built-up. In the second row, we give the strains ε_r obtained by equation 33.16. In the third row, we give the experimentally found values of ε_r. We see that the theoretical and

TABLE 6.
Potential differences and strains near the first resonance at the center of a plate

V [mv]	$\varepsilon_r^{(h)} \times 10^{-6}$	$\varepsilon_r^{(p)} \times 10^{-6}$
260	0.53	0.547
475	0.98	1.00
542	1.14	1.14

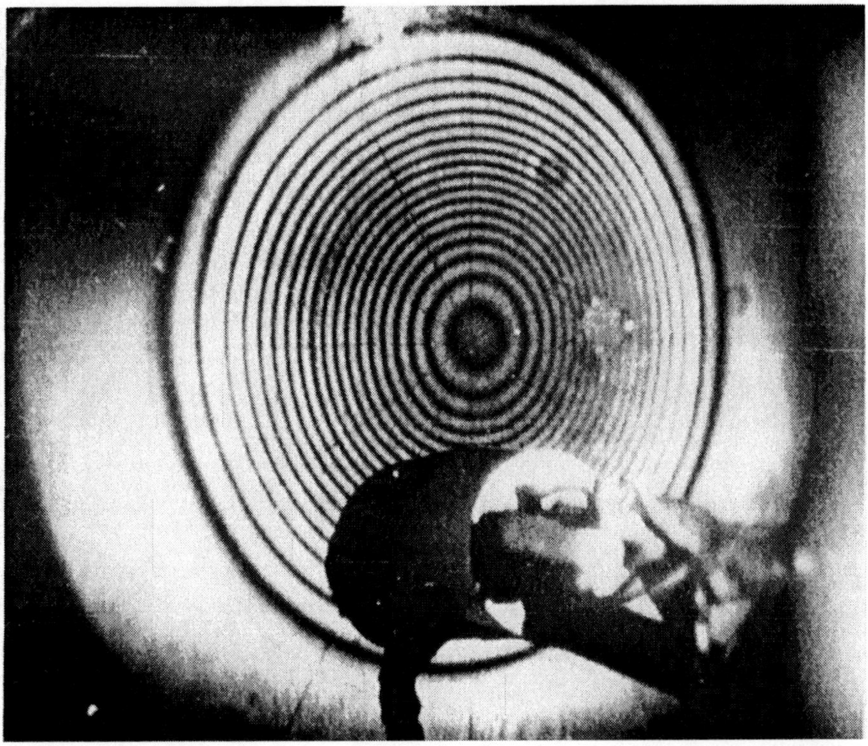

FIGURE 29. Hologram of a vibrating plate near the first resonance

experimental values of the strain coincide for the greatest build-up; and at the lower levels, we have discrepancies of about 2%. Figure 29 shows the first resonance axisymmetric form.

For the rectangular piezoceramic gauge glued in the center of the circular plate, we found the constant p on the same vibration form. We measured the electrical potential and experimentally found the respective strains. We then computed p from one of the formulas in 33.24. For our gauge, p turned out

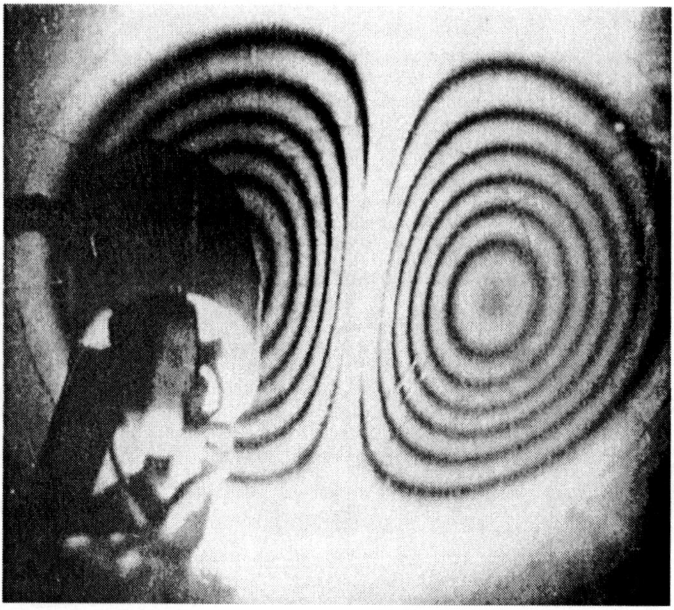

FIGURE 30. Hologram of a vibrating plate near the second resonance

TABLE 7.
Potential differences and strains near the second resonance in the region of the highest amplitude of a plate

V_1 [mv]	V_2 [mv]	$\varepsilon_r^{(h)} \times 10^{-6}$	$\varepsilon_r^{(p)} \times 10^{-6}$	$\varepsilon_\varphi^{(h)} \times 10^{-6}$	$\varepsilon_\varphi^{(p)} \times 10^{-6}$
90	200	1,25	1,25	0,54	0,40
155	340	1,87	2,10	0,80	0,68
200	420	2,50	2,60	1,10	0,90

to be 1.15. At the second resonance (the second resonance vibration form is shown in Figure 30), we measured the electric signals using two rectangular gauges glued at the antinode ($r = 0.04$ m) in the circular and radial directions. Table 7 presents the experimental and theoretical results (obtained by formulas 33.25 for $p = 1.15$) for the strains at three levels of resonance build-up. The resonance frequency ω was 1340 Hz, and the complex conductivity Y was $(1 + i0.25) \times 10^{-6}\Omega^{-1}$. The two upper rows of Table 7 contain the electrical potentials for the first and second gauges; the third and fifth rows contain the strains ε_r and ε_φ computed by equations 33.25; and the fourth and sixth rows contain the respective experimentally obtained strain values. Note that the theoretical and experimental results agree.

We performed similar experiments and computations for higher resonance frequencies. The discrepancies between the experimental and theoretical strain values became more pronounced. This is explained by greater experimental difficulties and faster variation of the stressed-strained state to the coordinates. On the whole, the measurement of dynamic strains using piezoelectric gauges seems rather promising.

Part II. Asymptotic Method as Applied to Electroelastic Shell Theory

The theories of piezoceramic shells constructed in Chapter 2 were originally obtained using the asymptotic method [78–85, 87–89]. Though not as simple and visual as the method of hypotheses, the asymptotic reduction method has a number of merits. Firstly, the theories constructed by the asymptotic method are mathematically well-grounded; secondly, the error introduced when passing from the three-dimensional equations and boundary conditions to the two-dimensional ones is known. Thirdly, the method gives a qualitative picture of the total electroelastic state of the shell without solving the problem. Fourthly, it allows to construct an iterative process that in the roughest approximation leads to the two-dimensional theories given in Part I.

When using the asymptotic methods we commonly deal with quantities of the same order at every stage of the iterative process. This is achieved by asymptotically extending the scale with respect to the independent variables. The new dimensionless variables are introduced so that the differentiation with respect to them would not result in an increase or decrease of the needed functions. Besides, for every solution class the quantities are expanded in asymptotic series of the form

$$P = \eta^r \sum_{s=0}^{\infty} \eta^s P_s$$

in the small parameter η, where r assumes a certain value for each quantity. In the asymptotic integration of one differential equation the choice of r is a simple matter. For a system of differential equations the choice of the exponent r is similar. The technique of asymptotic integration for a system of differential equations can be found, say, in Cole's work [16]. The asymptotics of the needed quantities for the system of differential equations is chosen as follows.

Suppose that we have a system of differential equations with small parameter and n unknowns $y_{(i)}$, $i = 1, 2, \ldots, n$. We represent every needed function in the form

$$y_{(i)} = \eta^{x_i} y_{(i)*}.$$

Here all $y_{(i)*}$ have the same asymptotic order and x_i are the degrees of the small parameter found from the condition for the limiting equations and boundary conditions to be consistent. We can use the known analytical and numerical solutions and experimental results to check the correctness of the chosen asymptotics.

We will use the asymptotic method for constructing electroelastic shell theories for different directions of preliminary polarization. Two iterative processes for integrating three-dimensional electroelasticity equations will be used in order to pass from the three-dimensional to two-dimensional problems. The first iterative process allows us to construct the internal electroelastic state which in the first approximation can be described by the theory of electroelastic shells. It turned out that there exist special electroelastic states that have no analogue in the theory of nonelectrical shells.The second iterative process is used to construct a quickly decreasing electroelastic state; i.e., boundary layer, localized near the edge. The boundary layer and the internal electroelastic state interact at the edge provided the boundary conditions are satisfied. The investigation of this interaction gives the boundary conditions for the theory of piezoceramic shells and boundary layer.

We do not need to solve boundary layer problems in the theory of nonelectrical shells because the boundary layer corrections are inessential there. In the theory of piezoelectric shells the boundary layer corrections may be small, but may also be comparable to the principal terms. We must find them by solving some boundary layer problems.

The asymptotic analysis of three-dimensional electroelastic equations allowed us to qualitatively investigate the electroelastic state of piezoceramic shells and find out in what cases the two-dimensional problem will be mechanical or connected electromechanical, classical or nonclassical.

We met some difficulties when applying the asymptotic method to electroelasticity. The use of reduction technique was complicated by the lack of numerical or analytical standard solutions that help in choosing the asymptotics. This problem did not exist when using the asymptotic method in the theory of nonelectrical shells where the thoroughly verified Kirchhoff-Love theory served the purpose.

Chapter 6

CONSTRUCTING EQUATIONS IN THE THEORY OF PIEZOCERAMIC SHELLS

A complete system of equations in the theory of nonelectrical shells consists of the equilibrium equations, geometrical relations, and constitutive relations. The equilibrium equations and relations between strains and displacements are the same in all shell theories because they are mathematically strict. The differences lie in the elasticity relations. We also observe this situation in the theory of piezoceramic shells. The first iteration process used to construct the equations of the internal electroelastic state leads to the equations of the electroelastic shell theory in which the equilibrium equations and strain-displacement formulas are the same as in the theory of nonelectrical shells. Therefore, we will only construct and discuss two-dimensional electroelasticity relations and electrostatic equations.

34 SHELLS WITH THICKNESS POLARIZATION

The three-dimensional constitutive relations, electrostatic equations, and conditions on the shell faces can be found in Section 3.

For the electroelasticity equations, we will now a scale extension typical of the asymptotic methods, with respect to the coordinates

$$\alpha_i = \eta^s R\xi_i, \qquad \gamma = \eta^1 R\zeta, \qquad \eta = h/R. \tag{34.1}$$

Here, h is the shell thickness; R is some characteristic dimension of the shell; and s is the index of variability of the internal electroelastic shell. The coordinates ξ_i and ζ are chosen in such a way that differentiation with them does not result in any substantial increase or decrease in the sought-for quantities.

Except for a shell with faces completely covered with electrodes, the following asymptotics

$$\begin{aligned}
&\frac{v_3}{R} = \eta^c v_{3*}, && \frac{v_i}{R} = \eta^s v_{i*}, && s_{11}^E \tau_{ii} = \tau_{ii*} \\
&s_{11}^E \tau_{ij} = \tau_{ij*}, && s_{11}^E \tau_{33} = \eta^{1-c}\tau_{33*}, && s_{11}^E \tau_{i3} = \eta^{1-s}\tau_{i3*} \\
&d_{31}E_3 = E_{3*}, && d_{31}\frac{\psi}{R} = \eta^1\psi_*, && d_{31}E_i = \eta^{1-s}E_{i*} \\
&\frac{d_{31}}{\varepsilon_{11}^T} D_i = \eta^{1-s} D_{i*}, && \frac{d_{31}}{\varepsilon_{33}^T} D_3 = D_{3*} \\
&s_{11}^E q_3^{\pm} = \eta^{1-c} q_{3*}^{\pm}, && s_{11}^E q_i^{\pm} = \eta^{1-s} q_{i*}^{\pm}.
\end{aligned} \tag{34.2}$$

Here, quantity c is

$$c = \begin{cases} 0 & \text{for } 0 \leq s < 1/2 \\ -1 + 2s & \text{for } 1/2 \leq s < 1. \end{cases}$$

For a shell without electrodes on the faces, we will take the asymptotics for D_3 in the form

$$\frac{d_{31}}{\varepsilon_{33}^T} D_3 = \eta^{2-2s} D_{3*}.$$

All quantities with asterisks in 34.2 are of the same order and are dimensionless.

Substituting the asymptotics 34.2 and 34.1 in the equations of the electroelasticity, we obtain the following equations

$$\begin{aligned}
\tau_{ii*} &= \frac{1}{1-\nu^2}\left(\frac{a_j}{a_i} e_{i*} + \nu e_{j*}\right) + \eta^{1-c} \frac{\nu_1}{1-\nu} \frac{\tau_{33*}}{a_i} - \frac{1}{1-\nu} \frac{E_{3*}}{a_i} \\
\tau_{ij*} &= \frac{1}{\sigma_1}\left(\frac{a_i}{a_j} m_{i*} + m_{j*}\right) \\
\frac{\partial v_{3*}}{\partial \zeta} &= -\eta^{1-c} \nu_1 \left(\frac{\tau_{11*}}{a_2} + \frac{\tau_{22*}}{a_1}\right) + \eta^{2-2c} \sigma_3 \frac{\tau_{33*}}{a_1 a_2} + \eta^{1-c} \frac{d_{33}}{d_{31}} \frac{E_{3*}}{a_1 a_1} \\
\frac{\partial v_{i*}}{\partial \zeta} + \nu^{1-2s+c} \frac{g_{i*}}{a_i} &= \eta^{2-2s} \sigma_2 \frac{\tau_{i3*}}{aj} + \eta^{2-2s} \frac{d_{15}}{d_{31}} \frac{E_{i*}}{a_i} \\
D_{i*} &= \frac{a_j}{a_i} E_{i*} + t_1 \frac{\tau_{i3*}}{a_j} \\
D_{3*} &= E_{3*} + t_2(\tau_{11*} a_1 + \tau_{22*} a_2) + \eta^{1-c}\, t_3 \tau_{33*} \qquad (34.3) \\
E_{3*} &= -a_1 a_2 \frac{\partial \psi_*}{\partial \zeta}, \qquad E_{i*} = -\frac{1}{A_i} \frac{\partial \psi}{\partial \xi_i},
\end{aligned}$$

$$\frac{\partial D_{3*}}{\partial \zeta} + \eta^{2-2s} \frac{\varepsilon_{11}^T}{\varepsilon_{33}^T} \left[\frac{1}{A_1} \frac{\partial D_{1*}}{\partial \xi_1} + \frac{1}{A_2} \frac{\partial D_{2*}}{\partial \xi_2} + \eta^s R\left(k_1 D_{2*} + k_2 D_{1*}\right)\right] = 0 \qquad (34.4)$$

$$\begin{aligned}
e_{i*} &= \frac{1}{A_i} \frac{\partial v_{i*}}{\partial \xi_i} + \eta^s R k_i v_{j*} + \eta^c \frac{R}{R_i} v_{3*} \\
m_{i*} &= \frac{1}{A_j} \frac{\partial v_{i*}}{\partial \xi_j} - \eta^s R k_j v_{j*} \\
g_{j*} &= \frac{1}{A_i} \frac{\partial v_{3*}}{\partial \xi_i} - \eta^{2s-c} \frac{R}{R_i} v_{i*}, \\
a_i &= 1 + \eta_\zeta^1 \frac{R}{R_i} \qquad (34.5)
\end{aligned}$$

$$\psi_*|_{\zeta=\pm 1} = \pm V_*, \qquad \frac{\tau_{i3*}}{a_i}|_{\zeta=\pm 1} = \pm q_{i*}^{\pm},$$
$$\frac{\tau_{3*}}{a_1 a_2}|_{\zeta=\pm 1} = \pm q_{3*}^{\pm} \tag{34.6}$$

$$\nu_1 = -\frac{s_{13}^E}{s_{11}^E}, \qquad \sigma_1 = \frac{s_{66}^E}{s_{11}^E}, \qquad \sigma_2 = \frac{s_{44}^E}{s_{11}^E},$$
$$\sigma_3 = \frac{s_{33}^E}{s_{11}^E}, \qquad t_1 = \frac{d_{15}d_{31}}{\varepsilon_{11}^T s_{11}^E}, \qquad t_2 = \frac{d_{31}^2}{\varepsilon_{33}^T s_{11}^E},$$
$$t_3 = \frac{d_{33}d_{31}}{\varepsilon_{33}^T s_{11}^E}, \qquad t_4 = \frac{d_{15}^2}{\varepsilon_{33}^T s_{11}^E}, \qquad t_5 = \frac{d_{31}d_{15}}{\varepsilon_{33}^T s_{11}^E}, \qquad t_6 = \frac{d_{33}d_{15}}{\varepsilon_{33}^T s_{11}^E}. \tag{34.7}$$

In order to reduce approximately the three-dimensional electroelasticity equations to two-dimensional ones, we integrate equations 34.3 to 34.5 to ζ. Then all the needed quantities (displacements, stresses, components of the field's electric intensity vector, components of the electric induction vector, and electrical potential) will be represented as polynomials in ζ. The remainder of the series presented are those whose greatest terms are small quantities of order ε where

$$\varepsilon = O(\eta^{2-2s}). \tag{34.8}$$

We will neglect the terms of order ε in the resultant asymptotic expansions and write the latter as

$$\begin{aligned}
\tau_{ii*} &= \tau_{ii,0} + \eta^{1-2s+c}\zeta\tau_{ii,1}(\tau_{ij*}, v_{i*}, e_{i*}, m_{i*}, E_{3*})\\
v_{3*} &= v_{3,0} + \eta^{1-c}\zeta v_{3,1}(g_{i*})\\
\tau_{i3*} &= \tau_{i3,0} + \zeta\tau_{i3,1} + \eta^{1-2s+c}\zeta^2\tau_{i3,2}, (E_{i*}, D_{i*}, \psi_*)\\
\tau_{33*} &= \tau_{33,0} + \zeta\tau_{33,1} + \eta^{1-2s+c}\zeta^2\tau_{33,2} + \eta^{2-4s+2c}\zeta^3\tau_{33,3}\\
D_{3*} &= D_{3,0}.
\end{aligned} \tag{34.9}$$

In the parentheses on the right of the formulas, we enumerate the quantities whose asymptotic expansions have the same form. The quantity D_{3*} for a shell with faces completely covered with electrodes is independent of ζ. The formula for D_{3*} for a shell without electrodes on the faces looks like

$$D_{3*} = D_{3,0} + \eta^{1-2s+c}\zeta D_{3,1} + \zeta^2 D_{3,2} + \eta^{1-2s+c}\zeta^3 D_{3,3}.$$

We substitute formulas 34.9 into equations 34.3 to 34.6. By equating the coefficients at the same powers of ζ, we get a system of 50 differential equations with 50 unknowns:

$$
\begin{aligned}
\tau_{ii,0} &= \frac{1}{1-\nu^2}\left(e_{i,0}+\nu e_{j,0}\right)+\eta^{1-c}\frac{\nu_1}{1-\nu}\tau_{33,0}-\frac{1}{(1-\nu)}E_{3,0}\\
\tau_{ii,1} &= \frac{1}{1-\nu^2}\left[e_{i,1}+\nu e_{j,1}+\eta^{2s-c}\left(\frac{R}{R_j}-\frac{R}{R_i}\right)e_{i,0}\right]\\
&\quad -\eta^{2s-2c}\frac{\nu_1}{1-\nu}\tau_{33,1}-\frac{1}{(1-\nu)}\left(E_{3,1}-\eta^{2s-c}\frac{R}{R_i}E_{3,0}\right)\\
\tau_{ij,0} &= \frac{1}{\sigma_1}(m_{i,0}+m_{j,0})\\
\tau_{ij,1} &= \frac{1}{\sigma_1}\left[m_{i,1}+m_{j,1}+\eta^{2s-c}\left(\frac{R}{R_i}-\frac{R}{R_j}\right)m_{i,0}\right]\\
v_{3,1} &= -\nu_1(\tau_{11,0}+\tau_{22,0})+\frac{d_{33}}{d_{31}}E_{3,0},\qquad v_{i,1}=-g_{i,0}\\
D_{i,0} &= E_{i,0}+t_1\tau_{i3,0}\\
D_{i,1} &= E_{i,1}+t_1\tau_{i3,1}+\eta^{1}\left(\frac{R}{R_j}-\frac{R}{R_i}\right)E_{i,0}\\
D_{i,2} &= E_{i,2}+t_1\tau_{i3,2}+\eta^{2s-c}\left(\frac{R}{R_j}-\frac{R}{R_i}\right)E_{i,1}\\
D_{3,1} &= E_{3,1}+t_2\left[\tau_{11,1}+\tau_{22,1}+\eta^{2s-c}R\left(\frac{\tau_{11,0}}{R_1}+\frac{\tau_{22,0}}{R_2}\right)\right]\\
&\quad +\eta^{2s-2c}t_3\tau_{33,1}\\
D_{3,0} &= E_{3,0}+t_2(\tau_{11,0}+\tau_{22,0})+\eta^{1-c}t_3\tau_{33,0}\\
E_{3,0} &= -\psi_{,1},\qquad E_{3,1}=-2\psi_{,2}-\eta^{2s-c}\psi_{,1}\left(\frac{R}{R_1}+\frac{R}{R_2}\right)\\
E_{ik} &= -\frac{1}{A_i}\frac{\partial\psi_{,k}}{\partial\xi_i}\qquad k=0,1,2\\
\tau_{i3,0}+\eta^{1-2s+c}\tau_{i3,2} &= \frac{1}{2}\left[q_{i*}^{+}-q_{i*}^{-}+\eta^{1}\frac{R}{R_j}l(q_{i*}^{+}+q_{i*}^{-}\right]\\
\tau_{i3,1} &= \frac{1}{2}\left[q_{i*}^{+}+q_{i*}^{-}+\eta^{1}\frac{R}{R_j}(q_{i*}^{+}-q_{i*}^{-})\right]\\
\tau_{33,0}+\eta^{1-2s+c}\tau_{33,2} &= \frac{1}{2}\left[q_{3*}^{+}-q_{3*}^{-}+\eta^{1}\left(\frac{R}{R_1}+\frac{R}{R_2}\right)(q_{3*}^{+}+q_{3*}^{-})\right]\\
\tau_{33,1}+\eta^{2-4s+2c}\tau_{33,3} &= \frac{1}{2}\left[q_{3*}^{+}+q_{3*}^{-}+\eta^{1}\left(\frac{R}{R_1}+\frac{R}{R_2}\right)(q_{3*}^{+}-q_{3*}^{-})\right].
\end{aligned}
\tag{34.10}
$$

The quantities $e_{i,0}, g_{i,0}, m_{i,0}$, and $m_{i,1}$ are the same as those in equations 34.5 where the asterisk should be replaced by a zero or a one, respectively. We find $e_{i,1}$ and $g_{i,1}$ from the formulas

$$e_{i,1} = \frac{1}{A_1}\frac{\partial v_{i,1}}{\partial \xi_1} + \eta^s k_i R v_{j,1} + \eta^{2s-c} R \frac{v_{3,1}}{R_i}$$

$$g_{i,1} = \frac{1}{A_i}\frac{\partial v_{3,1}}{\partial \xi_i} - \eta^c R \frac{v_{i,1}}{R_i}. \tag{34.11}$$

If we add to equations 34.10 and 34.11 equivalent forms of equilibrium equations and geometrical relations, we will obtain a system where the number of equations is equal to the number of unknowns. The system can be integrated without resorting to the traditional notions of shell theory, i.e., forces, moments, displacements, and strains of the middle surface. Nevertheless, the transition is necessary if we are to make a comparison with the theory of nonelectrical shells.

We apply the notation of shell theory to our equations. The displacements u_i and w of the middle surface are related (through equations 34.1, 34.2, and 34.9) to the three-dimensional displacements v_i and v_3 as

$$u_i = v_i|_{\zeta=0} = \eta^s R v_{i,0}, \; w = -v_3|_{\zeta=0} = -\eta^c R v_{3,0}. \tag{34.12}$$

We take into account equations 34.1, 34.2, and 34.9 and express the forces and moments through the stresses

$$T_i = \int_h^h \tau_{ii} d\gamma = \frac{2h}{s_{11}^E}\tau_{ii,0}, \qquad S_{ij} = \int_h^h \tau_{ij} d\gamma = 2h\frac{\tau_{ij,0}}{s_{11}^E}$$

$$G_i = -\int_{-h}^h \tau_{ii}\gamma d\gamma = -\frac{2h^2}{3}\eta^{1-2s+c}\frac{\tau_{ii,1}}{s_{11}^E}$$

$$H_{ij} = \int_{-h}^h \tau_{ij}\gamma d\gamma = \frac{2h^2}{3}\eta^{1-2s+c}\frac{\tau_{ij,1}}{s_{11}^E}. \tag{34.13}$$

Leaving out the simple, but cumbersome, transformations, we write some intermediate results

$$e_{i,0} = \varepsilon_i, \qquad m_{i,0} = \omega_j, \qquad g_{i,0} = \eta^{s-c}\gamma_i,$$

$$v_{i,1} = -\eta^{s-c}\gamma_i$$

$$v_{3,1} = -\frac{\nu}{1-\nu}(e_{i,0} + e_{j,0}) + \left(\frac{d_{33}}{d_{31}} + \frac{2\nu_1}{1-\nu}\right)E_{3,0}. \tag{34.14}$$

We transform formulas 34.13 using equations 34.10 to 34.12 and 34.14 to obtain the electroelasticity relations

$$T_i = \frac{2h}{s_{11}^E(1-\nu^2)}(\varepsilon_i + \nu\varepsilon_j) - \frac{2hd_{31}}{s_{11}^E(1-\nu)}E_3^{(0)} - \left\{\left[\frac{hs_{13}^E}{s_{11}^E(1-\nu)}(q_3^+ - q_3^-)\right]\right\}$$

$$S_{ij} = \frac{2h}{s_{66}^E}\omega, \qquad H_{ij} = \frac{4h^3}{3s_{66}^E}\left(\tau - \frac{\omega}{2R_j}\right)$$

$$G_i = -\frac{2h^3}{3s_{11}^E(1-\nu^2)}(\kappa_i + \nu\kappa_j) + \frac{2h^3}{3(1-\nu)}\frac{d_{31}}{s_{11}^E}E_3^{(1)}$$
$$+\left\{\frac{2h^3}{3s_{11}^E(1-\nu^2)}\left[\left(\frac{1}{R_i} - \frac{1}{R_j}\right)\varepsilon_i - \frac{s_{13}^E}{s_{11}^E(1-\nu)}\left(\frac{1}{R_i} + \frac{\nu}{R_j}\right)(\varepsilon_i + \varepsilon_j)\right.\right.$$
$$-\left(\left(d_{33} - \frac{2s_{13}^E d_{31}}{(1-\nu)s_{11}^E}\right)\left(\frac{1}{R_i} + \frac{\nu}{R_j}\right)E_3^{(0)} - \frac{d_{31}(1+\nu)}{R_i}\right)E_3^{(0)}$$
$$\left.\left. + \frac{s_{13}^E(1+\nu)}{2h}(q_3^+ + q_3^-)\right]\right\}. \tag{34.15}$$

The electrical quantities vary with respect to γ according to the laws

$$\begin{aligned}
E_3 &= E_3^{(0)} + \gamma E_3^{(1)} \\
E_i &= E_i^{(0)} + \gamma E_i^{(1)} + \gamma^2 E_i^{(2)} \\
D_3 &= D_3^{(0)} + \gamma D_3^{(1)} + \gamma^2 D_3^{(2)} + \gamma^3 D_3^{(3)} \\
D_i &= D_i^{(0)} + \gamma D_i^{(1)} + \gamma^2 D_i^{(2)} \\
\psi &= \psi^{(0)} + \gamma\psi^{(1)} + \gamma^2\psi^{(2)}.
\end{aligned} \tag{34.16}$$

Here, the electrical quantities with superscripts in parentheses are coefficients at the powers γ in the formulas derived from equation 34.9.

The formulas for $E_3^{(0)}, \ldots, \psi^{(2)}$ can have different forms depending on the electrical conditions on the shell's faces. For example, if the faces are covered with electrodes and their electrical potential is defined, we find the electrical quantities as

$$\begin{aligned}
\psi^{(1)} &= \frac{V}{h} \\
E_3^{(0)} &= -\frac{V}{h} \\
\psi^{(0)} &= -h^2\psi^{(2)} \\
E_i^{(1)} &= 0 \\
E_i^{(0)} &= -h^2 E_i^{(2)} \\
D_3^{(0)} &= \varepsilon_{33}^T E_3^{(0)} + \frac{d_{31}}{2h}(T_1 + T_2) + \left\{\frac{d_{33}}{2}(q_3^+ - q_3^-)\right\}
\end{aligned}$$

$$\frac{2h^3}{3}E_3^{(1)} = \frac{d_{31}}{\varepsilon_{33}^T}(G_1 + G_2) - \left\{\frac{h^2}{3}\left[\frac{d_{31}}{\varepsilon_{33}^T}\left(\frac{T_1}{R_1} + \frac{T_2}{R_2}\right) + \frac{d_{33}}{\varepsilon_{33}^T}\left(q_3^+ + q_3^-\right)\right]\right\} \tag{34.17}$$

$$D_i^{(0)} = \varepsilon_{11}^T E_i^{(0)} - \frac{d_{15}}{4h}\left[3N_i + h(q_i^+ - q_i^-)\right] - \left\{\frac{d_{15}h}{4R_j}(q_i^+ + q_i^-)\right\}$$

$$D_i^{(1)} = \varepsilon_{11}^T E_i^{(1)} + \frac{d_{15}}{2h}(q_i^+ + q_i^-) + \left\{\varepsilon_{11}^T\left(\frac{1}{R_j} - \frac{1}{R_i}\right)E_i^{(0)} + \frac{d_{15}}{2R_j}(q_i^+ - q_i^-)\right\}$$

$$D_i^{(2)} = \varepsilon_{11}^T E_i^{(2)} + \frac{3d_{15}}{4h^3}\left[N_i + h(q_i^+ - q_i^-)\right] + \left\{\frac{3d_{15}}{4hR_j}(q_i^+ + q_i^-)\right\}$$

$$nD_3^{(n)} = -\frac{1}{A_1}\frac{\partial D_1^{(n-1)}}{\partial \xi_1} - \frac{1}{A_2}\frac{\partial D_2^{(n-1)}}{\partial \xi_2} - k_1 D_2^{(n-1)} - k_2 D_1^{(n-1)}, \quad (n = 1, 2, 3)$$

$$\psi^{(2)} = -\frac{1}{2}E_3^{(1)} + \left\{\left(\frac{1}{R_1} + \frac{1}{R_2}\right)\frac{E_3^{(0)}}{2}\Big]\right\}. \tag{34.18}$$

If the faces have no electrodes, we should replace equations 34.17 (with equations 34.18 remaining as is) by

$$E_3^{(0)} = -\psi^{(1)} = -\frac{d_{31}}{2h\varepsilon_{33}^T}(T_1 + T_2) - \left\{\frac{d_{33}}{2\varepsilon_{33}^T}(q_3^+ - q_3^-)\right\}$$

$$D_3^{(0)} = -h^2 D_3^{(2)}, \qquad D_3^{(1)} = -h^2 D_3^{(3)}. \tag{34.19}$$

We use equations 34.19 and 34.18, for shells without electrodes on the faces, to obtain the following differential relation, which will be needed for further investigation:

$$\left(\frac{\partial}{\partial\alpha_1}\frac{A_2}{A_1}\frac{\partial}{\partial\alpha_1} + \frac{\partial}{\partial\alpha_2}\frac{A_1}{A_2}\frac{\partial}{\partial\alpha_2}\right)\left(\psi^{(0)} + \frac{h^2}{3}\psi^{(2)}\right)$$
$$+ \frac{d_{15}}{2h\varepsilon_{33}^T}\left(\frac{\partial}{\partial\alpha_1}A_2N_1 + \frac{\partial}{\partial\alpha_2}A_1N_2\right)$$
$$-\left\{\frac{\partial}{\partial\alpha_1}A_2\left(\frac{1}{R_2} - \frac{1}{R_1}\right)E_1^{(1)} + \frac{\partial}{\partial\alpha_2}A_1\left(\frac{1}{R_1} - \frac{1}{R_2}\right)E_2^{(1)}\right\} = 0. \tag{34.20}$$

The terms in the braces are small. They are on the order of $O(\eta^1)$ compared to the principal terms.

Just as in nonelectrical shells [22], the accuracy of equation 34.8 is maximal for hypotheses of the Kirchhoff-Love type. For more accurate two-dimensional equations, the order of the system of the differential equations will be increased, and we will have to allow for the transverse shears and other small quantities usually neglected in this kind of theory. Note that refining the equations up to $O(\eta^{2-2s})$ makes sense only for the small variability indexes $0 < s < 1/2$; they become worthless as soon as s exceeds 1/2. For $s = 0$, the error of the refined equations is η^2. It is obvious that the boundary conditions should be refined too.

The equilibrium equations, electroelasticity relations (equation 34.15), equations 34.17 and 34.18 for shells with electrode-covered faces (equations 34.18 and 34.19 for faces without electrodes), and strain-displacement formulas form a complete system of order eight (ten for shell without faces' electrodes).

If the shell has electrode-covered faces and the electrical potential is given, the problem can be broken into mechanical and electrical parts. Then, the quantity $E_3^{(0)}$ in the electroelasticity relations for forces is found using the given electrical potential from the second formula of equations 34.17. We substitute its value into equations 34.15 and use the last equation in 34.17 to exclude $E_3^{(1)}$ from the electroelasticity relations for bending moments. The resultant electroelasticity relations will remind those of the theory of nonelectrical shells; the difference being in the sense of the coefficients at the strain components. By integrating the resultant constitutive relations together with the equilibrium equations and strain-displacement formulas (provided that the mechanical boundary conditions at the shell's edges are satisfied), we will find the forces, moments, and displacements.

Having solved the mechanical problem, we can find the electrical quantities by algebraic operations on the obtained forces and moments using equations 34.17 and 34.18.

Whether the problem can be divided into mechanical and electrical for faces not covered with electrodes, the decision is based on the electrical boundary conditions on the edges. This question will be considered below.

In Sections 7 and 8, we discussed the theory of shells with thickness polarization by constructing appropriate hypotheses. Now that we have the asymptotic formulas (equations 34.3, 34.4, and 34.9), we can estimate the error of each hypothesis. It is clear from equations 34.3 that, by neglecting τ_{33} in the first and last equations of equations 34.3, we introduce the error $O(\eta^{1-c})$. In a similar way, we can find the errors for the other hypotheses and hence for the constructed equations. A more exact theory may be based on other hypotheses where we keep the terms $O(\eta^{1-c})$ in equations 34.3 and 34.4 and neglect the terms $O(\eta^{2-2s})$. We will then obtain the laws 34.9 of variation of the sought-for quantities with thickness and the electroelasticity relations similar to those introduced in this section.

35 PURE MOMENT ELECTROELASTIC STATE OF SHELLS WITH THICKNESS POLARIZATION

We will construct electroelastic relations for a slowly varying stressed state arising in a shell with all free edges far from the edges [78].

The electroelastic relations will be exact within the quantities of order $\varepsilon = O(\eta^1 + \eta^{2-4s})$, where s is the variability index of the electroelastic state varying within $0 \leq s < 1/2$. For $s \geq 1/2$, the pure moment electroelastic state turns into the electroelastic state with great variability. This was considered in the previous section. We extend the scale along the coordinate lines (equation 34.1) and represent the required electroelastic state asymptotically, i.e.,

$$\begin{aligned}
&\frac{v_3}{R} = \eta^0 v_{3*}, && \frac{v_i}{R} = \eta^s v_{i*}, \\
&s_{11}^E \tau_{ii} = \eta^{1-2s}\tau_{ii*} && s_{11}^E \tau_{ij} = \eta^{1-2s}\tau_{ij*}, \\
&s_{11}^E \tau_{33} = \eta^{2-2s}\tau_{33*}, && s_{11}^E \tau_{i3} = \eta^{2-3s}\tau_{i3*} \\
&d_{31}E_3 = \eta^b E_{3*}, && d_{31}\frac{\psi}{R} = \eta^{1+b}\psi_* \\
&d_{31}E_i = \eta^{2-3s}E_{i*} && \frac{d_{31}}{\varepsilon_{11}^T}D_i = \eta^{2-2s}D_{i*}, \\
&\frac{d_{31}}{\varepsilon_{33}^T}D_3 = \eta^d D_{3*} && s_{11}^E(\tau_{ii,0}, \quad \tau_{ij,0}) = \eta^{1-2s}(\tau_{ii*}, \quad \tau_{ij*}).
\end{aligned} \tag{35.1}$$

In the case of completely electrode-covered faces, we should take $b = d = 0$. For the shells without electrodes on the faces, we should take $b = 1 - 2s$, $d = 3 - 4s$.

The chosen asymptotic representation can be substantiated by physical reasons and by the fact that, using formulas 35.1, we can construct a noncontradictory system of differential equations within appropriate approximation. The last relation in 35.1 shows that, for free shells, the stresses defining the normal and shearing forces are $O(\eta^{-1+2s})$ times less than the stresses defining the bending moments. Asymptotics 35.1 reflect the known phenomenon of the deformability of shells with unfixed edges when the dimensionless displacements v_3/R and the normal component of the electric field intensity $d_{31}E_3$ are η^{1-2s} times greater than the dimensionless stresses $s_{11}^E\tau_{ii}, s_{11}^E\tau_{ij}$.

We introduce equations 35.1 and 34.1 into equations 2.14 and 2.19 to obtain

$$\begin{aligned}
\eta^{1-2s}\tau_{ii*} &= \frac{1}{1-\nu^2}\left(\frac{a_j}{a_i}e_{i*} + \nu e_{j*}\right) + \eta^{2-2s}\frac{\nu_1}{1-\nu}\frac{\tau_{33*}}{a_i} - \eta^b\frac{1}{1-\nu}\frac{E_{3*}}{a_i} \\
\eta^{1-2s}\tau_{ij*} &= \frac{1}{\sigma_1}\left(\frac{a_i}{a_j}m_{i*} + m_{j*}\right)
\end{aligned} \tag{35.2}$$

$$\frac{\partial v_{3*}}{\partial \zeta} = -\eta^{2-2s}\nu_1\left(\frac{\tau_{11*}}{a_2} + \frac{\tau_{22*}}{a_1}\right) + \eta^{3-2s}\sigma_3\frac{\tau_{33*}}{a_1a_2} + \eta^{1+b}\frac{d_{33}}{d_{31}}E_{3*}$$

$$\frac{\partial v_{i*}}{\partial \zeta} + \eta^{1-2s}\frac{g_{i*}}{a_i} = \eta^{3-4s}\sigma_2\frac{\tau_{i3*}}{a_j} + \eta^{3-4s}\frac{d_{15}}{d_{31}}E_{i*} \tag{35.3}$$

$$D_{i*} = \frac{a_j}{a_i}E_{i*} + t_1\tau_{i3*}$$

$$\eta^d D_{3*} = \eta^b E_{3*} + \eta^{1-2s}t_2(\tau_{11*}a_1 + \tau_{22*}a_2) + \eta^{2-2s}t_3\tau_{33*}$$

$$\frac{\partial D_{3*}}{\partial \zeta} + \eta^{3-4s-d}\frac{\varepsilon_{11}^T}{\varepsilon_{33}^T}\frac{1}{A_1A_2}\left(\frac{\partial}{\partial \xi_1}A_2D_{1*} + \frac{\partial}{\partial \xi_2}A_1D_{2*}\right) = 0$$

$$E_{3*} = -a_1a_2\frac{\partial \psi_*}{\partial \zeta}, \qquad E_{i*} = -\frac{1}{A_i}\frac{\partial \psi_*}{\partial \xi_i}. \tag{35.4}$$

We add equations 34.5 and mechanical conditions 34.6 to equations 35.2 to 35.4.

Consider equations 35.3 and 35.4 as equations with independent variable ζ. We integrate them with respect to ζ up to η^{2-2s} and get

$$v_{3*} = v_{3,0} + \eta^1\frac{d_{33}}{d_{31}}\int_0^\zeta E_{3*}d\zeta$$

$$v_{i*} = v_{i,0} + \eta^{1-2s}\int_0^\zeta g_{i*}d\zeta$$

$$D_{3*} = D_{3,0}$$

$$E_{i*} = E_{i,0} - \int_0^\zeta \frac{1}{A_i}\frac{\partial \psi_*}{\partial \xi_i}d\zeta. \tag{35.5}$$

We continue the integration with respect to ζ of the electroelasticity equations. By neglecting the small terms (within our approximation) in the obtained relations, we get the formulas for the needed quantities:

$$\begin{aligned}
v_{3*} &= v_{3,0} + \eta^1\zeta v_{3,1}\\
v_{i*} &= v_{i,0} + \eta^{1-2s}\zeta v_{i,1}\\
g_{i*} &= g_{i,0},\\
\tau_{ii*} &= \eta^{1-2s}\tau_{ii,0} + \zeta\tau_{ii,1} + \eta^1\zeta^2\tau_{ii,2}, \qquad (\tau_{ij,*})\\
\tau_{i3*} &= \tau_{i3,0} + \eta^{1-2s}\zeta\tau_{i3,1} + \zeta^2\tau_{i3,2} + \eta^1\zeta^3\tau_{i3,3}, \qquad (\tau_{33*}, E_{i*}, D_{i*})\\
E_{3*} &= \eta^b E_{3,0} + \eta^{1-2s-b}\zeta E_{3,1} + r\zeta^2\eta^{2-2s-b}E_{3,2}\\
D_{3*} &= \eta^b D_{3,0} + r\eta^{2-2s-d+b}(\zeta D_{3,1} + \eta^{1-2s}\zeta^2 D_{3,2} + \zeta^3 D_{3,3} + \eta^1\zeta^4 D_{3,4})\\
\psi_* &= \eta^{1-2s-b}\psi_{,0} + \eta^b\zeta\psi_{,1} + \eta^{1-2s-b}\zeta^2\psi_{,2} + r\eta^{2-2s-b}\zeta^3\psi_{,3}.
\end{aligned} \tag{35.6}$$

Here, the number r is chosen to be zero if the shell faces are electrode-covered and to be one if there are no electrodes on the faces. In the parentheses on the right, we give the quantities whose expansions in ζ have similar forms. We substitute the expansions 35.6 into equations 35.2 to 35.4, equate the coefficients at the same powers of ζ, and use the notation accepted in the theory of shells.

We start with the formulas for shells with faces completely covered with electrodes. Instead of the electroelasticity relations for forces, just as in the theory for nonelectrical shells with purely moment stressed state, we get from the first two formulas 35.2 the approximate equations

$$\varepsilon_i = d_{31} E_3^{(0)}, \qquad \omega = 0. \tag{35.7}$$

The electroelasticity relations for moments are written as

$$G_i = -\frac{2h^3 B}{3}(\kappa_i + \sigma\kappa_j) - \frac{2h^3 B}{3}\left[d_{33}\left(\frac{1}{R_i} + \frac{\sigma}{R_j}\right) + d_{31}\left(\frac{\sigma}{R_i} + \frac{1}{R_j}\right)\right]E_3^{(0)}$$

$$H_{ij} = \frac{4h^3}{3s_{66}^E}\tau$$

$$E_3^{(0)} = -\frac{V}{h},$$

$$B = \frac{2 - (1-\nu)k_p^2}{2s_{11}^E(1-\nu^2)(1-k_p^2)}$$

$$\sigma = \frac{2\nu + (1-\nu)k_p^2}{2 - (1-\nu)k_p^2}, \qquad k_p^2 = \frac{2}{1-\nu}\frac{d_{31}^2}{s_{11}^E \varepsilon_{33}^T}. \tag{35.8}$$

The term in the square brackets is of the order η^{2s} compared to the principal terms. For small values of the variability index s, it may introduce an essential contribution; but as it grows, the contribution becomes much less.

For the electrical quantities, we will get

$$D_3^{(0)} = \varepsilon_{33}^T E_3^{(0)}, \qquad \psi^{(1)} = \frac{V}{h}, \qquad \psi^{(0)} = -h^2\psi^{(2)}$$

$$\frac{2h^2}{3}E_3^{(1)} = \frac{d_{31}}{\varepsilon_{33}^T}(G_1 + G_2)$$

$$\psi^{(2)} = -\frac{1}{2}E_3^{(1)} - \frac{1}{2}\psi^{(1)}\left(\frac{1}{R_1} + \frac{1}{R_2}\right) \tag{35.9}$$

$$D_i^{(0)} = \varepsilon_{11}^T E_i^{(0)} - \frac{d_{15}}{4h}\left[3N_i + h(q_i^+ - q_i^-)\right]$$

$$D_i^{(1)} = \varepsilon_{11}^T E_i^{(1)} - \varepsilon_{11}^T\left(\frac{1}{R_i} - \frac{1}{R_j}\right)E_i^{(0)}$$

$$D_i^{(2)} = \varepsilon_{11}^T E_1^{(2)} + \frac{3d_{15}}{4h^3}\left[N_i + h(q_i^+ - q_i^-)\right] - \varepsilon_{11}^T\left(\frac{1}{R_i} - \frac{1}{R_j}\right)E_i^{(1)}. \tag{35.10}$$

We do not write the formulas for strains and displacements here because they are the same as those in equations 7.19.

The forces T_i and S_{ij} are found from the equilibrium equations. With the forces known, we can find the sums of stresses (equations 35.6) using the formulas

$$\frac{T_i}{2h} = \tau_{ii}^{(0)} + \frac{h^2}{3}\tau_{ii}^{(2)}, \qquad \frac{S_{ij}}{2h} = \tau_{ij}^{(0)} + \frac{h^2}{3}\tau_{ij}^{(2)}. \tag{35.11}$$

We find $\tau_{ii}^{(2)}$ and $\tau_{ij}^{(2)}$ from the formulas

$$\begin{aligned}
\tau_{ii}^{(2)} &= \frac{1}{s_{11}^E(1-\nu^2)}\left[\frac{1}{2}\left(s_{13}^E - \frac{d_{31}d_{33}}{\varepsilon_{33}^T}\right)\left(\frac{1}{R_i} - \frac{1}{R_j}\right)(\tau_{ii}^{(1)} - \nu\tau_{jj}^{(1)})\right.\\
&\quad \left.-\left(\frac{1}{R_j} - \frac{1}{R_i}\right)\kappa_i + \frac{1}{2}k_p^2\left(\frac{\tau_{11}^{(1)}}{R_1} + \frac{\tau_{22}^{(1)}}{R_2}\right) + \frac{1}{2}k_p^2(\tau_{11}^{(2)} + \tau_{22}^{(2)})\right]\\
&\quad - \frac{d_{31}}{s_{11}^E(1-\nu^2)}\frac{1}{R_i}\left(\frac{1}{R_i} + \frac{V}{R_i}\right)E_3^{(0)}\\
\tau_{ij}^{(2)} &= \frac{1}{s_{66}^E}\left(\frac{1}{R_i} - \frac{1}{R_j}\right)\tau.
\end{aligned} \tag{35.12}$$

The quantities $\tau_{ii}^{(0)}$ and $\tau_{ij}^{(0)}$ are found from equations 35.11 after $T_i, S_{ij}, \tau_{ii}^{(2)}$, and $\tau_{ij}^{(2)}$ have been determined. As in the theory of nonelectrical shells with pure moment stressed-strained state, the transition formulas from forces to stresses differ from the Kirchhoff-Love formulas

$$\tau_{ii}^{(0)} = \frac{T_i}{2h}, \qquad \tau_{ij}^{(0)} = \frac{S_{ij}}{2h}.$$

For electrode-covered faces with given potential, we can divide the problem into mechanical and electrical. The complete system for the mechanical problem comprises equations 35.7 and 35.8, the equilibrium equations, and strain-displacement formulas. Having solved the mechanical problem, we use equations 35.9 and 35.10 to find the electrical quantities.

For a shell without electrodes on the faces, we should take $E_3^{(0)} = 0$ in equations 35.7, 35.8, and 35.12. Equations 35.10 do not change. We add the following differential equation for defining $\psi^{(0)}$:

$$\begin{aligned}
&\left(\frac{\partial}{\partial\alpha_1}\frac{A_2}{A_1}\frac{\partial}{\partial\alpha_1} + \frac{\partial}{\partial\alpha_2}\frac{A_1}{A_2}\frac{\partial}{\partial\alpha_2}\right)\left(\psi^{(0)} + \frac{h^3}{3}\psi^{(2)}\right)\\
&\quad + \frac{d_{15}}{2h\varepsilon_{11}^T}\left(\frac{\partial}{\partial\alpha_1}A_2N_1 + \frac{\partial}{\partial\alpha_2}A_1N_2\right) = 0.
\end{aligned} \tag{35.13}$$

We will show in the following section that the decision whether the problem can be divided into the mechanical and electrical parts is based on the type of electrical conditions on the shell's edge. This question will be discussed when we obtain the boundary conditions.

36 SHELLS WITH TANGENTIAL POLARIZATION (ELECTRODE-COVERED FACES)

For definiteness, we will assume, as before, that the shell is polarized along the α_2-lines. The constitutive relations 2.16 and 2.19 for this type of shell are given in Section 2.

We will investigate shells with and without electrodes on the faces separately. We start with shells with electrode-covered faces for which conditions 34.6 are specified. For simplicity, we assume that there is no mechanical surface load:

$$\tau_{i3}|_{\gamma=\pm h} = 0, \qquad \tau_{33}|_{\gamma=\pm h} = 0. \tag{36.1}$$

We represent the needed quantities of the three-dimensional elasticity theory in the form

$$\begin{aligned}
&\frac{v_1}{R} = \eta^{1-s+c} v_{i*}, && \frac{v_2}{R} = v_{2*},\\
&\left(\frac{v_2}{R}|_{\gamma=0}\right) = O(\eta^{1-s+c}) && \frac{v_3}{R} = \eta^{1-2s+2c} v_{3*},\\
&\frac{\tau_{ii}}{n_{22}} = \tau_{ii*}, && \frac{\tau_{ij}}{n_{22}} = \tau_{ij*}\\
&\frac{\tau_{i3}}{n_{22}} = \eta^{1-s}\tau_{i3*}, && \frac{\tau_{33}}{n_{22}} = \eta^{1-c}\tau_{33*},\\
&d_{15}E_3 = \eta^{-1}E_{3*} && d_{15}\frac{\psi}{R} = \psi_*,\\
&\frac{d_{15}}{\varepsilon_{33}^T} D_i = D_{i*}, && \frac{d_{15}}{\varepsilon_{33}^T} D_3 = \eta^{-1} D_{3*},\\
&d_{15}E_i = \eta^{2-2s} E_{i*}
\end{aligned}$$

$$c = \begin{cases} 0 & \text{for } 0 \le s < 1/2 \\ -1+2s & \text{for } 1/2 \le s < 1. \end{cases} \tag{36.2}$$

By formulas 34.1, we extend the scale in the electroelasticity relations 2.16 and 2.19 and substitute the sought-for quantities by the quantities with asterisks in equation 36.2 to get

$$\tau_{11*} = \frac{a_2}{a_1} q_1 e_{1*} + \eta^{1-2s+c} q_2 e_{2*} - \eta^{1-c} p_1 \frac{\tau_{33*}}{a_1} - \eta^{2-2s} r_1 E_{2*}$$

$$\tau_{22*} = \eta^{1-2s+c}\frac{a_1}{a_2} e_{2*} + q_2 e_{1*} - \eta^{1-c} P_2 \frac{\tau_{33*}}{a_2} - \eta^{2-2s} r_2 \frac{a_1}{a_2} E_{2*}$$

$$\tau_{12*} = q_3\left(\frac{a_1}{a_2}m_{1*} + \eta^{1-2s+c}m_{2*}\right) - \eta^{2-2s}r_3E_{1*}$$

$$\tau_{21*} = q_3\left(m_{1*} + \eta^{1-2s+c}\frac{a_2}{a_1}m_{2*}\right) - \eta^{2-2s}r_3\frac{a_2}{a_1}E_{1*}$$

$$\frac{\partial v_{3*}}{\partial\zeta} = \eta^{2s-2c}\left(\frac{1}{q_4}\frac{\tau_{11*}}{a_2} + \frac{1}{q_5}\frac{\tau_{22*}}{a_1}\right) + \eta^{1+2s-3c}\frac{1}{q_6}\frac{\tau_{33*}}{a_1a_2} + \eta^{2-2c}\frac{d_{31}}{d_{15}}\frac{1}{a_2}E_{2*}$$

$$\frac{\partial v_{1*}}{\partial\zeta} + \eta^{1-2s+c}\frac{g_{1*}}{a_1} = \eta^1\frac{1}{q_7}\frac{\tau_{13*}}{a_2}$$

$$\frac{\partial v_{2*}}{\partial\zeta} + \eta^{2-3s+2c}\frac{g_{2*}}{a_2} = \eta^{1-s}\frac{1}{q_3}\frac{\tau_{23*}}{a_1} + \frac{1}{a_1a_2}E_{3*} \tag{36.3}$$

$$D_{1*} = \eta^{2-2s}\frac{\varepsilon_{11}^T}{\varepsilon_{33}^T}\frac{a_2}{a_1}E_{1*} + d_1\tau_{21*}$$

$$D_{2*} = \eta^{2-2s}\frac{a_1}{a_2}E_{2*} + d_2\frac{a_1}{a_2}\tau_{11*} + d_3\tau_{22*} + \eta^{1-c}d_2\frac{\tau_{33*}}{a_2}$$

$$D_{3*} = \frac{\varepsilon_{11}^T}{\varepsilon_{33}^T}E_{3*} + \eta^{1-s}a_2\tau_{23*}$$

$$\frac{\partial D_{3*}}{\partial\xi} + \eta^{2-2s}\frac{1}{A_1A_2}\left(\frac{\partial}{\partial\xi_1}A_2D_{1*} + \frac{\partial}{\partial\xi_2}A_1D_{2*}\right) = 0$$

$$E_{3*} = -a_1a_2\frac{\partial\psi_*}{\partial\zeta}, \qquad \eta^{2-2s}E_{i*} = -\frac{1}{A_i}\frac{\partial\psi_*}{\partial\xi_i}. \tag{36.4}$$

All the quantities in equations 36.3 and 36.4 are on the same order. When deriving equations 36.3, we used the notation

$$e_{1*} = \eta^{1-2s+c}\frac{1}{A_1}\frac{\partial v_{1*}}{\partial\xi_1} + Rk_1v_{2*} + \eta^{1-2s+2c}\frac{R}{R_1}v_{3*}$$

$$e_{2*} = \frac{1}{A_2}\frac{\partial v_{2*}}{\partial\xi_2} + \eta^sRk_2v_{1*} + \eta^c\frac{R}{R_2}v_{3*}$$

$$m_{1*} = \eta^{1-2s+c}\frac{1}{A_2}\frac{\partial v_{1*}}{\partial\xi_2} - Rk_2v_{2*},$$

$$m_{2*} = \frac{1}{A_1}\frac{\partial v_{2*}}{\partial\xi_1} - \eta^sRk_1v_{1*}$$

$$g_{1*} = \frac{1}{A_1}\frac{\partial v_{3*}}{\partial\xi_1} + \eta^{2s-c}\frac{R}{R_1}v_{1*}$$

$$g_{2*} = \frac{1}{A_2}\frac{\partial v_{3*}}{\partial\xi_2} - \eta^{-1+3s-2c}\frac{R}{R_2}v_{2*}$$

$$
\begin{aligned}
q_1 &= \frac{n_{11}}{n_{22}}, \qquad q_2 = \frac{n_{12}}{n_{22}}, \qquad q_3 = \frac{1}{s_{44}^E n_{22}}, \qquad q_4 = \frac{1}{s_{12}^E n_{22}} \\
q_5 &= \frac{1}{s_{13}^E n_{22}}, \qquad q_6 = \frac{1}{s_{11}^E n_{22}}, \qquad q_7 = \frac{1}{s_{66}^E n_{22}} \\
r_1 &= \frac{c_1}{n_{22} d_{15}}, \qquad r_2 = \frac{c_2}{n_{22} d_{15}} \\
d_1 &= \frac{d_{15}^2 n_{22}}{\varepsilon_{33}^T}, \qquad d_2 = \frac{d_{31} d_{15} n_{22}}{\varepsilon_{33}^T}, \qquad d_3 = \frac{d_{33} d_{15} n_{22}}{\varepsilon_{33}^T}.
\end{aligned} \tag{36.5}
$$

When constructing the equations and boundary conditions of the shell theory in question, we will neglect small quantities of order ε where

$$\varepsilon = O(\eta^1 + \eta^{2-2s}). \tag{36.6}$$

In order to approximately reduce the three-dimensional equations to two-dimensional ones, we integrate equations 36.3 to 36.5 with respect to ζ. After integration all the needed quantities will be represented as polynomials in ζ. For example, we write the asymptotic expansion for the stress τ_{ii*} (which holds for τ_{ij*})

$$\tau_{ii*} = \eta^{1-2s+c}\tau_{ii,0} + \zeta\tau_{ii,1} + \eta^{1-c}\zeta^2\tau_{ii,2} + \eta^{2-2s}\zeta^3\tau_{ii,3} + \dots$$

Within the quantities $O(\varepsilon)$ defined by formula 36.6, we should have neglected all the terms starting with $\tau_{ii,2}$ in the formula above. But for small variability indexes s, the quantities $\tau_{ii,0}$ and $\tau_{ii,2}$ are comparable and give contributions of the same order into the force T_i. Therefore, we will keep three terms in the expansions of the principal stresses τ_{ii*} and τ_{ij*} and the quantities e_{i*} and m_{i*}. Taking this remark into account, we obtain the following formulas:

$$
\begin{aligned}
v_{1*} &= v_{1,0} + \eta^{1-2s+c}\zeta v_{1,1} \\
v_{2*} &= \eta^{1-s+c} v_{2,0} + \zeta v_{2,1} + \eta^1\zeta^2 v_{2,2} \\
v_{2*} &= \eta^{1-s+c} v_{2,0} + \zeta v_{2,1} + \eta^1\zeta^2 v_{2,2} \\
v_{3*} &= v_{3,0} + \eta^{1-c}\zeta v_{3,1} + \eta^{2s-2c}\zeta^2 v_{3,2} \\
g_{1*} &= g_{1,0}(E_{3*}, D_{3*}), \qquad g_{2*} = g_{2,0} + \eta^{-1+3s-2c}\zeta g_{2,1} \\
e_{1*} &= \eta^{1-2s+c} e_{1,0} + \zeta e_{1,1} + \eta^1\zeta^2 e_{1,2} \qquad (m_{1*}, \tau_{ij*}) \\
e_{2*} &= e_{2,0} + \zeta\eta^{1-2s+c} e_{2,1} + \eta^{2s-c}\zeta^2 e_{2,2} \qquad (m_{2*}) \\
\tau_{ii*} &= \eta^{1-2s+c}\tau_{ii,0} + \zeta\tau_{ii,1} + \eta^{1-c}\zeta^2\tau_{ii,2} \\
\tau_{i3*} &= \tau_{i3,0} + \zeta^2\tau_{i3,2} \\
\tau_{33*} &= \tau_{33,0} + \eta^{1-2s+c}\zeta\tau_{33,1} + \zeta^2\tau_{33,2} + \eta^{1-2s+c}\zeta^3\tau_{33,3} \\
D_{i*} &= \eta^{1-2s+c} D_{i,0} + \zeta D_{i,1}.
\end{aligned} \tag{36.7}
$$

We represent $v_{2,1}$ as a sum of two quantities:

$$v_{2,1} = v'_{2,1} + \eta^{2-3s+2c} v''_{2,1}, \qquad v'_{2,1} = Rd_{15}E_{3,0}, \tag{36.8}$$

where $v'_{2,1}$ does not depend on the variables ξ_i, ζ. Comparing formulas 36.7 to similar expressions in the theory of nonelectrical shells [15, 21], we will see that our expressions for the stresses τ_{ii*} and τ_{ij*} coincide with the respective formulas for the pure moment stressed state. The formulas for the displacements v_{2*} and v_{3*} have no analogs in the theory of nonelectrical shells. The greatest term in the displacement formulas is the rotation angle $v_{2,1}$; the largest principal stresses are $\tau_{ii,1}$ and $\tau_{ij,1}$ that define the moments. The displacement v_{3*} varies by a square law; and for the small variability $v_{3,2}$, it is comparable to $v_{3,0}$. This means that in our case using the Kirchhoff hypothesis, the normal has constant length and cannot be applied even in the roughest approximation.

The formulas for $e_{i,0}, \ldots, m_{2,1}$ are as follows

$$\begin{aligned}
e_{i,0} &= \frac{1}{A_i}\frac{\partial v_{i,0}}{\partial \xi_i} + \eta^s Rk_i v_{j,0} + \frac{\eta^c}{R_i} v_{3,0} \\
m_{i,0} &= \frac{1}{A_j}\frac{\partial v_{i,0}}{\partial \xi_j} - \eta^s Rk_j v_{j,0} \\
g_{i,0} &= \frac{1}{A_i}\frac{\partial v_{3,0}}{\partial \xi_i} - \eta^{2s} Rk_i v_{i,0} \\
g_{2,1} &= \eta^{2-3s+c}\frac{1}{A_2}\frac{\partial v_{3,1}}{\partial \xi_2} - \frac{R}{R_2} v_{2,1} \\
e_{1,1} &= \eta^{2-4s+2c}\frac{1}{A_1}\frac{\partial v_{1,1}}{\partial \xi_1} + Rk_1 v_{2,1} \\
e_{2,1} &= \frac{1}{A_2}\frac{\partial v''_{2,1}}{\partial \xi_2} + \eta^s Rk_2 v_{1,1} \\
e_{1,2} &= Rk_1 v_{2,2} + \frac{R}{R_1} v_{3,2} \\
e_{2,2} &= \frac{1}{A_2}\frac{\partial v_{2,2}}{\partial \xi_2} + \frac{R}{R_2} v_{3,2} \\
m_{1,1} &= \eta^{2-4s+2c}\frac{1}{A_2}\frac{\partial v_{1,1}}{\partial \xi_2} - Rk_2 v_{2,1}, \\
m_{1,2} &= -Rk_2 v_{2,2} \\
m_{2,1} &= \frac{1}{A_1}\frac{\partial v''_{2,1}}{\partial \xi_1} - \eta^s Rk_1 v_{1,1}, \\
m_{2,2} &= \frac{1}{A_1}\frac{\partial v_{2,2}}{\partial \xi_1}.
\end{aligned} \tag{36.9}$$

We substitute the expansions 36.7 into equations 36.3 and 36.4 and equate the coefficients at the same powers of ζ and get

$$\begin{aligned}
\tau_{11,0} &= q_1 e_{1,0} + q_2 e_{2,0} - \eta^{2s-2c} p_1 \tau_{33,0} \\
\tau_{22,0} &= e_{2,0} + q_2 e_{1,0} - \eta^{2s-2c} p_2 \tau_{33,0} \\
\tau_{11,1} &= q_1 e_{1,1} + \eta^{2-4s+2c} q_2 e_{2,1} \\
\tau_{22,1} &= \eta^{2-4s+2c} e_{2,1} + q_2 e_{1,1} \\
\tau_{ij,0} &= q_3(m_{i,0} + m_{j,0}), \qquad \tau_{ij,1} = q_3(m_{1,1} + \eta^{2-4s+2c} m_{2,1}) \\
\tau_{11,2} &= \eta^c \left[q_1 e_{1,2} + q_2 e_{2,2} + q_1 \left(\frac{R}{R_2} - \frac{R}{R_1} \right) e_{1,1} \right] - p_1 \tau_{33,2} \\
\tau_{22,2} &= \eta^c [e_{2,2} + q_2 e_{1,2}] - p_2 \tau_{33,2} \\
v_{3,1} &= \frac{1}{q_4} \tau_{11,0} + \frac{1}{q_5} \tau_{22,0} + \eta^{2s-2c} \frac{1}{q_6} \tau_{33,0} \\
v_{3,2} &= \frac{1}{2} \left(\frac{1}{q_4} \tau_{11,1} + \frac{1}{q_5} - \tau_{22,1} \right) \\
v_{1,1} &= -q_{1,0} + \eta^{2s-c} \frac{1}{q_7} \tau_{13,0}, \\
v''_{2,1} &= -q_{2,0} \\
v_{2,2} &= -\frac{1}{2} q_{2,1} - \frac{1}{2} \left(\frac{R}{R_1} + \frac{R}{R_2} \right) E_{3,0}.
\end{aligned} \tag{36.10}$$

Equations 36.8 to 36.10, together with the equations of motion written in the same form, constitute a complete system where the number of unknowns is equal to the number of equations. This verifies the validity of the asymptotic representation 36.2.

We pass in the obtained equations to the notation typical for shell theories. The displacements u_i and w of the middle surface of the shell are related to the three-dimensional displacements v_i and v_3 by equations 36.2 and 36.7 as

$$u_i = v_i|_{\zeta=0} = \eta^{1-s+c} v_{i,0}, \qquad w = -v_3|_{\zeta=0} = -\eta^{1-2s+2c} v_{3,0}. \tag{36.11}$$

The forces and moments are expressed through the stresses using equations 5.1, 5.2, 36.2, 36.10. We get

$$\begin{aligned}
T_i &= \int_{-h}^{+h} \tau_{ii} d\gamma = 2h n_{22} \left(\eta^{1-2s+c} \tau_{ii,0} + \frac{1}{3} \eta^{1-c} \tau_{ii,2} \right) \\
G_i &= -\int_{-h}^{+h} \tau_{ii} \gamma d\gamma = -\frac{2h^2}{3} n_{22} \tau_{ii,1} \\
S_{ij} &= \int_{-h}^{+h} \tau_{ij} d\gamma = 2h n_{22} \left(\eta^{1-2s+c} \tau_{ij,0} + \frac{1}{3} \eta^1 \tau_{ij,2} \right)
\end{aligned}$$

$$H_{ij} = \int_{-h}^{+h} \tau_{ij}\gamma d\gamma = \frac{2h^2}{3} n_{22}\tau_{ij,1}. \qquad (36.12)$$

Omitting the cumbersome, but simple, transformations, we write some intermediate results:

$$E_{3,0} = \eta^1 d_{15} E_3, \qquad e_{i,0} = \eta^{-1+2s-c}\varepsilon_i,$$

$$m_{i,0} = \eta^{-1+2s-c}\omega_j \qquad g_{i,0} = \eta^{-1+3s-2c}\gamma_i,$$

$$v_{1,1} = -\eta^{-1+3s-2c}\gamma_1 \qquad v_{2,1} = d_{15}\eta^1 E_3 - \eta^1\gamma_2,$$

$$v_{2,2} = -\frac{\eta^1 R}{2R_1} d_{15} E_3 \qquad v_{3,1} = \eta^{-1+2s-c}(p_1\varepsilon_1 + p_2\varepsilon_2),$$

$$v_{3,2} = \eta^1 \frac{1}{2} R k_1\, d_{15}\, p_1\, E_3 \qquad e_{1,1} = \eta^1 R(\kappa_1 + k_1 d_{15} E_3),$$

$$e_{2,1} = \eta^{-1+4s-2c} R\kappa_2 \qquad e_{1,2} = \frac{\eta^1 R^2 k_1 d_{15}}{2R_1}(-1 + p_1)E_3$$

$$e_{2,2} = \frac{\eta^1 R^2 k_1 d_{15}}{2}\left(\frac{1}{R_1} - \frac{p_1 - 1}{R_2}\right)E_3$$

$$m_{1,1} = \eta^1 R(\tau - k_2 d_{15} E_3), \qquad m_{2,1} = \eta^{-1+4s-2c} R\tau$$

$$m_{1,2} = \frac{\eta^1 R^2}{2R_1} k_2 d_{15} E_3, \qquad m_{2,2} = -\frac{\eta^1 R}{2} d_{15} E_3 \frac{1}{A_1}\frac{\partial}{\partial\alpha_1}\frac{1}{R_1}. \qquad (36.13)$$

From the third equilibrium equation 2.11 and the second condition in 36.1, we can get

$$n_{22}\left(\tau_{33,0} + \frac{1}{3}\tau_{33,2}\right) = -\frac{1}{3}\eta^1 R\left(\frac{n_{11}}{R_1} + \frac{n_{21}}{R_2}\right)k_1 d_{15} E_3. \qquad (36.14)$$

Into formulas 36.12, we substitute 36.10 that was transformed using equations 36.11, 36.13, and 36.14; and within the quantities $O(\eta^1 + \eta^{2-2s})$, we obtain the following electroelasticity relations:

$$T_i = 2h(n_{ii}\varepsilon_i + n_{ij}\varepsilon_j) + l_i \frac{h^3}{3} k_1 d_{15} E_3$$

$$G_i = -\frac{2h^3}{3}(n_{ii}\kappa_i + n_{ij}\kappa_j + n_{i1} k_1 d_{15} E_3)$$

$$S_{ij} = \frac{2h}{s_{44}^E}\omega + b_{ij}\frac{h^3 d_{15}}{3s_{44}^E} E_3$$

$$H_{ij} = \frac{4h^3}{3s_{44}^E}\tau - \frac{2h^3}{3s_{44}^E} k_2 d_{15} E_3 \qquad (36.15)$$

where

$$E_3 = -\frac{V}{h}, \qquad b_{ij} = \frac{A_2}{A_1}\frac{\partial}{\partial\alpha_1}\left(\frac{1}{A_2R_1}\right) - 2\left(\frac{1}{R_i} - \frac{1}{R_2}\right)k_2$$

$$l_i = n_{ii}\frac{p_1 - 1}{R_1} + n_{i2}\left(\frac{1}{R_1} + \frac{p_1 - 1}{R_2}\right) + 2n_{11}\left(\frac{1}{R_2} - \frac{1}{R_i}\right) + 2p_i\left(\frac{n_{11}}{R_1} + \frac{n_{21}}{R_2}\right).$$

For electroelastic states with great variability, the terms with coefficients l_i and b_{ij} can be neglected. Relations 36.15 together with the equilibrium equations and strain-displacement formulas, which coincide with the respective formulas from the theory of nonelectrical shells, form a complete system of eight-order differential equations. By formula 34.1, we return to the variable γ and can write the quantities we are seeking as polynomials in degrees of γ:

$$\begin{aligned}
v_1 &= v_1^{(0)} + \gamma v_1^{(1)},\\
v_2 &= v_2^{(0)} + \gamma v_2^{(1)} + \gamma^2 v_2^{(2)}\\
v_3 &= v_3^{(0)} + \gamma v_3^{(1)} + \gamma^2 v_3^{(2)},\\
\tau_{ii} &= \tau_{ii}^{(0)} + \gamma\tau_{ii}^{(1)} + \gamma^2\tau_{ii}^{(2)},\\
\tau_{i3} &= \tau_{i3}^{(0)} + \gamma^2\tau_{i3}^{(2)}\\
\tau_{ij} &= \tau_{ij}^{(0)} + \gamma\tau_{ij}^{(1)} + \gamma^2\tau_{ij}^{(2)},\\
\tau_{33} &= \tau_{33}^{(0)} + \gamma\tau_{33}^{(1)} + \gamma^2\tau_{33}^{(2)} + \gamma^3\tau_{33}^{(3)},\\
E_3 &= E_3^{(0)}, \qquad D_3 = D_3^{(0)}, \qquad D_i = D_i^{(0)} = \gamma D_i^{(1)}.
\end{aligned} \tag{36.16}$$

Having found the forces, moments, and displacements of the middle surface, we define the expansion coefficients in 36.16 by the formulas

$$\begin{aligned}
v_1^{(1)} &= -\gamma_1,\\
v_2^{(1)} &= d_{15}E_3 - \gamma_2,\\
v_2^{(2)} &= -\frac{d_{15}}{2R_2}E_3\\
v_3^{(1)} &= s_{12}^E\tau_{11}^{(0)} + s_{13}^E\tau_{22}^{(0)},\\
v_3^{(2)} &= \frac{k_1 d_{15}}{2}p_1E_3\\
\tau_{ii}^{(0)} &= \frac{T_i}{2h} - \frac{h^2}{3}\tau_{ii}^{(2)},\\
\tau_{ij}^{(0)} &= \frac{S_{ij}}{2h} - \frac{h^2}{3}\tau_{ij}^{(2)}\\
\tau_{ii}^{(2)} &= \frac{l_i k_i d_{15}E_3}{2} + \frac{3p_i}{h^2}\tau_{33}^{(0)},
\end{aligned}$$

$$\begin{aligned}
\tau_{ij}^{(2)} &= \frac{d_{15}b_{ij}}{2s_{44}^E}E_3\\
\tau_{ii}^{(1)} &= -\frac{3}{2h^3}G_i,\\
\tau_{ij}^{(1)} &= \frac{3}{2h^3}H_{ij}\\
\tau_{i3}^{(0)} &= -h^2\tau_{i3}^{(2)} = -\frac{3}{4h}N_i\\
\tau_{33}^{(0)} &= -h^2\tau_{33}^{(2)} = \frac{1}{2h}\left(\frac{1}{R_1}T_1 + \frac{1}{R_2}T_2\right)\\
&\quad + \frac{3}{4h}\left(\frac{1}{A_2}\frac{\partial N_1}{\partial \alpha_1} + \frac{1}{A_2}\frac{\partial N_2}{\partial \alpha_2} + k_2N_1 + k_1N_2\right) \qquad (36.17)
\end{aligned}$$

$$\begin{aligned}
D_1^{(0)} &= \frac{d_{15}}{s_{44}^E}\omega,\\
D_1^{(1)} &= d_{15}\tau_{21}^{(1)}\\
D_2^{(0)} &= d_{31}\tau_{11}^{(0)} + d_{33}\tau_{22}^{(0)} + d_{31}\tau_{33}^{(0)}\\
D_2^{(1)} &= d_{31}\tau_{11}^{(1)} + d_{33}\tau_{22}^{(1)},\\
D_3^{(0)} &= \varepsilon_{11}^T E_3^{(0)}. \qquad (36.18)
\end{aligned}$$

We should note that the problem consists of mechanical and electrical parts. After solving the mechanical problem, all the electrical quantities can be found from equations 36.18. We stress that the mechanical problem differs from the theories of Kirchhoff-Love in that the principal stresses and normal displacements vary by square laws; therefore, the transition from forces to stresses is done by more complicated formulas 36.17. Recall that in the Kirchhoff-Love theory, the transition formulas are as simple as

$$\tau_i^{(0)} = \frac{T_i}{2h}, \qquad \tau_{ij}^{(0)} = \frac{S_{ij}}{2h}.$$

37 SHELLS WITH TANGENTIAL POLARIZATION (FACES WITHOUT ELECTRODES)

Consider a shell whose faces are not covered with electrodes. Let a mechanical surface load be defined as in 3.1. The electrical conditions on the faces without electrodes are written as those in 3.4.

We choose the following asymptotic representation for the sought-for quantities of the electroelastic state:

$$v_i = \eta^s R v_{i*}, \qquad v_3 = \eta^c R v_{3*}, \qquad \tau_{ii} = n_{22}\tau_{ii*}$$
$$\tau_{ij} = n_{22}\tau_{ij*}, \qquad \tau_{33} = \eta^{1-c} n_{22}\tau_{i3*}, \qquad \tau_{i3} = \eta^{1-s} n_{22}\tau_{i3*}$$

$$d_{15}E_3 = \eta^{1-s}E_{3*}, \qquad d_{15}E_i = E_{i*}, \qquad \frac{d_{15}}{\varepsilon_{33}^T} D_i = D_{i*}$$

$$\frac{d_{15}}{\varepsilon_{33}^T} D_3 = \eta^{2-3s+c} D_{3*}, \qquad \frac{d_{15}}{R}\psi = \eta^s \psi_*. \tag{37.1}$$

Quantities s, η, and c have the same sense, as before.

We integrate the electroelasticity equations with respect to ζ within 36.6. Before the integration, we change the sought-for quantities by formulas 37.1 and the independent variables by formulas 34.1 to get the following laws of variation of the sought-for quantities to ζ:

$$\begin{aligned}
v_{i*} &= v_{i,0} + \eta^{1-2s+c}\zeta v_{i,1}\,(e_{i*}, m_{i,*}, \tau_{ii*}, \tau_{ij*}, D_{i*})\\
v_{3*} &= v_{3,0}\,(g_{i*}, E_{i*}, \psi_*)\\
\tau_{i3*} &= \tau_{i3,0} + \zeta\tau_{i3,1} + \eta^{1-2s+c}\tau_{i3,2}\,(E_{3*})\\
\tau_{33*} &= \tau_{33,0} + \zeta\tau_{33,1} + \eta^{1-2s+c}\zeta^2\tau_{33,2} + \eta^{2-4s+2c}\zeta^3\tau_{33,3}\\
D_{3*} &= D_{3,0} + \zeta^2 D_{3,2}.
\end{aligned} \tag{37.2}$$

Passing to the notation of the shell theory, we get the equations which will be divided into two groups. The first group will be comprised of

$$\begin{aligned}
T_i &= 2h(n_{ii}\varepsilon_i + n_{ij}\varepsilon_j) - 2hc_i E_2^{(0)}\\
S_{ij} &= \frac{2h}{s_{44}^E}(\omega - d_{15}E_1^{(0)})\\
G_i &= -\frac{2h^3}{3}(n_{ii}\kappa_i + n_{ij}\kappa_j),\\
H_{ij} &= \frac{4h^3}{3s_{44}^E}\tau\\
E_i^{(0)} &= -\frac{1}{A_i}\frac{\partial\psi^{(0)}}{\partial\alpha_i},\\
D_i^{(0)} &= \varepsilon_{11}^T E_1^{(0)} + \frac{d_{15}}{2h}S_{12}\\
D_2^{(0)} &= \varepsilon_{33}^T E_2^{(0)} + \frac{d_{31}}{2h}T_1 + \frac{d_{33}}{2h}T_2,
\end{aligned}$$

$$\frac{\partial(A_2 D_1^{(0)})}{\partial\alpha_1} + \frac{\partial(A_1 D_2^{(0)})}{\partial\alpha_2} = 0. \tag{37.3}$$

In addition to equations 37.3, we should include the equilibrium equations and strain-displacement formulas in the first group. We then get a closed system of tenth-order differential equations with respect to the unknown mechanical and electrical quantities. All other quantities will be included in the second group:

$$
\begin{aligned}
D_1^{(1)} &= \frac{3d_{15}}{2h^3} H_{12}, \\
D_2^{(1)} &= -\frac{3}{2h^3}(d_{15}G_1 + d_{33}G_2) \\
D_3^{(0)} &= -h^2 D_3^{(2)} = \frac{h^2}{2A_1A_2}\left[\frac{\partial(A_2 D_1^{(1)}}{\partial\alpha_1} + \frac{\partial(A_1 D_2^{(1)})}{\partial\alpha_2}\right] \\
\psi^{(k+1)} &= \frac{1}{k+1} E_3^{(k)}, \quad k = 0, 1 \\
E_3^{(0)} &= \frac{1}{\varepsilon_{11}^T} D_3^{(0)} + \frac{d_{15}}{4\varepsilon_{11}^T}\left[\frac{3N_2}{h} + (q_2^+ - q_2^-)\right] \\
E_3^{(1)} &= -\frac{d_{15}}{2h\varepsilon_{11}^T}(q_2^+ + q_2^-), \\
E_3^{(2)} &= \frac{1}{\varepsilon_{11}^T} D_3^{(2)} - \frac{3d_{15}}{4\varepsilon_{11}^T h^2}\left(\frac{N_2}{h} + q_2^+ - q_2^-\right).
\end{aligned}
\qquad (37.4)
$$

Having found the solution to the system of the first group, we use formulas 37.4 to find the electrical quantities that did not enter the first group. Note that the complete problem, generally speaking, cannot be divided into the mechanical and electrical problems with the exception of some special cases. Say, for an axially symmetric problem [26], that all the electrical quantities in equations 37.3 can be expressed in terms of the forces. This allows us to divide the problem into the mechanical and electrical parts. Here, the equations of the mechanical problem differ from those of the nonelectrical shell theory by the sense of the coefficients before the strain components in the constitutive relations.

Chapter 7

THE THEORY OF ELECTROELASTIC BOUNDARY LAYER

We will assume that just as for nonelectrical shells [22, 43] the complete electroelastic state can be represented as a sum of two electroelastic states: the inner electrostatic state and the boundary layer. The inner electroelastic state varies relatively slowly along the coordinate lines of the middle surface and is described by the equations of piezoceramic shell theory. The boundary layer damps down quickly in the direction perpendicular to the edge and is described by three-dimensional electroelasticity equations.

In the theory of nonelectrical shells, the boundary layer plays a secondary role and is rarely resorted to for a strict calculation of the inner stressed-strained state. The corrections due to allowing for the boundary layer are rather small. We will show that this situation can also be observed in the theory of piezoceramic shells but with some exceptions: the calculation should be started with the boundary layer because it defines the greatest stresses. Also, the corrections due to the boundary layer under the conditions of the shell theory introduce qualitative changes into the description of the internal electroelastic state [81, 82].

For constructing the boundary conditions in the theory of electroelastic shells, we should use the Saint Venant principle generalized to the case of electroelasticity [83]. In what follows the generalized Saint Venant principle is obtained from the solutions of three-dimensional electroelastic problems.

38 THE BOUNDARY LAYER OF A SHELL WITH THICKNESS POLARIZATION

We assume that the investigated edge of a shell coincides with the surface $\alpha_1 = \alpha_{10}$. As in Part I, we introduce the electroelasticity equations by using equations 2.12 and 2.13, the nonsymmetric tensor $\tau_{\rho\mu}$ instead of the symmetric stress tensor $\sigma_{\rho\mu}$, and the vectors $\mathbf{D}$ and $\mathbf{E}$ instead of the electric induction vector $\mathcal{D}$ and electric strength vector $\mathcal{E}$.

We asymptotically extend the scale near the edge

$$\alpha_1 - \alpha_{10} = R\eta^1\xi_1, \qquad \alpha_2 = R\eta^S\xi_2, \qquad \gamma = R\eta^1\zeta \tag{38.1}$$

This means that the sought-for electroelastic state has the same great variability with respect to the variables α_1 and γ and much less variability with respect to the variable α_2, which is equal to the variability of the inner electroelastic state.

The equations for a piezoceramic preliminarily polarized in the thickness direction will have the form

$$\eta^{-1}\frac{1}{R}\frac{1}{A_1}\frac{\partial\tau_{11}}{\partial\xi_1}+\eta^{S}\frac{1}{R}\frac{1}{A_2}\frac{\partial\tau_{12}}{\partial\xi_2}+\eta^{-1}\frac{a_1}{R}\frac{\partial\tau_{13}}{\partial\zeta}$$
$$+k_2(\tau_{11}-\tau_{22})+k_1(\tau_{12}+\tau_{21})+\frac{2}{R_1}\tau_{13}=0$$
$$\eta^{-1}\frac{1}{R}\frac{1}{A_1}\frac{\partial\tau_{21}}{\partial\xi_1}+\eta^{S}\frac{1}{R}\frac{1}{A_2}\frac{\partial\tau_{22}}{\partial\xi_2}+\eta^{-1}\frac{a_2}{R}\frac{\partial\tau_{23}}{\partial\zeta}$$
$$+k_1(\tau_{22}-\tau_{11})+k_2(\tau_{12}+\tau_{21})+\frac{2}{R_2}\tau_{23}=0$$
$$\eta^{-1}\frac{1}{R}\frac{1}{A_1}\frac{\partial\tau_{31}}{\partial\xi_1}+\eta^{-S}\frac{1}{R}\frac{1}{A_2}\frac{\partial\tau_{32}}{\partial\xi_2}+\eta^{-1}\frac{1}{R}\frac{\partial\tau_{33}}{\partial\zeta}+k_2\tau_{13}+k_1\tau_{23}-\frac{\tau_{11}}{R_1}-\frac{\tau_{22}}{R_2}=0$$
$$a_2\left(\eta^{-1}\frac{1}{R}\frac{1}{A_1}\frac{\partial v_1}{\partial\xi_1}+k_1v_2+\frac{v_3}{R_1}\right)-[s_{11}^{E}a_1\tau_{11}+s_{12}^{E}a_2\tau_{22}+s_{13}^{E}\tau_{33}]-d_{31}E_3=0$$
$$a_1\left(\eta^{-s}\frac{1}{R}\frac{1}{A_2}\frac{\partial v_2}{\partial\xi_2}+k_2v_1+\frac{v_3}{R_2}\right)+[s_{12}^{E}a_1\tau_{11}+s_{11}^{E}a_2\tau_{22}+s_{13}^{E}\tau_{33}]-d_{31}E_3=0$$
$$\eta^{-1}\frac{a_1a_2}{R}\frac{\partial v_3}{\partial\zeta}-[s_{13}^{E}a_1\tau_{11}+s_{13}^{E}a_2\tau_{22}+s_{33}^{E}\tau_{33}]-d_{33}E_3=0$$
$$\eta^{-s}\frac{a_1}{R}\frac{1}{A_2}\frac{\partial v_1}{\partial\xi_2}+\eta^{-1}\frac{a_2}{R}\frac{1}{A_1}\frac{\partial v_2}{\partial\xi_1}-k_1a_2v_1-k_2a_1v_2-s_{66}^{E}a_2\tau_{12}=0$$
$$\eta^{-s}\frac{a_1}{R}\frac{1}{A_2}\frac{\partial v_1}{\partial\xi_2}+\eta^{-1}\frac{a_2}{R}\frac{1}{A_1}\frac{\partial v_2}{\partial\xi_1}-k_1a_2v_1-k_2a_1v_2-s_{66}^{E}a_2\tau_{21}=0$$
$$a_2\left[\eta^{-1}\frac{a_1}{R}\frac{\partial v_1}{\partial\zeta}+\eta^{-1}\frac{1}{R}\frac{1}{A_1}\frac{\partial v_3}{\partial\xi_1}-\frac{v_1}{R_1}\right]-s_{44}^{E}a_1\tau_{13}-d_{15}a_2E_1=0$$
$$a_1\left[\eta^{-1}\frac{a_2}{R}\frac{\partial v_2}{\partial\zeta}+\eta^{s}\frac{1}{R}\frac{1}{A_2}\frac{\partial v_3}{\partial\xi_2}-\frac{v_2}{R_2}\right]-s_{44}^{E}a_2\tau_{23}-d_{15}a_1E_2=0$$
$$(38.2)$$

$$a_1D_1=\varepsilon_{11}^{T}a_2E_1+d_{15}a_1\tau_{13}$$
$$a_2D_2=\varepsilon_{11}^{T}a_1E_2+d_{15}a_2\tau_{23}$$
$$D_3=\varepsilon_{33}^{T}E_3+d_{31}(a_1\tau_{11}+a_2\tau_{22})+d_{33}\tau_{33} \qquad (38.3)$$

$$\eta^{-1}\frac{1}{R}\frac{\partial D_3}{\partial\zeta}+\eta^{-1}\frac{1}{R}\frac{1}{A_1}\frac{\partial D_1}{\partial\xi_1}+\eta^{-s}\frac{1}{R}\frac{1}{A_2}\frac{\partial D_2}{\partial\xi_2}+k_1D_2+k_2D_1=0$$
$$E_1=-\eta^{-1}\frac{1}{R}\frac{1}{A_1}\frac{\partial\psi}{\partial\xi_1},\qquad E_2=-\eta^{-s}\frac{1}{R}\frac{1}{A_2}\frac{\partial\psi}{\partial\xi_2}$$

$$E_3 = -\eta^{-1}\frac{a_1 a_2}{R}\frac{\partial\psi}{\partial\zeta}, \qquad a_i = 1 + \eta^{-1}\zeta\frac{R}{R_i}. \tag{38.4}$$

The equations of the electroelastic boundary layer, like those of the boundary layer of nonelectrical shells, can be divided into the equations for the plane and antiplane boundary layers. In the first approximation, the equations of the plane boundary layer are the equations of a plane electroelastic problem. The equations of the antiplane boundary layer coincide within the coefficients with the equations of the antiplane problem of the theory of elasticity.

We assume that the asymptotics of the plane and antiplane layers leads to a noncontradictory iterative process:

$$\left(s_{11}^E\tau_{21}^k, s_{11}^E\tau_{12}^k, s_{11}^E\tau_{23}^k, s_{11}^E\tau_{32}^k, \frac{v_2^k}{h}, \frac{d_{31}}{\varepsilon_{11}^T}D_2^k\right) = \eta^r(\tau_{21*}^k, \tau_{12*}^k, \tau_{23*}^k, \tau_{32*}^k, v_{2*}^k, D_{2*}^k) \tag{38.5}$$

$$\left(s_{11}^E\tau_{11}^k, s_{11}^E\tau_{22}^k, s_{11}^E\tau_{33}^k, s_{11}^E\tau_{13}^k, s_{11}^E\tau_{31}^k, \frac{v_1^k}{h}, \frac{v_3^k}{h}, d_{31}\frac{\psi^k}{h}, d_{31}E_1^k, d_{31}E_3^k, \frac{d_{31}}{\varepsilon_{11}^T}D_1^k, \frac{d_{31}}{\varepsilon_{33}^T}D_3^k\right)$$
$$= \eta^{1-s-r}(\tau_{11*}^k, \tau_{22*}^k, \tau_{33*}^k, \tau_{13*}^k, \tau_{31*}^k, v_{1*}^k, v_{3*}^k, \psi_*^k, E_{1*}^k, E_{3*}^k, D_{1*}^k, D_{3*}^k)$$
$$d_{31}E_2^k = \eta^{2-2s}E_{2*}^k. \tag{38.6}$$

As usual, the degrees of η are chosen so that the dimensionless quantities with asterisks are of the same order.

Formulas 38.5 and 38.6 unite two asymptotics for the plane and antiplane boundary layers. The superscript k should be replaced by a if the quantity belongs to the antiplane boundary layer and by b if it belongs to the plane boundary layer. For the antiplane problem, we choose

$$r = 0. \tag{38.7}$$

For the plane one, we choose

$$r = 1 - s. \tag{38.8}$$

We change the variables 38.1 in the three-dimensional electroelasticity equations, taking into account the asymptotics 38.5 to 38.8, and break the resultant system into two subsystems. In the first subsystem, the quantities 38.5 and 38.7 of the antiplane layer are principal. In the second subsystem, we find the quantities 38.6 and 38.8 of the plane layer. We write out the equations of the antiplane and plane boundary layers uniting them for brevity:

$$\frac{1}{A_{10}}\frac{\partial\tau_{21}^k}{\partial\xi_1} + \frac{\partial\tau_{23*}^k}{\partial\zeta} + X_2^k + \eta^1 X_2^{k1} = 0$$

$$\frac{1}{A_{10}}\frac{\partial v_{2*}^k}{\partial \xi_1} - \sigma_1 \tau_{12*}^k + W_{12}^k + \eta^1 W_{12}^{k1} = 0$$

$$\frac{1}{A_{10}}\frac{\partial v_{2*}^k}{\partial \xi_1} - \sigma_1 \tau_{21*}^k + W_{21}^k + \eta^1 W_{21}^{k1} = 0$$

$$\frac{\partial v_{2*}^k}{\partial \zeta} - \sigma_2 \tau_{23*}^k + W_{23}^k + \eta^1 W_{23}^{k1} = 0 \quad (38.9)$$

$$D_{2*}^k - t_1 \tau_{23*}^k + K_{23}^k + \eta^1 K_{23}^{k1} = 0 \quad (38.10)$$

$$\frac{1}{A_{10}}\frac{\partial \tau_{11*}^k}{\partial \xi_1} + \frac{\partial \tau_{13*}^k}{\partial \zeta} + X_1^k + \eta^1 X_1^{k1} = 0$$

$$\frac{1}{A_{10}}\frac{\partial \tau_{13*}^k}{\partial \xi_1} + \frac{\partial \tau_{33*}^k}{\partial \zeta_1} + X_3^k + \eta^1 X_3^{k1} = 0$$

$$\frac{1}{A_{10}}\frac{\partial v_{1*}^k}{\partial \xi_1} - (\tau_{11*}^k - \nu \tau_{22*}^k - \nu_1 \tau_{33*}^k) - E_{3*}^k + W_1^k + \eta^1 W_1^{k1} = 0$$

$$\nu \tau_{11*}^k - \tau_{22*}^k + \nu_1 \tau_{33*}^k - E_{3*}^k + W_2^k + W_2^{k1} = 0$$

$$\frac{\partial v_{3*}^k}{\partial \zeta} - (\sigma_3 \tau_{33*}^k - \nu_1 \tau_{11*}^k - \nu_1 \tau_{22*}^k) - \frac{d_{33}}{d_{31}} E_{3*}^k + \eta^1 W_3^{k1} = 0$$

$$\frac{\partial v_1^k}{\partial \zeta} + \frac{1}{A_{10}}\frac{\partial v_{3*}^k}{\partial \xi_1} - \sigma_2 \tau_{13*}^k - \frac{d_{15}}{d_{31}} E_{1*}^k + \eta^1 W_{13}^{k1} = 0$$

$$D_{1*}^k - E_{1*}^k - t_1 \tau_{13*}^k + \eta^1 K_1^{k1} = 0$$

$$D_{3*}^k - E_{3*}^k - t_2(\tau_{13*}^k + \tau_{22*}^k) - t_3 \tau_{33*}^k + \eta^1 K_3^{k1} = 0$$

$$\frac{1}{A_{10}}\frac{\partial D_{1*}^k}{\partial \xi_1} + \frac{\varepsilon_{33}^T}{\varepsilon_{11}^T}\frac{\partial D_{3*}^k}{\partial \zeta} + M^k + \eta^1 M^{k1} = 0$$

$$E_{1*}^k + \frac{1}{A_{10}}\frac{\partial \psi_*^k}{\partial \xi_1} + \eta^1 L_1^{k1} = 0$$

$$E_3^k + \frac{\partial \psi_*}{\partial \zeta} + \eta^1 L_3^{k1} = 0. \quad (38.11)$$

In the formulas above, we used the notation 34.7.

The notations $X_2^k, \ldots, L_3^{k1}$ follow from the formulas

$$X_2^a = W_{12}^a = W_{21}^a = W_{23}^a = K_{23}^a = 0$$

$$X_1^a = \frac{1}{A_{20}}\frac{\partial \tau_{12*}^a}{\partial \xi_2} + \eta^s R k_{10}(\tau_{12*}^a + \tau_{21*}^a)$$

$$X_3^a = \frac{1}{A_{20}}\frac{\partial \tau_{32*}^a}{\partial \xi_2} + \eta^s R k_{10} \tau_{23*}^a$$

$$W_1^a = \eta^s R k_{10} v_{2*}^a,$$

$$W_2^a = \frac{1}{A_{20}}\frac{\partial v_{2*}^a}{\partial \xi_2}$$

$$
\begin{aligned}
M^a &= \frac{1}{A_{20}}\frac{\partial D^a_{2*}}{\partial \xi_2} + R\eta^s k_{10} D^a_{2*},\\
X^{a1}_1 &= \left(\frac{1}{A_1}\right)'\xi_1\frac{\partial \tau^a_{11*}}{\partial \xi_1} + \left(\frac{1}{A_2}\right)'\xi_1\frac{\partial \tau^a_{12*}}{\partial \xi_2} + \frac{R}{R_2}\zeta\frac{\partial \tau^a_{13*}}{\partial \zeta} + Rk_2(\tau^a_{11*} - \tau^a_{22*})\\
&\quad + \eta^s R(k_1)' R\xi_1(\tau^a_{12*} + \tau^a_{21*}) + \frac{2R}{R_1}\tau^a_{13*}\\
X^{a1}_3 &= \left(\frac{1}{A_1}\right)'\xi_1\frac{\partial \tau^a_{31*}}{\partial \xi_1} + \left(\frac{1}{A_2}\right)'\xi_1\frac{\partial \tau^a_{32*}}{\partial \xi_2} + Rk_2\tau^a_{23*} + \eta^s R(k_1)'\xi_1\tau^a_{23*}\\
&\quad - \frac{R}{R_1}\tau^a_{11*} - \frac{R}{R_2}\tau^a_{22*}\\
K^1_1 &= \zeta\frac{R}{R_1}D^a_{1*} - \zeta\frac{R}{\mathcal{R}_2}E^a_{1*} - \zeta t_1\frac{R}{R_1}\tau^a_{13*}\\
K^{\alpha 1}_3 &= -\zeta t_2\frac{R}{R_1}\tau^a_{11*} - \zeta t_2\tau^a_{22*}\\
M^{a1} &= k_2 D^a_{1*} + \left(\frac{1}{A_1}\right)'\xi_1\frac{\partial D^a_{1*}}{\partial \xi_1} - \left(\frac{1}{A_2}\right)'\xi_1\frac{\partial D^a_{2*}}{\partial \xi_2} + \eta^s(k_1)' R\xi_1 D^a_2\\
L^{a1}_1 &= \left(\frac{1}{A_1}\right)'\xi_1\frac{\partial \psi^a_*}{\partial \xi_1},\\
L^{a1}_3 &= \xi_1\left(\frac{R}{R_1} + \frac{R}{R_2}\right)\frac{\partial \psi^a_*}{\partial \zeta} \qquad (38.12)\\
X^b_1 &= X^b_3 = W^b_1 = W^b_2 = M^b = 0,\\
K^b_{23} &= -E^b_{2*}\\
X^b_2 &= \frac{1}{A_{20}}\frac{\partial \tau^b_{22*}}{\partial \xi_2} + \eta^s Rk_{10}(\tau^b_{22*} - \tau^b_{11*})\\
W^b_{12} &= W^b_{21} = \frac{1}{A_{20}}\frac{\partial v^b_{1*}}{\partial \xi_2} - \eta^s Rk_{10}v^b_{1*}\\
W^b_{23} &= \frac{1}{A_{20}}\frac{\partial v^b_{3*}}{\partial \xi_2}\\
X^{b1}_2 &= \left(\frac{1}{A_1}\right)'\xi_1\frac{\partial \tau^b_{21*}}{\partial \xi} + \left(\frac{1}{A_2}\right)'\xi_1\frac{\partial \tau^b_{22*}}{\partial \xi_2}\\
&\quad + \frac{R}{R_2}\zeta\frac{\partial \tau^b_{23*}}{\partial \zeta} + \eta^s R(k_1)'\xi_1(\tau^b_{22*} - \tau^b_{11*})\\
&\quad + Rk_2(\tau^b_{12*} + \tau^b_{21*}) + \frac{2R}{R_2}\tau^b_{23*}\\
X^{b1}_1 &= \left(\frac{1}{A}\right)'\xi_1\frac{\partial \tau^b_{11*}}{\partial \xi_1} + \frac{R}{R_1}\zeta\frac{\partial \tau^b_{13*}}{\partial \zeta}Rk_2(\tau^b_{11*} - \tau^b_{22*}) + \frac{2R}{R_1}\tau^b_{13*}
\end{aligned}
$$

$$
\begin{aligned}
X_3^{b1} &= \left(\frac{1}{A_1}\right)' \xi_1 \frac{\partial \tau_{31*}^b}{\partial \xi_1} + Rk_2\tau_{13*}^b - \frac{R}{R_1}\tau_{11*}^b - \frac{R}{R_2}\tau_{22*}^b \\
K_1^{b1} &= \zeta\frac{R}{R_1}D_{1*}^b - \zeta\frac{R}{R_2}E_{1*}^b - t_1\zeta\frac{R}{R_1}\zeta_{13*}^b \\
K_3^{b1} &= -\zeta t_2\tau_{11*}^b - \zeta t_2\tau_{22*}^b \\
M^{b1} &= \left(\frac{1}{A_1}\right)' \xi_1 \frac{\partial D_{1*}^b}{\partial \xi_1} + k_2RD_{1*}^b.
\end{aligned} \tag{38.13}
$$

The quantities $A_1, A_2, 1/A_1, 1/A_2, 1/R_1, 1/R_2, k_1$, and k_2 that characterize the geometry of the shell's middle surface are expanded in power series in the variable α_1 in the neighborhood of $\alpha_1 = \alpha_{10}$. For example, for A_1, we can take into account formulas 38.1 and write a series

$$
\begin{aligned}
A_1 &= A_{10} + \sum_{n=1}^{\infty}(\alpha_1 - \alpha_{10})^n (A_1)_n \\
&= A_{10} + \sum_{n-1}^{\infty} \xi_1^n R^n \eta^n (A_1)_n = A_{10} + \eta^1 \xi_1 (A_1)' \\
(A_1)' &= \sum_{n-1}^{\infty} \xi_1^{n-1} R^n \eta^{n-1} (A_1)_n.
\end{aligned} \tag{38.14}
$$

Here, A_{10} and $(A_1)_n$ are coefficients of the Taylor series.

For quantities with superscript 1, formulas 38.12 and 38.13 are written up to $O(\eta^1)$.

For defining an approximate solution of type a, we neglect the small terms with factor η^1 in equations 38.9 and obtain a system for the antiplane problem of elasticity theory:

$$
\begin{aligned}
&\frac{1}{A_{10}}\frac{\partial \tau_{21*}^a}{\partial \xi_1} + \frac{\partial \tau_{23*}^a}{\partial \zeta} = 0 \\
&\frac{1}{A_{10}}\frac{\partial v_{2*}^a}{\partial \xi_1} = \sigma_1 \tau_{21*}^a, \qquad \tau_{21*}^a = \tau_{12*}^a \\
&\frac{\partial v_{2*}^a}{\partial \zeta} = \sigma_2 \tau_{23*}^a.
\end{aligned} \tag{38.15}
$$

We solve this system and find the required quantities with superscript a. We can find the rest of the quantities in this group by integrating system 38.11 by the terms with factor η^1 and treating the quantities found when solving the antiplane problem as known quantities.

By using the same two stages, we approximately define the quantities with superscript b. In the first stage, we find the quantities $\tau_{11*}^b, \tau_{22*}^b, \tau_{33*}^b, \tau_{13*}^b, V_{1*}^b, V_{3*}^b$,

$\psi_*^b, E_{1*}^b, E_{3*}^b, D_{1*}^b$ and D_{3*}^b from system 38.11, by neglecting the terms $O(\eta^1)$. We thus get the system of equations of the plane problem of electroelasticity theory. In the second stage, the rest of the quantities are found from system 38.9, where we also neglect the terms $o(\eta^1)$ and treat the quantities found in the first stage as known quantities. We write the equations of the plane problem:

$$\frac{1}{A_{10}}\frac{\partial \tau_{11*}^b}{\partial \xi_1} + \frac{\partial \tau_{13*}^b}{\partial \zeta} = 0$$

$$\frac{1}{A_{10}}\frac{\partial \tau_{31*}^b}{\partial \xi_1} + \frac{\partial \tau_{33*}^b}{\partial \zeta} = 0$$

$$\frac{1}{A_{10}}\frac{\partial v_{1*}^b}{\partial \xi_1} - (\tau_{11*}^b - \nu\tau_{22*}^b - \nu_1\tau_{33*}^b) - E_{3*}^b = 0$$

$$\nu\tau_{11*}^b - \tau_{22*}^b + \nu_1\tau_{33*}^b - E_{3*}^b = 0$$

$$\frac{\partial v_{3*}^b}{\partial \zeta} - (\sigma_3\tau_{33*}^b - \nu_1\tau_{11*}^b - \nu_1\tau_{22*}^b) - \frac{d_{33}}{d_{31}}E_{3*}^b = 0$$

$$\frac{\partial v_{1*}^b}{\partial \zeta} + \frac{1}{A_{10}}\frac{\partial v_{3*}^b}{\partial \xi} - \sigma_2\tau_{13*}^b - \frac{d_{15}}{d_{13}}E_{1*}^b = 0$$

$$D_{1*}^b - E_{1*}^b - t_1\tau_{13*}^b = 0$$

$$D_{3*}^b - E_{3*}^b - t_2(\tau_{11*}^b + \tau_{22*}^b) - t_3\tau_{33*}^b = 0$$

$$\frac{1}{A_{10}}\frac{\partial D_{1*}^b}{\partial \xi_1} + \frac{\varepsilon_{33}^T}{\varepsilon_{11}^T}\frac{\partial D_{3*}^b}{\partial \zeta} = 0$$

$$E_{1*}^b = -\frac{1}{A_{20}}\frac{\partial \psi_*^b}{\partial \xi_2}$$

$$E_{3*}^b = -\frac{\partial \psi_*^b}{\partial \zeta}. \tag{38.16}$$

We will call systems 38.15 and 38.16 used to find the greatest quantities the principal systems and the sought-for quantities in them the principal quantities. The systems used at the second stage to find smaller quantities will be called auxiliary.

Since the mechanical and electrical loads are allowed for while integrating the equations of the inner electroelastic state, the following conditions on the faces are valid for the antiplane and plane boundary layers:

$$\tau_{23*}^a \mid_{\zeta=\pm 1} = 0 \tag{38.17}$$

$$\tau_{33*}^b \mid_{\zeta=\pm 1} = 0$$

$$\tau_{13*}^b \mid_{\zeta=\pm 1} = 0 \tag{38.18}$$

$$\psi_*^b \mid_{\zeta=\pm 1} = 0 \tag{38.19}$$

$$D_{3*}^b \mid_{\zeta=\pm 1} = 0. \tag{38.20}$$

The condition 38.19 is valid for electrode-covered surfaces. The condition 38.20 is valid for surfaces without electrodes.

39 BOUNDARY LAYER AT THE EDGE $\alpha_1 = \alpha_{10}$ (PRELIMINARY POLARIZATION ALONG THE α_2-LINES)

Since the directions α_1 and α_2 are not equally important for preliminary polarization along the α_2-lines, we will consider the edges $\alpha_1 = \alpha_{10}$ and $\alpha_2 = \alpha_{20}$, separately. Let one of the coordinate lines of $\alpha_1 = \alpha_{10}$ be the boundary contour of the shell. We write constitutive relations for a piezoceramic polarized along the α_2-lines

$$\left(1+\frac{\gamma}{R_2}\right)\left(\frac{1}{A_1}\frac{\partial v_1}{\partial \alpha_1}+k_1 v_2+\frac{v_3}{R_1}\right)$$
$$-\left[\left(1+\frac{\gamma}{R_1}\right)s_{11}^E\tau_{11}+\left(1+\frac{\gamma}{R_2}\right)s_{13}^E\tau_{22}+s_{12}^E\tau_{33}\right]-d_{31}\left(1+\frac{\gamma}{R_1}\right)E_2=0$$

$$\left(1+\frac{\gamma}{R_1}\right)\left(\frac{1}{A_2}\frac{\partial v_2}{\partial \alpha_2}+k_2 v_1+\frac{v_3}{R_2}\right)$$
$$-\left[\left(1+\frac{\gamma}{R_1}\right)s_{13}^E\tau_{11}+\left(1+\frac{\gamma}{R_2}\right)s_{33}^E\tau_{22}+s_{13}^E\tau_{33}\right]-d_{33}\left(1+\frac{\gamma}{R_1}\right)E_2=0$$

$$\left(1+\frac{\gamma}{R_1}\right)\left(1+\frac{\gamma}{R_2}\right)\frac{\partial v_3}{\partial \gamma}-\left[\left(1+\frac{\gamma}{R_1}\right)s_{12}^E\tau_{11}+\left(1+\frac{\gamma}{R_2}\right)s_{13}^E\tau_{22}+s_{11}^E\tau_{33}\right]$$
$$-d_{31}\left(1+\frac{\gamma}{R_1}\right)E_2=0$$

$$\left[\left(1+\frac{\gamma}{R_1}\right)\frac{1}{A_2}\frac{\partial v_1}{\partial \alpha_2}+\left(1+\frac{\gamma}{R_2}\right)\frac{1}{A_1}\frac{\partial v_2}{\partial \alpha_1}-k_1\left(1+\frac{\gamma}{R_2}\right)v_1\right.$$
$$\left.-k_2\left(1+\frac{\gamma}{R_1}\right)v_2\right]-s_{44}^E\left(1+\frac{\gamma}{R_1}\right)\tau_{21}-d_{15}\left(1+\frac{\gamma}{R_2}\right)E_1=0$$

$$\left[\left(1+\frac{\gamma}{R_1}\right)\frac{1}{A_2}\frac{\partial v_1}{\partial \alpha_2}+\left(1+\frac{\gamma}{R_2}\right)\frac{1}{A_1}\frac{\partial v_2}{\partial \alpha_1}-k_1\left(1+\frac{\gamma}{R_2}\right)v_1\right.$$
$$\left.-k_2\left(1+\frac{\gamma}{R_1}\right)v_2\right]-s_{44}^E\left(1+\frac{\gamma}{R_2}\right)\tau_{12}-d_{15}\left(1+\frac{\gamma}{R_2}\right)E_1=0$$

$$\left(1+\frac{\gamma}{R_2}\right)\left[\left(1+\frac{\gamma}{R_1}\right)\frac{\partial v_1}{\partial \gamma}+\frac{1}{A_1}\frac{\partial v_3}{\partial \alpha_1}-\frac{v_1}{R_1}\right]-s_{66}^E\left(1+\frac{\gamma}{R_1}\right)\tau_{13}=0$$

$$\left(1+\frac{\gamma}{R_1}\right)\left[\left(1+\frac{\gamma}{R_2}\right)\frac{\partial v_2}{\partial \gamma}+\frac{1}{A_2}\frac{\partial v_3}{\partial \alpha_2}-\frac{v_2}{R_2}\right]-s_{66}^E\left(1+\frac{\gamma}{R_2}\right)\tau_{23}-d_{15}E_3=0$$

$$\left(1+\frac{\gamma}{R_1}\right)D_1=\varepsilon_{11}^T\left(1+\frac{\gamma}{R_2}\right)E_1+d_{15}\left(1+\frac{\gamma}{R_1}\right)\tau_{21}$$

$$D_3=\varepsilon_{11}^T E_3+d_{15}\left(1+\frac{\gamma}{R_2}\right)\tau_{23}$$

$$\left(1+\frac{\gamma}{R_2}\right)D_2=\varepsilon_{33}^T\left(1+\frac{\gamma}{R_1}\right)E_2+d_{31}\left(1+\frac{\gamma}{R_1}\right)\tau_{11}$$

$$+d_{33}\left(1+\frac{\gamma}{R_2}\right)\tau_{22}+d_{31}\tau_{33}. \tag{39.1}$$

In the constitutive relations 39.1, we change variables by formulas 38.1. The equilibrium equations and electrostatic equations, where the variables have been changed by 38.1, can be found in the previous section.

We asymptotically represent the quantities of the antiplane and plane boundary layers as follows:

$$\left(s_{11}^E\tau_{12}^K, s_{11}^E\tau_{21}^K, \frac{v_2^k}{h}, \frac{d_{15}\psi^k}{h}, d_{15}E_1^k, d_{15}E_3^k, \frac{d_{15}}{\varepsilon_{33}^T}D_1^k, \frac{d_{15}}{\varepsilon_{33}^T}D_3^k\right)$$

$$=\eta^r\left(\tau_{12*}^k, \tau_{21*}^k, v_{2*}^k, \Psi_*^k, E_{1*}^k, E_{3*}^k, D_{1*}^k, D_{3*}^k\right) \tag{39.2}$$

$$d_{15}E_2^k=\eta^{1-s+r}E_{2*}^k$$

$$\left(s_{11}^E\tau_{11}^k, s_{11}^E\tau_{22}^k, s_{11}^E\tau_{33}^k, s_{11}^k\tau_{13}, \frac{v_1^k}{h}, \frac{v_3^k}{h}, \frac{d_{15}}{\varepsilon_{33}^T}D_2^k\right)$$

$$=\eta^{1-s-r}\left(\tau_{11*}^k, \tau_{22*}^k, \tau_{33*}^k, \tau_{13*}^k, v_{1*}^k, v_{3*}^k, D_{2*}^k\right). \tag{39.3}$$

In order to find the asymptotics of the antiplane boundary layer from equations 39.2 and 39.3, we should take

$$k=a, \qquad r=0. \tag{39.4}$$

For the plane boundary layer, we should take

$$k=b, \qquad r=1-s. \tag{39.5}$$

We substitute equations 39.2 to 39.5 into equations 38.2, 38.4, and 39.1. We then divide the complete system into two subsystems and unit them for brevity.

In the first subsystem, where the principal quantities are those of the antiplane problem, we have

$$\frac{1}{A_{10}}\frac{\partial \tau_{21*}^k}{\partial \xi_1} + \frac{\partial \tau_{23*}^k}{\partial \zeta} = -X_2^k - \eta^1 X_2^{k1}$$

$$\frac{1}{A_{10}}\frac{\partial v_{2*}^k}{\partial \xi_1} - E_{1*}^k - \sigma_2 \tau_{21*}^k = -P_{21}^k - \eta^1 P_{21}^{k1}$$

$$\frac{1}{A_{10}}\frac{\partial v_{2*}^k}{\partial \xi_1} - E_{1*}^k - \sigma_2 \tau_{12*}^k = -P_{12}^k - \eta^1 P_{12}^{k1}$$

$$\frac{\partial v_{2*}^k}{\partial \zeta} - E_{3*}^k - \sigma_1 \tau_{23*}^k = -p_{23}^k - \eta^1 P_{23}^{k1}$$

$$D_{1*}^k - \frac{\varepsilon_{11}^T}{\varepsilon_{33}^T} E_{1*}^k - t_4 \tau_{21*}^k = \eta^1 L_1^{k1}$$

$$D_{3*}^k - \frac{\varepsilon_{11}^T}{\varepsilon_{33}^T} E_{3*}^k - t_4 \tau_{23*}^k = \eta^1 L_3^{k1}$$

$$\frac{\partial D_{3*}^k}{\partial \zeta} + \frac{1}{A_{10}}\frac{\partial D_{1*}^k}{\partial \xi_1} = -M^k - \eta^1 M^{k1}$$

$$E_3^k + \frac{\partial \psi_*^k}{\partial \zeta} = -\eta^1 K_3^{k1}$$

$$E_{1*}^k + \frac{1}{A_{10}}\frac{\partial \psi_*^k}{\partial \xi_1} = -\eta^1 K_1^{k1}$$

$$E_{2*}^k + \frac{1}{A_{20}}\frac{\partial \psi_*^k}{\partial \xi_2} = -\eta^1 K_2^{k1}. \tag{39.6}$$

For subsystem 39.6, the conditions should hold on the faces covered with electrodes

$$\psi_*^k \mid_{\zeta=\pm 1} = 0, \qquad \tau_{23*}^k \mid_{\zeta=\pm 1} = 0 \tag{39.7}$$

and on the faces without electrodes

$$D_{3*}^k \mid_{\zeta=\pm 1} = 0, \qquad \tau_{23*}^k \mid_{\zeta=\pm 1} = 0. \tag{39.8}$$

For the second subsystem, where the quantities of the plane boundary layer are principal, we have

$$\frac{1}{A_{10}}\frac{\partial \tau_{11*}^k}{\partial \xi_1} + \frac{\partial \tau_{13*}^k}{\partial \zeta} = -X_1^k - \eta^1 X_1^{k1}$$

$$\frac{1}{A_{10}}\frac{\partial \tau_{31*}^k}{\partial \xi_1} + \frac{\partial \tau_{33*}^k}{\partial \zeta} = -X_3^k - \eta^1 X_3^{k1}$$

$$\frac{1}{A_{10}}\frac{\partial v_{1*}^k}{\partial \xi_1} - \tau_{11*}^k + \nu_1 \tau_{22*}^k + \nu \tau_{33*}^k = -P_1^k - \eta^1 P_1^{k1}$$

$$\frac{\partial v_{3*}^k}{\partial \zeta} + \nu \tau_{11*}^k + \nu_1 \tau_{22*}^k - \tau_{33*}^k = -P_3^k - \eta^1 P_3^{k1}$$
$$\frac{\partial v_{1*}^k}{\partial \zeta} + \frac{1}{A_{10}} \frac{\partial v_{3*}^k}{\partial \xi_1} - \sigma_1 \tau_{13*}^k = -\eta^1 P_{13}^{k1}$$
$$-\nu_1 \left(\tau_{11*}^k + \tau_{33*}^k \right) + \sigma_3 \tau_{22*}^k = P_2^k + \eta^1 P_2^{k1}$$
$$D_{2*}^k - t_5 \tau_{11*}^k - t_6 \tau_{22*}^k - t_5 \tau_{33*}^k = -L_2^k - \eta^1 L_2^{k1}. \quad (39.9)$$

The mechanical conditions

$$\tau_{33*}^k \mid_{\zeta=\pm 1} = 0, \qquad \tau_{13*}^k \mid_{\zeta=\pm 1} = 0 \quad (39.10)$$

should hold when we solve equations 39.9.

In equations 39.6 and 39.9, we have introduced the notation

$$X_1^a = \frac{1}{A_{20}} \frac{\partial \tau_{12*}^a}{\partial \xi_2} + \eta^s R k_{10} \left(\tau_{12*}^a + \tau_{21*}^a \right)$$
$$X_3^a = \frac{1}{A_{20}} \frac{\partial \tau_{23*}^a}{\partial \xi_2} + \eta^s R k_{10} \tau_{23*}^a$$
$$P_1^a = -\frac{d_{31}}{d_{15}} E_{2*}^a + \eta^s R k_{10} v_{2*}^a$$
$$P_2^a = -\frac{d_{33}}{d_{15}} E_{2*}^a + \frac{1}{A_{20}} \frac{\partial v_{2*}^a}{\partial \xi_2}$$
$$P_3^a = -\frac{d_{31}}{d_{15}} E_{2*}^a$$
$$L_2^a = E_{2*}^a$$
$$X_2^a = P_{12}^a = P_{23}^a = M^a = 0$$
$$X_1^{a1} = \left(\frac{1}{A_1} \right)' \xi_1 \frac{\partial \tau_{11*}^a}{\partial \xi_1} + \left(\frac{1}{A_2} \right)' \xi_1 \frac{\partial \tau_{12*}^a}{\partial \xi_2} + \frac{R}{R_2} \zeta \frac{\partial \tau_{13*}^a}{\partial \zeta} + R k_2 \left(\tau_{11*}^a - \tau_{22*}^a \right)$$
$$+ \eta_s R (k_1)' \xi_1 \left(\tau_{12*}^a + \tau_{21*}^a \right) + \frac{2R}{R_1} \tau_{13*}^a$$
$$X_2^{a1} = \left(\frac{1}{A_1} \right)' \xi_1 \frac{\partial \tau_{21*}^a}{\partial \xi_1} + \eta^{1-2s} \frac{1}{A_2} \frac{\partial \tau_{22*}^a}{\partial \xi_2} + \frac{R}{R_2} \zeta \frac{\partial \tau_{23*}^a}{\partial \zeta} + \eta^{1-s} R k_1 \left(\tau_{22*}^a - \tau_{11*}^a \right)$$
$$+ R k_2 \left(\tau_{12*}^a + \tau_{21*}^a \right) + \frac{2R}{R_2} \tau_{23*}^a$$
$$X_3^{a1} = \left(\frac{1}{A_1} \right)' \xi_1 \frac{\partial \tau_{31*}^a}{\partial \xi_1} + \left(\frac{1}{A_2} \right)' \xi_1 \frac{\partial \tau_{32*}^a}{\partial \xi_2} + R k_2 \tau_{13*}^a + \eta^s R \left(k_1 \right)' \xi_1 \tau_{23*}^a$$
$$- \frac{R}{R_1} \tau_{11*}^a - \frac{R}{R_2} \tau_{22*}^a$$

$$X_2^b = \frac{1}{A_{20}}\frac{\partial \tau_{22*}}{\partial \xi_2} + \eta^s R k_{10}(\tau_{22*} - \tau_{11*})$$

$$P_{23}^b = \frac{1}{A_{20}}\frac{\partial v_{3*}^b}{\partial \xi_2}$$

$$M^b = \frac{1}{A_{20}}\frac{\partial D_{2*}^b}{\partial \xi_2} + \eta^s R k_{10} D_{2*}^b$$

$$P_{12}^b = P_{21}^b = \frac{1}{A_{20}}\frac{\partial v_{1*}^b}{\partial \xi_2} - \eta^s R k_{10} v_{1*}^b$$

$$X_1^b = X_3^b = P_1^b = P_2^b = P_3^b = L_2^b = 0$$

$$X_2^{b1} = \left(\frac{1}{A_1}\right)' \xi_1 \frac{\partial \tau_{21*}^b}{\partial \xi_1} + \left(\frac{1}{A_2}\right)' \xi_1 \frac{\partial \tau_{22*}^b}{\partial \xi_2} + \frac{R}{R_2}\zeta \frac{\partial \tau_{23*}^b}{\partial \zeta}$$
$$+ \eta^s R\left(k_1\right)' \xi_1 \left(\tau_{22*}^b - \tau_{11*}^b\right) + R k_2 \left(\tau_{12*}^b + \tau_{21*}^b\right) + \frac{2R}{R_2}\tau_{23*}^b.$$

Of the quantities with superscript 1, we have only written the ones we will need for further investigation. For the coefficients of the first and second quadratic forms, we use expansions in the type of power series of the type 38.14.

In order to get the equations of the principal and auxiliary problems for the antiplane boundary layer, we should neglect the terms $O(\eta^1)$ in equations 39.6 and 39.9. The principal subsystem coincides with the homogeneous equations of the antiplane electroelasticity problem. For this problem to have a solution, the conditions 39.7 or 39.8 should hold. The second (auxiliary) subsystem is used to find the other smaller quantities of the antiplane boundary layer after the principal problem has been solved. The mechanical conditions 39.10 on the shell faces should hold when we integrate the auxiliary subsystem.

The plane boundary layer can be calculated in the same way. We neglect the terms with $O(\eta^1)$. We use equations 39.9 to find the homogeneous equations of the principal problem of the plane boundary layer and equations 39.6 to find the equations of the auxiliary problem.

40 BOUNDARY LAYER NEAR THE EDGE $\alpha_2 = \alpha_{20}$ (PRELIMINARY POLARIZATION ALONG THE α_2-LINES

We change the variables

$$\alpha_1 = R\eta^s \xi_1, \qquad \alpha_2 - \alpha_{20} = R\eta^1 \xi_2, \qquad \gamma = R\eta^1 \zeta. \tag{40.1}$$

For the quantities of the plane and antiplane boundary layers, we choose the asymptotic representation

$$\left(s_{11}^E\tau_{12}^k, s_{11}^E\tau_{13}^k, s_{11}^E\tau_{21}^k, \frac{v_1^k}{h}, \frac{d_{15}}{\varepsilon_{33}^T}D_1^k\right) = \eta^r\left(\tau_{12*}^K, \tau_{21*}^k, \tau_{13*}^k, v_{1*}^k, D_{1*}^k\right) \quad (40.2)$$

$$\left(s_{11}^E\tau_{11}^k, s_{11}^E\tau_{22}^k, s_{11}^E\tau_{33}^k, s_{11}^E\tau_{23}^k, \frac{v_2^k}{h}, \frac{v_3^k}{h}, \frac{d_{15}}{h}\psi^k, d_{15}E_2^k, d_{15}E_3^k, \frac{d_{15}}{\varepsilon_{33}^T}D_2^k, \frac{d_{15}}{\varepsilon_{33}^T}D_3^k\right)$$
$$= \eta^{1-s-r}\left(\tau_{11*}^k, \tau_{22*}^k, \tau_{33*}^k, \tau_{23*}^k, v_{2*}^k, v_{3*}^k, \psi_*^k, E_{2*}^k, E_{3*}^k, D_{2*}^k, D_{3*}^k\right)$$

$$d_{15}E_1^k = \eta^{2-2s-r}E_{1*}^k. \quad (40.3)$$

In order to get the asymptotics of the antiplane boundary layer, we should assume that in equations 40.2 and 40.3

$$k = a, \qquad r = 0. \quad (40.4)$$

For the asymptotics of the plane boundary layer, we should take

$$k = b \qquad r = 1 - s. \quad (40.5)$$

As in the previous cases, we use equations 40.2 to 40.5 to get, from the three-dimensional electroelasticity equations 38.2, 38.4, and 39.1, the two subsystems

$$\frac{1}{A_{20}}\frac{\partial\tau_{12*}^k}{\partial\xi_2} + \frac{\partial\tau_{13*}^k}{\partial\zeta} = -X_1^k - \eta^1 X_1^{k1}$$
$$\frac{1}{A_{20}}\frac{\partial v_{1*}^k}{\partial\xi_2} - \sigma_2\tau_{21*}^k = -P_{21}^k - P_{21}^{k1}$$
$$\frac{1}{A_{20}}\frac{\partial v_{1*}^k}{\partial\xi_2} - \sigma_2\tau_{12*}^k = -P_{12}^k - \eta^1 p_{12}^{K1}$$
$$\frac{\partial v_{1*}^k}{\partial\zeta} - \sigma_1\tau_{13*}^y = -P_{13}^k - \eta^1 P_{13}^{k1}$$
$$D_{1*}^k - t_4\tau_{21*}^k = L_1^k + \eta^1 L_1^{k1} \quad (40.6)$$
$$\tau_{13*}^k\,|_{\zeta=\pm1} = 0 \quad (40.7)$$
$$\frac{1}{A_{20}}\frac{\partial\tau_{22*}^k}{\partial\xi_2} + \frac{\partial\tau_{23*}^k}{\partial\zeta} = -X_2^k - \eta^1 X_2^{k1}$$
$$\frac{1}{A_{22}}\frac{\partial\tau_{23*}^k}{\partial\xi_2} + \frac{\partial\tau_{33*}^k}{\partial\zeta} = -X_2^k - \eta^1 X_3^{k1}$$
$$\tau_{11*}^k - \nu_1\tau_{22*}^k - \nu\tau_{33*}^k + \frac{d_{31}}{d_{15}}E_{2*}^k = P_1^k + \eta^1 P_1^{k1}$$

$$\frac{1}{A_{20}}\frac{\partial v_{2*}^k}{\partial \xi_2} + \nu_1\tau_{11*}^k - \sigma_3\tau_{22*}^k + \nu_1\tau_{33*}^k - \frac{d_{33}}{d_{15}}E_{2*}^k = -P_2^k - \eta^1 P_2^{k1}$$
$$\frac{\partial v_{3*}^k}{\partial \zeta} + \nu\tau_{11*}^k + \nu_1\tau_{22*}^k - \tau_{33*}^k - \frac{d_{31}}{d_{15}}E_{2*}^k = -\eta^1 P_3^{k1}$$
$$\frac{\partial v_{2*}^k}{\partial \zeta} + \frac{1}{A_{20}}\frac{\partial v_{3*}^k}{\partial \xi_2} - \sigma_1\tau_{23*}^k - E_{3*}^k = -\eta^1 P_{23*}^{k1}$$
$$D_{3*}^k - \frac{\varepsilon_{11}^T}{\varepsilon_{33}^T}E_{3*}^k - t_4\tau_{23*}^k = \eta^1 L_3^{k1}$$
$$D_{2*}^k - E_{2*}^k - t_5\tau_{11*}^k - t_6\tau_{22*}^k - t_5\tau_{33*}^k = \eta^1 L_2^{k1}$$
$$\frac{\partial D_{3*}^k}{\partial \zeta} + \frac{1}{A_{20}}\frac{\partial D_{2*}^k}{\partial \xi_2} = -M^k - \eta^1 M^{k1}$$
$$E_{3*}^k + \frac{\partial \psi_*^k}{\partial \zeta} = -\eta^1 K_3^{k1}$$
$$E_{1*}^k + \frac{1}{A_{10}}\frac{\partial \psi_*^k}{\partial \xi_1} = -\eta^1 K_1^{k1}$$
$$E_{2*}^k + \frac{1}{A_{20}}\frac{\partial \psi_*^k}{\partial \xi_2} = -\eta^1 K_2^{k1} \tag{40.8}$$

$$\tau_{33*}^k\,|_{\zeta=\pm1} = 0 \qquad \tau_{23*}^k\,|_{\zeta=\pm1} = 0 \qquad \psi_*^k\,|_{\zeta=\pm1} = 0 \tag{40.9}$$
$$\tau_{33*}^k\,|_{\zeta=\pm1} = 0 \qquad \tau_{23*}^k\,|_{\zeta=\pm1} = 0 \qquad D_{3*}^k\,|_{\zeta=\pm1} = 0. \tag{40.10}$$

Here, we used the notation

$$X_2^a = \frac{1}{A_{10}}\frac{\partial \tau_{21*}^a}{\partial \xi_1} + \eta^s R k_{10}\left(\tau_{12*}^a + \tau_{21*}^a\right)$$
$$X_3^a = \frac{1}{A_{10}}\frac{\partial \tau_{31*}^a}{\partial \xi_1} + \eta^s R k_{10}\tau_{13*}^a$$
$$P_1^a = \frac{1}{A_{10}}\frac{\partial v_{1*}^a}{\partial \xi_1} \qquad P_2^a = \eta^s k_{20} v_{1*}^a$$
$$M^a = \frac{1}{A_{10}}\frac{\partial D_{1*}^a}{\partial \xi_1} + \eta^s R k_{20} D_{1*}^a$$
$$X_1^a = P_{21}^a = P_{12}^a = P_{13}^a = L_1^a = 0$$
$$X_1^{a1} = \eta^{1-2s}\frac{1}{A_1}\frac{\partial \tau_{11*}^a}{\partial \xi_1} + \left(\frac{1}{A_1}\right)'\xi_2\frac{\partial \tau_{12*}^a}{\partial \xi_2} + \frac{R}{R_2}\zeta\frac{\partial \tau_{13*}^a}{\partial \zeta}$$
$$+\eta^{1-s}Rk_2\left(\tau_{11*}^a - \tau_{22*}^a\right) + Rk_1\left(\tau_{12*}^a + \tau_{21*}^a\right) + \frac{2R}{R_1}\tau_{13*} \tag{40.11}$$
$$X_1^b = \frac{1}{A_{10}}\frac{\partial \tau_{11*}^b}{\partial \xi_1} + \eta^s R k_{20}\left(\tau_{11*}^b - \tau_{22*}^b\right)$$

$$
\begin{aligned}
P_{12}^{b} &= P_{21}^{b} = \frac{1}{A_{10}}\frac{\partial v_{2*}^{b}}{\partial \xi_1} - \eta^{s} R k_{20} v_{2*}^{b} - E_{1*}^{b} \\
P_{13}^{b} &= \frac{1}{A_{10}}\frac{\partial v_{3*}^{b}}{\partial \xi_1} \\
L_{1}^{b} &= \frac{\varepsilon_{11}^{T}}{\varepsilon_{33}^{T}} E_{1*} \\
X_{2}^{b} &= X_{3}^{b} = P_{1}^{b} = P_{2}^{b} = K^{b} = 0 \\
X_{2}^{b1} &= \left(\frac{1}{A_2}\right)' \xi_2 \frac{\partial \tau_{22*}^{b}}{\partial \xi_2} + \eta^{1-2s}\frac{1}{A_1}\frac{\partial \tau_{21*}^{b}}{\partial \xi_1} + \frac{R}{R_2}\zeta\frac{\partial \tau_{23*}^{b}}{\partial \zeta} \\
&\quad + Rk_1\left(\tau_{22*}^{b} - \tau_{11*}^{b}\right) + \eta^{1-s} Rk_2\left(\tau_{12*}^{b} + \tau_{21*}^{b}\right) + \frac{2R}{R_2}\tau_{23*}^{b} \\
X_{3}^{b1} &= \left(\frac{1}{A_2}\right)' \xi_2 \frac{\partial \tau_{32*}^{b}}{\partial \xi_2} + \eta^{1-2s}\frac{1}{A_1}\frac{\partial \tau_{31*}^{b}}{\partial \xi_1} + Rk_1\tau_{23*}^{b} \\
&\quad + \eta^{1-s} Rk_2 \tau_{13*}^{b} - \frac{R}{R_1}\tau_{11}^{b} - \frac{R}{R_2}\tau_{22*}^{b}.
\end{aligned} \tag{40.12}
$$

By neglecting the small terms of the order $O(\eta^{1})$, we reduce the subsystem 40.6 to the homogeneous equations of the antiplane problem of the anisotropic elasticity theory. The subsystem 40.8 is transformed by neglecting the small terms into the system of homogeneous equations of the plane problem of electroelasticity theory. When solving the antiplane problem, we must meet the conditions 40.7 on the shell faces. For the plane problem, the three conditions 40.9 or 40.10 should hold.

41 THE SAINT VENANT PRINCIPLE GENERALIZED TO ELECTROELASTICITY

The equations of piezoceramic shells and the respective boundary layer equations are strongly dependent on the direction of preliminary polarization. For concreteness, we consider the boundary layer near the edge $\alpha_1 = \alpha_{10}$ in a piezoceramic shell with preliminary polarization along the α_2-lines and faces covered with electrodes.

We saw in the previous sections that an approximate computation of the boundary layer is reduced to solving the plane and antiplane problems for a piezoceramic half-band with homogeneous conditions on the faces.

We assume that arbitrary mechanical and electrical loads

$$\tau_{11} = f_1(\zeta), \qquad \tau_{13} = f_2(\zeta) \tag{41.1}$$

$$\tau_{12} = f_3(\zeta), \qquad \psi = \phi_1(\zeta), \qquad D_{1*} = \phi_2(\zeta) \tag{41.2}$$

are given at the edge of the half-band $\alpha_1 = \alpha_{10}$. On the half-band faces without electrodes $\zeta = 1$ and $\zeta = -1$, the homogeneous conditions

$$\tau_{23} = 0, \qquad D_3 = 0, \qquad \tau_{13} = 0, \qquad \tau_{33} = 0 \tag{41.3}$$

are specified.

At the edge $\alpha_1 = \alpha_{10}$, the equations of the plane electroelastic boundary layer coincide with those of the plane boundary layer in electroelasticity theory within physical constants. Therefore, we will have the usual Saint Venant conditions for the plane electroelastic boundary layer

$$\int_{-h}^{+h} \tau_{11}(\alpha_{10})d\gamma = 0, \qquad \int_{-h}^{+h} \tau_{13}(\alpha_{10})d\gamma = 0, \qquad \int_{-h}^{+h} \gamma\tau_{11}(\alpha_{10})d\gamma = 0. \tag{41.4}$$

These conditions imply that the edge mechanical load satisfying conditions 41.4 gives a solution that damps down in the narrow edge zone and cannot be found from the equations of shell theory.

Let us find the damping conditions for the antiplane layer. The equations of the antiplane problem in dimensionless coordinates will have the form

$$\begin{aligned}
&\frac{\partial\tau_{21}}{\partial\xi} + \frac{\partial\tau_{23}}{\partial\zeta} = 0\\
&\frac{\partial D_1}{\partial\xi} + \frac{\partial D_3}{\partial\zeta} = 0\\
&\frac{\partial v_2}{\partial\xi} = s_{44}^E\tau_{21} - d_{15}\frac{\partial\psi}{\partial\xi}\\
&\frac{\partial v_2}{\partial\zeta} = s_{66}^E\tau_{23} - d_{15}\frac{\partial\psi}{\partial\zeta}\\
&D_1 = d_{15}\tau_{21} - \varepsilon_{11}^T\frac{\partial\psi}{\partial\xi}\\
&D_3 = d_{15}\tau_{23} - \varepsilon_{11}^T\frac{\partial\psi}{\partial\zeta}
\end{aligned} \tag{41.5}$$

where

$$A_1\xi_1 = \xi.$$

System 41.5 can be reduced to two equations with respect to v_2 and ψ:

$$\begin{aligned}
&k^2\frac{\partial^2 v_2}{\partial\xi^2} + \frac{\partial^2 v_2}{\partial\zeta^2} = (1-k^2)d_{15}\frac{\partial^2\psi}{\partial\xi^2}\\
&\frac{\partial^2\psi}{\partial\xi^2} + \frac{\partial^2\psi}{\partial\zeta^2} = 0\\
&k^2 = \frac{s_{66}^E}{s_{44}^E}.
\end{aligned} \tag{41.6}$$

When solving the system of equations 41.6, the first two conditions in 41.3 should be met on the faces of the half-band and the conditions 41.2 should be met at the edges. This is the second condition in 41.2 for the electrode-covered edge. For the edge without electrodes, this should be the third condition in 41.2.

We take the Laplace integral transform with respect to the variable ξ in the system of equations 41.6 and get

$$U = \int_0^\infty v_2\left(\xi,\zeta\right)e^{-p\xi}d\xi$$

$$\Phi = \int_0^\infty \psi\left(\xi,\zeta\right)e^{-p\xi}d\xi$$

$$\frac{d^2U}{d\zeta^2} + k^2p^2U = k\Omega_2 - \Omega_1 + \left(1-k^2\right)d_{15}p\Phi$$

$$\frac{d^2\Phi}{d\zeta^2} + p^2\Phi = \Omega_1$$

$$\Omega_1 = \left(\frac{\partial\psi}{\partial\xi} + p\psi\right)|_{\xi=0}$$

$$\Omega_2 = k\left(\frac{\partial v_2}{\partial\xi} + pv_2\right)|_{\xi=0} + kd_{15}\Omega. \qquad (41.7)$$

In the new notation, the conditions on the faces without electrodes $\zeta = 1, \zeta = -1$ look like

$$\frac{dU}{d\zeta} = 0, \qquad \frac{d\Phi}{d\zeta} = 0. \qquad (41.8)$$

We integrate equations 41.7. Their solutions are

$$\Phi = c_1\cos p\zeta + c_2\sin p\zeta + a_1(p,\zeta)\cos p\zeta + b_1(p,\zeta)\sin p\zeta$$

$$U = c_3\cos kp\zeta + c_4\sin kp\zeta - d_{15}\left\{\left(c_1 + a_1(p,\zeta)\right)\cos p\zeta\right.$$

$$\left.+\left(c_2 + b_1(p,\zeta)\right)\sin p\zeta\right\} + a_2(p,\zeta)\cos kp\zeta + b_2(p,\zeta)\sin kp\zeta$$

$$a_1(p,\zeta) = -\frac{1}{p}\int_{\zeta_0}^{\zeta}\Omega_1\sin p\zeta d\zeta$$

$$b_1(p,\zeta) = \frac{1}{p}\int_{\zeta_1}^{\zeta}\Omega_1\cos p\zeta d\zeta$$

$$a_2(p,\zeta) = -\frac{1}{p}\int_{\zeta_0}^{\zeta}\Omega_2\sin kp\zeta d\zeta$$

$$b_2(p,\zeta) = \frac{1}{p}\int_{\zeta_1}^{\zeta} \Omega_2 \cos kp\zeta d\zeta. \tag{41.9}$$

The constants $c_1, \ldots, c_4$ are found in conditions 41.8.

We assume that the functions 41.2 specified at the edge are even functions of ζ. We set $\zeta_0 = -1$ and $\zeta_1 = 0$ to get

$$c_2 = c_4 = 0, \qquad a_1 = a_2 = 0, \qquad c_1 = b_1(p,1)\frac{\cos p}{\sin p}$$

$$c_3 = \frac{1}{k \sin kp}\left[-d_{15}b_1(p.1)\cos p + d_{15}c_1 \sin p + kb_2(p.1)\cos kp\right]. \tag{41.10}$$

We substitute 41.10 into 41.9 and find the residues of the functions U and Φ. Then, by using the inversion theorem, we get the final formulas for v_2 and ψ. For these formulas to have no exponentially increasing terms, we equate the residues to zero at the points $p = 0$ and get the following Saint Venant-type damping conditions:

$$\int_{-h}^{+h} \tau_{12}(\alpha_{10})d\gamma = 0, \qquad \int_{-h}^{+h} \psi(\alpha_{10})d\gamma = 0, \qquad \int_{-h}^{+h} D_1(\alpha_{10})d\gamma = 0. \tag{41.11}$$

By using an analogy to mechanical load, we will call the electrical load satisfying the last two conditions in 41.11 a self-balanced electrical load. We should meet the second condition in 41.11 on electrode-covered edges, and we should meet the third condition for edges without electrodes.

Suppose that the functions 41.2 specified at the edge are odd functions. We set $\zeta_0 = 0, \zeta_1 = -1$ and get

$$c_1 = c_3 = 0, \qquad b_1(p,1) = b_2(p,2) = 0$$

$$c_2 = \frac{a_1(p,1)\sin p}{\cos p}, \qquad c_4 = \frac{a_2(p,1)\sin kp}{\cos kp}.$$

It is clear from these formulas that the functions U and Φ have residues only at the points where $\cos p_n = 0$ and $\cos kp_m = 0$. The roots of these equations are the numbers $p_n, -p_n, p_m$, and $-p_m$, where

$$p_n = \frac{\pi}{2}(2n-1), \qquad p_m = \frac{\pi}{2k}(2m-1), \qquad n, m = 1, 2, 3\ldots$$

At the points $p = 0$, the sought-for functions have no residues; therefore, the solution for the antiplane boundary layer will attenuate for any edge load 41.2 at odds with ζ.

Using the inversion theorem, we find that

$$\psi = \sum_{n-1}^{\infty}(res_{p_n}\Phi \cdot e^{p_n\xi} + res_{-p_n}\Phi \cdot e^{p_n\xi})$$

$$v_2 = \sum_{n-1}^{\infty}(res_{p_n}U \cdot e^{p_n\xi} + res_{-p_n}U \cdot e^{-p_n\xi}) + \sum_{n=1}^{\infty}(res_{p_m}U \cdot e^{p_m\xi} + res_{-p_m}U \cdot e^{p_m\xi}).$$

The residues at the points p_n and p_m give solutions that increase with the distance from the edge. We equate them to zero (the resultant solution has enough arbitrary integration functions to satisfy these conditions) and get

$$a_1(p_n, 1) = -\frac{1}{p_n}\int_0^1 \left(\frac{\partial\psi}{\partial\xi} + p_n\psi\right)|_{\xi=0} \sin p_n\zeta d\zeta = 0$$

$$a_2(p_m, 1) = -\frac{k}{p_m}\int_0^1 \left[\frac{\partial v_2}{\partial\xi} + d_{15}\frac{\partial\psi}{\partial\xi} + p_m(v_2 + d_{15}\psi)\right]|_{\xi=0} \sin kp_m\zeta = 0. \tag{41.12}$$

Using formulas 41.12 for ψ and v_2, we write an exponentially decreasing solution

$$\psi = \sum_{n=1}^{\infty} res_{-p_n}\Phi \cdot e^{-p_n\xi}$$

$$v_2 = \sum_{n-1}^{\infty} res_{-p_n}U \cdot e^{-p_n\xi} + \sum_{m=1}^{\infty} res_{-p_m}U \cdot e^{-p_m\xi}).$$

Here,

$$res_{-p_n}\Phi = a_1(-p_n, 1) \cdot \sin p_n\zeta$$
$$res_{-p_m}U = k^2 a_2(-p_m, 1) \cdot \sin kp_m\zeta$$
$$res_{-p_n}U = d_{15}a_1(-p_n, 1) \cdot \sin p_n\zeta. \tag{41.13}$$

If we are given the value of the electrical potential ψ at the edge by taking into account formulas 41.12 and discard the unknown functions $(\partial\psi/\partial\xi)\,|_{\xi=0}$ and $(v_2 + d_{15}\psi)\,|_{\xi=0}$ in equations 41.13, we find the final formulas

$$\psi = \sum_{n-1}^{\infty} c^{(n)} \sin p_n\zeta \cdot e^{-p_n\xi}$$

$$v_2 = -d_{15}\psi - \sum_{m=1}^{\infty} b^{(m)} \sin p_m\zeta \cdot e^{-p_m\zeta}$$

$$c^{(n)} = 2\int_0^1 \phi_1(\zeta)\sin p_n\zeta d\zeta$$

$$b^{(m)} = \frac{2s_{44}^E}{p_n}\int_0^\infty f_3(\zeta)\sin kp_m\zeta d\zeta.$$

If we know the component of the electric induction vector D_1 normal to the edge surface, the formula for $c^{(n)}$ should be replaced by

$$c^{(n)} = -\frac{2}{p_n\varepsilon_{11}^T}\int_0^1 [d_{15}f_3(\zeta) - \phi_2(\zeta)]\sin p_n\zeta d\zeta.$$

Thus, for piezoceramic shells with faces without electrodes, the Saint Venant conditions generalized to electroelasticity for mechanical stresses specified at the edges retains the form accepted in elasticity theory (the nonself-balanced edge load generate a solution exponentially decreasing at the edge). For electrical quantities, the attenuation conditions are

$$\int_{-h}^{+h} \psi(\alpha_{10})d\gamma = 0 \tag{41.14}$$

$$\int_{-h}^{+h} D_1(\alpha_{10})d\gamma = 0 \tag{41.15}$$

for the electrode-covered edge and the edge without electrodes, respectively.

Conditions 41.14 were obtained for the edge $\alpha_1 = \alpha_{10}$ for a shell preliminarily polarized along the α_2-lines. By analogy, the obtained attenuation conditions are valid for the edge $\alpha_2 = \alpha_{20}$ and for shells with thickness polarization. We have also obtained solutions for respective problems of the boundary layer, but we do not give them here because they are very cumbersome. The system of equations for the plane electroelastic problem has the sixth order (in elasticity theory, the order is four) and the algebraic transformations are very complicated.

If the shell's faces are covered with electrodes, we should replace the second condition in 41.3 by $\psi = 0$ for $\gamma = h, \gamma = -h$. We can show by using 41.9, that in this case, the attenuation conditions for a mechanical edge load completely coincide with Saint Venant's conditions in elasticity theory. Any electrical edge load, self-balanced or not, causes an electroelastic state exponentially decreasing with the distance from the edge. This result is fully consistent with the fact that the equations of the theory of piezoceramic shells with electrode-covered faces coincide with the equations for nonelectrical shells up to the constants that characterize the physical properties of the material. They do not have arbitrary integration functions for satisfying the electrical conditions at the edges.

Chapter 8

INTERACTION BETWEEN THE INTERNAL ELECTROELASTIC STATE AND THE BOUNDARY LAYER

The boundary layer and the inner electroelastic state are responsible for the three-dimensional boundary conditions at the edge. It follows from the Saint Venant principle generalized to electroelasticity that the nonself-balanced part of the edge mechanical and electrical load generates a penetrating solution described by the equations of the inner electroelastic state. The self-balanced part of the edge load generates a quickly decreasing electroelastic state at the edge, described by the boundary layer equations.

The boundary layer and the inner electroelastic state interact at the edge when the boundary conditions are observed. As a result of this interaction, the quantities of the boundary layer enter the two-dimensional boundary conditions of the theory of piezoceramic shells. The effect exerted by the boundary layer on the internal electroelastic state through the boundary conditions can be even stronger than in the theory of nonelectrical shells; and in some cases, which will be investigated later, the internal electroelastic state cannot be reliably defined without allowing for the influence of the boundary layer [81–82].

As a rule, the stressed-strained state at a distance from the edge is of principal interest in the theory of nonelectrical shells; therefore, we can restrict ourselves to the calculation of the inner stressed-strained state. It is only necessary to consider the boundary layer when (1) we have to find the stresses at the edge and (2) we have to calculate the stressed-strained state very accurately and allow for the corrections due to the boundary layer effect within the boundary conditions of the shell theory.

When designing a piezoceramic shell, we can sometimes define only the inner electroelastic state. But we will see that there are cases when the calculation should be started with the boundary layer. For example, in a piezoceramic shell polarized along the α_2-lines and having electrode-covered faces near a rigidly fixed edge $\alpha_1=$ constant, there appears a boundary layer more intensive than the internal electroelastic state. Besides, the corrections due to the boundary layer can have the same effect within the boundary conditions as the principal terms of the boundary conditions.

In the theory of nonelectrical shells, the number of auxiliary problems to be solved in order to construct refined boundary conditions of shell theory is not large [22]. For piezoceramic shells, the number of auxiliary problems is much greater, and the problems are more complicated, because the material is anisotropic and electroelastic. We will restrict ourselves to some interesting problems for the antiplane and plane boundary layers.

42 TWO-DIMENSIONAL ELECTROELASTIC STATE WITH GREAT VARIABILITY IN THE DIRECTION NORMAL TO THE EDGE

The internal electroelastic state is not always uniform where variability is concerned. In the presence of distortion lines, the electroelastic state consists of the principal and auxiliary components. In statics, the internal electroelastic state with small variability can be represented as a sum of the principal electroelastic state (membrane or pure moment state) and the simple edge effects localized at the edges and quickly decreasing with the distance from the edge. In dynamics, the auxiliary state can decrease at the edge or it can extend through the entire shell depending on the vibration frequency. The variability in the direction normal to the edge is much greater than the variability of the principal electroelastic state and essentially less than the variability of the boundary layer. As before, we will call the problem of determining the principal electroelastic state, the principal problem; and we will call the problem of finding the auxiliary electroelastic state, the auxiliary problem.

In Chapter 3, we gave the asymptotic representation of the inner electroelastic state under the assumption that the variability along the coordinates of the middle surface was constant. The asymptotic representation of the needed quantities of the auxiliary electroelastic state can be found similarly, if we assume that the variability index s varies with respect to the variable α_j and the index q varies with respect to α_i (the edge $\alpha_i = \alpha_{i0}$). We make an asymptotic change of the variables

$$\alpha_j = \eta^s R\xi_j, \qquad \alpha_i = \eta^q R\xi_i. \tag{42.1}$$

The variability index s coincides with that of the principal electroelastic state ($0 \leq s < 1/2$), while in statics $q = 1/2$, and in dynamics $1/2 \leq q < 1$. We write the asymptotic expansions with respect to ζ for the sought-for quantities (the edge $\alpha_i = \alpha_{i0}$)

$$\begin{aligned}
\frac{v_3}{R} &= v_{3*} = v_{3,0} + \eta^1\zeta v_{3,1} + \eta^1\zeta^2 v_{3,2}\\
\frac{v_i}{R} &= \eta^q v_{i*} = \eta^q\left(v_{i,0} + \eta^{1-2q}\zeta v_{v_i,1}\right)\\
\frac{v_j}{R} &= \eta^{2q-s} v_{j*} = \eta^{2q-s}\left(v_{j,0} + \eta^{1-2q}\zeta v_{j,1}\right)\\
\tau_{ij}s_{11}^E &= \eta^{q-1}\tau_{ij*} = \eta^{q-1}\left(\tau_{ij,0} + \zeta\tau_{ij,1}\right)\\
\tau_{jj}s_{11}^E &= \eta^{1-2q}\tau_{jj*} = \eta^{1-2q}\left(\eta^{2q-1}\tau_{jj,0} + \zeta\tau_{jj,1}\right)
\end{aligned}$$

$$\tau_{ii}s_{11}^E = \eta^{1-2q}\tau_{ii*} = \eta^{1-2q}\left[\left(\eta^{-1+3q}+\eta^{1-s}\right)\tau_{ii,0}+\zeta\tau_{ii,1}\right]$$
$$\tau_{i3}s_{11}^E = \eta^{2-3q}\tau_{i3*} = \eta^{2-3q}\left(\tau_{i3,0}+\eta^{2-2q}\zeta\tau_{i3,1}+\eta^{2-2q}\zeta^3\tau_{i3,3}\right). \quad (42.2)$$

For shells with tangential polarization, the factor s_{11}^E should be replaced by $1/n_{11}$ in order to make the stresses dimensionless.

The form of the asymptotic expansions of the electrical quantities is dependent on the direction of the preliminary polarization and on the type of electrical conditions on the shell faces; therefore, we will write them separately for every case.

Shells with tangential polarization

(a) with electrode-covered surfaces

$$\frac{d_{31}}{\varepsilon_{11}^T}D_i = \eta^{2-3q}D_{i*} = \tau^{2-3q}\left(D_{i,0}+\eta^{2-2q}\zeta D_{i,1}+\zeta^2 D_{i,2}+\eta^{2-2q}\zeta^3 D_{i,3}\right) \quad (42.3)$$

(b) without electrodes on the faces

$$\frac{d_{31}}{\varepsilon_{11}^T}D_i = \eta^{2-3q}D_{i*} = \eta^{2-3q}\left(D_{i,0}+\eta^{2q-1}\zeta D_{i,1}+\zeta^2 D_{i,2}\right)$$
$$\frac{d_{31}}{R}\psi = \eta^{2-2q}\psi_* = \eta^{2-2q}\left(\psi_{,0}+\eta^{2q-1}\zeta\psi_{,1}+\zeta^2\psi_{,2}\right). \quad (42.4)$$

Shells with polarization along the α_2-lines with faces covered by electrodes

$$\frac{d_{15}}{\varepsilon_{33}^T}D_1 = \eta^{q-1}D_{1*} = \eta^{q-1}\left(D_{1,0}+\zeta D_{1,1}\right), \qquad (\alpha_1 = \alpha_{10})$$
$$\frac{d_{15}}{\varepsilon_{33}^T}D_2 = D_{2*} = D_{2,0}+\eta^{1-2q}\zeta D_{2,1} \quad (\alpha_2 = \alpha_{20}).$$

We have only written the formulas required to derive the boundary conditions.

43 BOUNDARY CONDITIONS IN THE THEORY OF SHELLS WITH THICKNESS POLARIZATION (ELECTRODE-COVERED FACES)

We will construct the boundary conditions with the accuracy accepted for the electroelasticity relations in Chapter 6, i.e.,

$$\varepsilon = 0(\eta^{2-2s}).$$

An exclusion is made for some boundary conditions whose error is not specially specified. As before, we will consider only two variants of the electrical boundary conditions on the faces: (1) when the electrical potential on the electrode-covered faces is specified and (2) when the faces are not covered with electrodes.

Consider a rigidly fixed edge $\alpha_1 = \alpha_{10}$ of a shell with electrode-covered faces. The three-dimensional boundary conditions are

$$v_1 = 0, \qquad v_2 = 0, \qquad v_3 = 0, \qquad D_1 = 0 \quad (\alpha_1 = \alpha_{10}). \tag{43.1}$$

We represent every quantity in the boundary conditions as a sum of three terms: the first term is found from the equations of the internal electroelastic state, and the second and third terms are found from the antiplane and plane problems, respectively. Then, if the variability index s of the internal electroelastic state is smaller than 1/2, every quantity of the internal electroelastic state is represented as a sum of two terms. We find one of these from the equations of the principal electroelastic state (the principal problem), and we find the other from the equations of the quickly varying electroelastic state (the auxiliary problem). In statics, this is a membrane or pure moment stressed-strained state and simple edge effects. The asymptotic representations given below are suitable for the edge that does not coincide with the asymptotic line of the middle surface.

Taking into account formulas 34.2, 34.9, 38.5 to 38.8, and 43.1, we write

$$\begin{aligned}
&\eta^s\left(v_{1,0}^{(p)} + \eta^{1-2s+c}\zeta v_{1,1}^{(p)}\right) + \eta^{\delta+q}\left(v_{1,0}^{(a)} + \eta^{1-2q}\zeta v_{1,1}^{(a)}\right)\\
&\qquad + \eta^{2-s+\alpha} v_{1*}^{a} + \eta^{1+\beta} v_{1*}^{b} = 0\\
&\eta^s\left(v_{2,0}^{(p)} + \eta^{1-2s+c}\zeta v_{2,1}^{(p)}\right) + \eta^{\delta+2q-s}\left(v_{2,0}^{(a)} + \eta^{1-2q}\zeta V_{\alpha}2, 1^{(a)}\right)\\
&\qquad + \eta^{1+\alpha} v_{2*}^{a} + \eta^{2-s+\beta} v_{2*}^{b} = 0\\
&\eta^c\left(v_{3,0}^{(p)} + \eta^{1-c}\zeta v_{3,1}^{(p)}\right) + \eta^{\delta}\left(v_{3,0}^{(a)} + \eta^{1}\zeta v_{3,1}^{(a)} + \eta^{1}\zeta^2 v_{3,2}^{(a)}\right)\\
&\qquad + \eta^{2-s+\alpha} v_{3*}^{a} + \eta^{1+\beta} v_{3*}^{b} = 0\\
&\eta^{1-s}\left(D_{1,0}^{(p)} + \zeta D_{1,1}^{(p)} + \eta^{1-2s+c}\zeta^2 D_{1,2}^{(p)}\right) + \eta^{\delta+2-3q}\left(D_{1,0}^{(a)} + \eta^{2q-1}\zeta D_{1,1}^{(a)}\right)\\
&\qquad + \zeta^{(a)} D_{1,2}^{a} + \eta^{1-2+\alpha} D_{1*}^{a} + \eta^{\beta} D_{1*}^{b} = 0.
\end{aligned} \tag{43.2}$$

The superscripts (p) and (a) are used to denote the quantities of the principal and auxiliary problems, respectively. If the internal electroelastic state has $s \geq 1/2$, we should use $c = -1 + 2s$ in 43.2 and delete the quantities with the superscript (a). If $0 \leq s \leq 1/2$, we have $c = 0$. The principal electroelastic state is found from the nonhomogeneous equations, while the solutions to the auxiliary problem and the boundary layer are found from the homogeneous equations. This is the reason for the scale factors η^{δ}, η^{α}, and η^{β} before the quantities of the auxiliary problem, the antiplane, and the plane boundary layers, respectively. The values of δ, α, and β are chosen so that the problems obtained under the initial approximation for any of the required electroelastic states would not be contradictory. This requirement is satisfied for

$$\delta = 0, \qquad \alpha = 1 - s, \qquad \beta = 0. \tag{43.3}$$

We substitute 43.3 into 43.2:

$$\begin{aligned}
v_{1,0} + \eta^{1-q-s(0)}\zeta v_{1,1} + \eta^{3-3s} v_{1*}^{a} + \eta^{1-s} v_{1*}^{b} &= 0 \\
v_{2,0} + \eta^{1-2s+c}\zeta v_{2,1} + \eta^{2-2s}\left(v_{2*}^{a} + v_{2*}^{b}\right) &= 0 \\
v_{3,0} + \eta^{1-c}\zeta v_{3,1} + \eta^{1-c}\zeta^{2} v_{3,2} + \eta^{3-2s-c} v_{3*}^{a} + \eta^{1-c} v_{3*}^{b} &= 0 \\
D_{1,0} + \eta^{2q-1(o)}\zeta D_{1,1} + \zeta^{2} D_{1,2} + \eta^{3q-2s(1-s)} D_{1*}^{a} + \eta^{3q-2(-1+s)} D_{1*}^{b} &= 0.
\end{aligned} \tag{43.4}$$

In formulas 43.4, we use double subscripts for η; the left one refers to the case when the variability of the inner electroelastic state is small ($s < 1/2, c = 0$) and the right one in the parentheses refers to greater variability ($s \geq 1/2, c = -1 + 2s$).

We use the following notation for the sums of quantities in the principal and auxiliary problems:

$$\begin{aligned}
v_{1,0} &= v_{1,0}^{(p)} + \eta^{q-s} v_{1,0}^{(a)} \\
v_{1,1} &= \eta^{q-s} v_{1,1}^{(p)} + v_{1,1}^{(a)} \\
v_{2,0} &= v_{2,0}^{(p)} + \eta^{2q-2s} v_{2,0}^{(a)} \\
v_{2,1} &= v_{2,1}^{(p)} + v_{2,1}^{(a)} \\
v_{3,0} &= v_{3,0}^{(p)} + v_{3,0}^{(a)} \\
v_{3,1} &= v_{3,1}^{(p)} + v_{3,1}^{(a)} \\
v_{3,2} &= v_{3,2}^{(a)} \\
D_{1,0} &= \eta^{3q-1-s} D_{1,0}^{(p)} + D_{1,0}^{(a)} \\
D_{1,1} &= \eta^{q-s} D_{1,1}^{(p)} + D_{1,1}^{(a)} \\
D_{1,2} &= \eta^{3q-3s} D_{1,2}^{(p)} + D_{1,2}^{(a)}.
\end{aligned}$$

By deleting the small terms within our approximation, we get from the second condition in 43.4

$$v_{2,0} = 0, \quad (\alpha_1 = \alpha_{10}). \tag{43.5}$$

Since the problem is linear, we consider the symmetrical and antisymmetrical parts of the plane boundary layer separately. The symmetrical part of the plane boundary layer corresponds to a bending of the half-band, and the antisymmetrical part corresponds to an extension of the half-band.

For the symmetrical part of the electroelastic half-band, the edge conditions have the form ($\alpha_1 = \alpha_{10}$)

$$\begin{aligned} \eta^{-q(-1+s)}\zeta v_{1,1} + v_{1*}^b &= 0 \\ \eta^{-1+c} v_{3,0} + \eta^0 \zeta^2 v_{3,2} + v_{3*}^b &= 0 \\ \eta^{2-3q(1-s)} D_{1,0} + \eta^{2-3q(1-s)} \zeta^2 D_{1,2} + D_{1*}^b &= 0. \end{aligned} \tag{43.6}$$

The homogeneous boundary conditions 38.18 and 38.19 should hold for the faces. Since the problem is linear, we can represent its solution as a linear combination of the solutions to six auxiliary problems I–VI with the following conditions at the edge:

$$\begin{array}{llll} \text{I.} & v_{1*}^b + \zeta = 0, & v_{3*}^b = 0, & D_{1*}^b = 0 \\ \text{II.} & v_{1*}^b = 0, & v_{3*}^b + 1 = 0, & D_{1*}^b = 0 \\ \text{III.} & v_{1*}^b = 0, & v_{3*}^b + \zeta^2 = 0, & D_{1*}^b = 0 \\ \text{IV.} & v_{1*}^b = 0, & v_{3*}^b = 0, & D_{1*}^b = 0 \\ \text{V.} & v_{1*}^b = 0, & v_{3*}^b = 0, & D_{1*}^b + 1 = 0 \\ \text{VI.} & v_{1*}^b = 0, & v_{3*}^b = 0, & D_{1*}^b + \zeta^2 = 0 \end{array}$$

The solutions to problems I–VI should be multiplied by

$$\eta^{-q(-1+s)} v_{1,1}, \qquad \eta^{-1+c} v_{3,0}, \qquad \eta^0 v_{3,2}, \qquad \eta^1,$$
$$\eta^{2-3q(1-s)} D_{1,0}, \qquad \eta^{2-3q(1-s)} D_{1,2},$$

respectively. Problem IV consists of the integration of the nonhomogeneous system of equations 38.11, whose free terms have the order of η^1 and are given by formulas 38.13. The solutions to problems I–VI multiplied by the respective factors are added together to give the solution of the plane problem under the conditions 43.6.

When solving problems I–VI, we require that the displacements at infinity be equal to zero. According to the Saint Venant principle generalized to electroelasticity, a fastly decreasing boundary layer in a shell with electrode-covered faces is obtained if the vertical resultant of the vertical forces and bending moments at the edge (we treat the direction coinciding with the direction of the axis ζ as vertical) is equal to zero. We use $N_1, \ldots, N_6$ to denote the vertical resultant forces in problems I–VI and $M_1, \ldots, M_6$ to denote the resultant bending moments; we write the equilibrium conditions for the half-band with the edge unloaded at infinity:

$$\begin{aligned}
&\eta^{-q(-1+s)}v_{1,1}N_1 + \eta^{-1+c}v_{3,0}N_2 + v_{3,2}N_3 \\
&\quad +\eta^{2-3q(1-s)}(D_{1,0}N_4 + D_{1,2}N_5) + \eta^1 N_6 = 0 \\
&\eta^{-q(-1+s)}v_{1,1}M_1 + \eta^{-1+c}v_{3,0}M_2 + v_{3,2}M_3 \\
&\quad +\eta^{2-3q(1-s)}(D_{1,0}M_4 + D_{1,2}M_5) + \eta^1 M_6 = 0.
\end{aligned}$$

We neglect the terms $O(\eta^{2-2s})$ as compared to the principal terms and get

$$\begin{aligned}
\eta^{-q(-1+s)}v_{1,1}N_1 + \eta^{-1+c}v_{3,0}N_2 + v_{3,2}N_3 &= 0 \\
\eta^{-q(-1+s)}v_{1,1}M_1 + \eta^{-1+c}v_{3,0}M_2 + v_{3,2}M_3 &= 0.
\end{aligned}$$

We solve this algebraic system with respect to $v_{3,0}$ and $v_{1,1}$ and obtain

$$\eta^{1-q-s(0)}v_{1,1} = \eta^{1-s}\rho_1 v_{3,2}, \qquad v_{3,0} = \eta^{1-c}\rho_2 v_{3,2}$$

where

$$\rho_1 = \Delta_1/\Delta, \qquad \rho_2 = \Delta_2/\Delta.$$

$$\Delta_1 = \begin{vmatrix} N_2 & N_3 \\ M_2 & M_3 \end{vmatrix}, \qquad \Delta_2 = \begin{vmatrix} N_3 & N_1 \\ M_3 & M_1 \end{vmatrix}, \qquad \Delta = \begin{vmatrix} N_1 & N_2 \\ M_1 & M_2 \end{vmatrix}.$$

we find ρ_1 and ρ_2 by solving problems I–III.

For the antisymmetrical part of the plane boundary layer, the edge conditions have the form

$$\begin{aligned}
v_{1,0} + \eta^{1-s}v_{1*}^b &= 0 \\
\eta^{1-c}\zeta v_{3,1} + \eta^{1-c}v_{3*}^b &= 0 \\
\eta^{2q-1(0)}\zeta D_{1,1} + \eta^{3q-2(-1+s)}D_{1*}^b &= 0.
\end{aligned}$$

We introduce auxiliary problems VII–X with edge conditions

$$\begin{array}{llll}
\text{VII.} & v_{1*}^b + 1 = 0, & v_{3*}^b = 0, & D_{1*}^b = 0 \\
\text{VIII.} & v_{1*}^b = 0, & v_{3*}^b + \zeta = 0, & D_{1*}^b = 0 \\
\text{IX.} & v_{1*}^b = 0, & v_{3*}^b = 0, & D_{1*}^b + \zeta = 0 \\
\text{X.} & v_{1*}^b = 0, & v_{3*}^b = 0, & D_{1*}^b = 0.
\end{array}$$

As before, we assume that all stresses vanish at infinity and then, by the equilibrium conditions for the half-band, the horizontal resultant directed along the α_1-lines should vanish at the edge (Γ_i is the horizontal resultant of the ith problem, $i = 7, \ldots, 10$):

$$\eta^{-1+s}v_{1,0}\Gamma_7 + v_{3,1}\Gamma_8 + \eta^{1-q(1-s)}D_{1,1}\Gamma_9 + \eta^1\Gamma_{10} = 0.$$

We neglect the small terms and rewrite this formula within the quantities $O(\eta^{2-s-q(2-2s)})$

$$v_{1,0} + \eta^{1-s} m_1 v_{3,1} = 0, \qquad m_1 = \Gamma_7/\Gamma_8,$$

where the numbers Γ_7, Γ_8 are found by solving problems VII–VIII. Thus, after eliminating the quantities of the boundary layer in the boundary conditions, we get

$$v_{1,0} + m_1 \eta^{1-s} v_{3,1} = 0, \qquad v_{2,0} = 0, \qquad v_{3,0} = \eta^{1-c} \rho_2 v_{3,2}$$
$$\eta^{1-q-s(0)} v_{1,1} = \eta^{1-s} \rho_1 v_{3,2}.$$

We find the constants m_1, ρ_1, and ρ_2 that enter the terms allowing for the boundary layer corrections by solving the five auxiliary problems I–III, VII, and VIII.

In the resultant boundary conditions, we turn to the notation typical for shell theory by the formulas of Section 7.

At the *rigidly fixed edge* $\alpha_1 = \alpha_{10}$ of a shell with electrode-covered faces, we get the boundary conditions

$$u_1 + m_1 h \left[\frac{s_{13}^E}{s_{11}^E (1-\nu)} \frac{1}{A_1} \frac{\partial u}{\partial \alpha_1} - \left(d_{31} - \frac{2 s_{13}^E d_{31}}{s_{11}^E (1-\nu)} \right) \frac{V}{h} \right] = 0. \quad [\eta^{1-s}]$$

$$u_2 = 0$$
$$\gamma_1 + \rho_1 h a \kappa_1 = 0, \quad [\eta^{1-s}]$$
$$w + h^2 \rho_2 \kappa_1 = 0, \quad [\eta^{1-c}]$$
$$a = \frac{1+\sigma}{2} B \left(s_{13}^E - \frac{d_{33} d_{31}}{\varepsilon_{33}^T} \right). \tag{43.7}$$

Here, and below, the brackets on the right of every formula give the order of the greatest correction term allowing for the effect of the boundary layer compared to the principal terms of the same condition.

Compare the refined boundary conditions at the rigidly fixed edge of an electroelastic shell with the appropriate boundary conditions obtained by A.L. Gol'denveiser [22] with the same accuracy in the theory of nonelectrical shells; the latter have the form

$$u_1 - \frac{\nu m h}{1-\nu} \frac{1}{A_1} \frac{\partial u}{\partial \alpha_1} = 0, \qquad u_2 = 0, \qquad w = 0, \qquad \gamma_1 = 0. \tag{43.8}$$

If we ignore the term of purely electrical nature (which is proportional to the electrical potential) and restrict ourselves to the static case as in [22], we will observe that in the last two conditions 43.8 there are no correction terms containing ρ_1 and ρ_2. They have been lost because it was not taken into account the simple edge effect when representing the boundary quantities as a sum of quantities corresponding to stressed states with different variability in formulas like 43.2. This caused an error of order η^{1-s} in the formulas accurate to $O(\eta^{2-2s})$.

We will not derive the rest of the boundary conditions. They have been obtained by the technique we have just described within the quantities $O(\eta^{2-2s})$.

Free edge $\alpha_1 = \alpha_{10}$ *without electrodes:*

$$\begin{aligned} T_1 &= 0, \\ S_{21} &= 0 \\ G_1 + 3J\frac{h}{A_2}\frac{\partial H_{21}}{\partial \alpha_2} &= 0 \qquad [\eta^{1-s}], \\ N_1 - \frac{1}{A_2}\frac{\partial H_{21}}{\partial \alpha_2} &= 0. \end{aligned} \tag{43.9}$$

The number J is found from the solution to the antiplane problem 38.15 with the homogeneous conditions 38.17 on the faces and the edge condition

$$\tau_{12*}^a + \zeta = 0$$

by the formula

$$J = \int_{-1}^{+1} \zeta d\zeta \int_{-\infty}^{0} \tau_{12*}^a A_{10} d\xi_1. \tag{43.10}$$

Hinge-supported edge $\alpha_1 = \alpha_{10}$ *without electrodes:*

$$\begin{aligned} &u_2 = 0, \qquad w = 0, \qquad G_1 = 0 \\ &T_1 + n_2 h k_2 \left\{ \frac{1}{2}\frac{s_{13}^E}{s_{11}^E} T_2 - \frac{d_{33}}{s_{11}^E} V \right\} = 0, \quad [\eta^1]. \end{aligned} \tag{43.11}$$

We find the number n_2 by the formula

$$n_2 = \int_{-1}^{+1} d\zeta \int_{-\infty}^{0} \tau_{2*}^b A_{10} d\xi_1$$

from the solution of the plane problem 38.16 with the homogeneous conditions 38.18 and 38.19 on the faces and the edge conditions

$$\tau_{11*}^b = 0, \qquad v_{3*}^b + \zeta = 0, \qquad D_{1*}^b = 0.$$

44 BOUNDARY CONDITIONS FOR A SHELL WITH THICKNESS POLARIZATION (FACES WITHOUT ELECTRODES)

Consider a free electrode-covered edge $\alpha_1 = \alpha_{10}$. We will assume that the shell has other edges which are well fixed so that the shell is in a membrane electroelastic state. The three-dimensional boundary conditions at the edge are

$$\tau_1 = 0, \qquad \tau_{21} = 0, \qquad \tau_{31} = 0, \qquad \psi = V. \tag{44.1}$$

As before, we represent the total electroelastic state as a sum

$$\tau_{11,0}^{(p)} + \eta^{1-2s+c}\zeta\tau_{11,1}^{(p)} + \eta^{\delta+1-2q}\left[\left(\eta^{-1+3q} + \eta^{1-s}\right)\tau_{11,0}^{(a)} + \zeta\tau_{11,1}^{(a)}\right] + \eta^{1-s+\alpha}\tau_{11*}^{a} + \eta^{\beta}\tau_{11*}^{b} = 0$$

$$\tau_{21,0}^{(p)} + \eta^{1-2s+c}\zeta\tau_{21,1}^{(p)} + \eta^{\delta+q-1}\left(\tau_{21,0}^{(a)} + \zeta\tau_{21,1}^{(a)}\right) + \eta^{\alpha}\tau_{21*}^{a} + \eta^{1-s+\beta}\tau_{21*}^{b} = 0$$

$$\eta^{1-s}\left(\tau_{31,0}^{(p)} + \zeta\tau_{31,1}^{(p)} + \eta^{1-2s+c}\zeta^2\tau_{31,2}^{(p)}\right) + \eta^{\delta+2-3q}\left(\tau_{13,0}^{(a)} + \eta^{2-2q}\zeta\tau_{13,1}^{(a)} + \zeta^2\tau_{13,2}^{(a)} + \eta^{2-2q}\zeta^3\tau_{13,3}^{(a)}\right) + \eta^{1-s+\alpha}\tau_{31*}^{a} + \eta^{\beta}\tau_{31*}^{b} = 0$$

$$\eta^1\left(\psi_{,0}^{(p)} + \zeta\psi_1^{(p)} + \eta^{1-2s+c}\zeta^2\psi_{,2}^{(p)}\right) + \eta^{\delta+2-2q}\left(\psi_{,0}^{(a)} + \eta^{2q-1}\zeta\psi_{,1}^{(a)} + \zeta^2\psi_{,2}^{(a)}\right) + \eta^{2-s+\alpha}\psi_*^a + \eta^{1+\beta}\psi_*^b = \eta^1 V_*. \tag{44.2}$$

The problem is physically and mathematically noncontradictory if

$$\alpha = 1 - 2s + c, \qquad \beta = 0, \qquad \delta = 2q - 2s. \tag{44.3}$$

We allow for 44.3 and introduce the notation

$$\tau_{11,0}^{(p)} + \eta^{1-2s}(\eta^{-1+3q} + \eta^{1-s})\tau_{11,0}^{(a)} = \tau_{11,0}; \ldots \psi_{,2}^{(p)} + \eta^{-c}\psi_{,2}^{(a)} = \psi_{,2}.$$

Then the conditions 44.2 can be written in the form

$$\begin{aligned}
\tau_{11,0} + \eta^{1-2s+c}\zeta\tau_{11,1} + \eta^{2-3s+c}\tau_{11*}^{a} + \tau_{11*}^{b} &= 0\\
\tau_{21,0} + \eta^{1-2s+c}\zeta\tau_{21,1} + \eta^{1-2s+c}\tau_{21*}^{a} + \eta^{1-s}\tau_{21*}^{b} &= 0\\
\tau_{31,0} + \zeta\tau_{31,1} + \eta^{1-2s+c}\zeta^2\tau_{31,2} + \eta^{1-2s+c}\tau_{31*}^{a} + \eta^{-1+s}\tau_{31*}^{b} &= 0\\
\psi_{,0} + \zeta\psi_{,1} + \eta^{1-2s+c}\zeta^2\psi_{,2} + \eta^{2-3s+c}\psi_*^a + \psi_*^b &= V_*.
\end{aligned} \tag{44.4}$$

When deriving the boundary conditions of the shell theory, we use the generalized Saint Venant principle formulated in Section 41. We integrate the conditions 44.4 with respect to ζ, then multiply the first condition by ζ and also integrate it with respect to ζ to obtain

$$
\begin{aligned}
2\tau_{11,0} + \int_{-1}^{+1} \left(\tau_{11*}^{b} + \eta^{2-3s+c}\tau_{11*}^{a}\right) |_{\xi_1=0}\, d\zeta &= 0 \\
\eta^{1-2s+c}\frac{2}{3}\tau_{11,1} + \int_{-1}^{+1} (\tau_{11*}^{b} + \eta^{2-3s+c}\tau_{11*}^{a}) |_{\xi_1=0}\, \zeta d\zeta &= 0 \\
2\tau_{21,0} + \int_{-1}^{+1} (\eta^{1-s}\tau_{21*}^{b} + \eta^{1-2s+c}\tau_{21*}^{a}) |_{\xi_1=0}\, d\zeta &= 0 \\
2\tau_{31,0} + \eta^{1-2s+c}\frac{2}{3}\tau_{31,2} + \int_{-1}^{+1} (\eta^{1-2s+c}\tau_{31*}^{a} + \eta^{-1+s}\tau_{31*}^{b}) |_{\xi_1=0}\, d\zeta &= 0 \\
2\psi_{,0} + \eta^{1-2s+c}\frac{2}{3}\psi_{,2} + \int_{-1}^{+1} (\eta^{2-3s+c}\psi_{*}^{a} + \psi_{*}^{b}) |_{\xi_1=0}\, d\zeta &= 2V_{*}.
\end{aligned}
\tag{44.5}
$$

We use the equilibrium equations 38.2 from the boundary layer theory which allows for the homogeneous conditions on the faces 38.17, 38.18, and 38.20 and the damping conditions 41.4 and 41.14, to transform the integrals in the formulas 44.5:

$$
\begin{aligned}
&\int_{-1}^{+1} (\tau_{11*}^{b} + \eta^{2-3s+c}\tau_{11}^{a})|_{\xi_1=0} d\zeta \\
&= A_{10}\int_{-1}^{+1} d\zeta \int_{-\infty}^{\circ} \left(\frac{1}{A_{10}}\frac{\partial \tau_{11*}^{b}}{\partial \xi_1} + \eta^{2-3s+c}\frac{1}{A_{10}}\frac{\partial \tau_{11*}^{a}}{\partial \xi_1}\right) d\xi_1 \\
&= -A_{10}\int_{-1}^{+1} d\zeta \int_{-\infty}^{\circ} \left[-\frac{\partial \tau_{13*}^{b}}{\partial \zeta} - \eta^{2-3s+c}\left(\frac{\partial \tau_{13*}^{a}}{\partial \zeta} + X_1^{a}\right)\right] d\xi_1 \\
&= -\eta^{2-3s+c}\frac{A_{10}}{A_{20}}\frac{\partial}{\partial \xi_2}\int_{-1}^{+1} d\zeta \int_{-\infty}^{\circ} \tau_{12*}^{a} d\xi_1 = 0
\end{aligned}
$$

$$
\begin{aligned}
&\int_{-1}^{+1} (\eta^{1-s}\tau_{21*}^{b} + \eta^{1-2s+c}\tau_{21*}^{a}) |_{\xi_1=0}\, d\zeta \\
&= \int_{-1}^{+1} d\zeta \int_{-\infty}^{0} \left[\eta^{1-s}\frac{1}{A_{10}} \quad \frac{\partial \tau_{21*}^{b}}{\partial \xi_1} + \eta^{1-2s+c}\frac{1}{A_{10}}\frac{\partial \tau_{21*}^{a}}{\partial \xi_1}\right] A_{10} d\xi_1 \\
&= \int_{-1}^{+1} d\zeta \int_{-\infty}^{\circ} \left\{\eta^{1-s}\left[-\frac{\partial \tau_{23*}^{b}}{\partial \zeta} - X_2^{b}\right] - \eta^{1-2s+c}\frac{\partial \tau_{23*}^{a}}{\partial \zeta}\right\} A_{10} d\xi_1 \\
&= -\eta^{1-s}\int_{+1}^{+1} d\zeta \int_{0}^{-\infty} \frac{1}{A_{20}}\frac{\partial \tau_{22*}^{b}}{\partial \xi_2} A_{10} d\xi_1
\end{aligned}
$$

$$= -\eta^{1-s}\frac{1}{A_{20}}\frac{\partial}{\partial\xi_2}\int_{-1}^{+1}d\zeta\int_{-\infty}^{0}\tau_{22*}^{b}A_{10}d\xi_1. \tag{44.6}$$

In order to calculate the last integral which will be denoted

$$n_1 = \int_{-1}^{+1}d\zeta\int_{-\infty}^{0}\tau_{22*}^{b}A_{10}d\xi_1 \tag{44.7}$$

we should solve an auxiliary plane electroelasticity problem with the conditions at the edge $\xi_1 = 0$

$$\tau_{11*}^{b} = 0, \qquad \tau_{13*}^{b} = 0, \qquad \psi_*^{b} + \zeta = 0. \tag{44.8}$$

Allowing for equations 44.7 and 44.8, we rewrite the last expression in the second equation 44.6 as

$$\int_{-1}^{+1}\left(\eta^{1-s}\tau_{21*}^{b} + \eta^{1-2s+c}\tau_{21*}^{a}\right)|_{\xi_1=0}\,d\zeta = -\eta^{1-s}n_1\frac{1}{A_{20}}\frac{\partial\psi_{,1}}{\partial\xi_2}.$$

When transforming the following integrals, we will omit the calculations coinciding with those made in [22]:

$$\int_{-1}^{+1}\left(\eta^{1-2s+c}\tau_{31*}^{a} + \eta^{1-s}\tau_{31*}^{b}\right)|_{\xi_1=0}\,d\zeta$$

$$= \int_{-1}^{+1}d\zeta\int_{-\infty}^{0}\left(\eta^{1-2s+c}\frac{1}{A_{10}}\frac{\partial\tau_{31*}^{a}}{\partial\xi_1} + \eta^{-1+s}\frac{1}{A_{10}}\frac{\partial\tau_{31*}^{b}}{\partial\xi_1}\right)A_1 d\xi_1$$

$$= -\int_{-1}^{+1}d\zeta\int_{-\infty}^{0}\left[\eta^{s}\frac{R}{R_{20}}\tau_{22*}^{b} + \eta^{1-2s+c}\left(\frac{1}{A_{20}}\frac{\partial\tau_{32*}^{a}}{\partial\xi_2} + \eta^{s}Rk_{10}\tau_{23*}^{a}\right)\right]A_{10}d\xi_1$$

$$= \eta^{s}\frac{R}{R_{20}}n_1\psi_{,1} + \eta^{1-2s+c}\frac{2}{3}\frac{1}{A_{20}}\frac{\partial\tau_{21,1}}{\partial\xi_2}$$

$$\int_{-1}^{+1}\left(\tau_{11*}^{b} + \eta^{2-3s+c}\tau_{11*}^{a}\right)|_{\xi_1=0}\,d\zeta = -\eta^{2-3s+c}2J\frac{1}{A_{20}}\frac{\partial\tau_{21,1}}{\partial\xi_2}.$$

Here, for computing J by formula 43.10, we should solve an auxiliary antiplane problem with the following conditions at the edge $\xi_1 = 0$:

$$\tau_{12*}^{a} + \zeta = 0$$

$$\int_{-1}^{+1}\left(\eta^{2-3s+c}\psi_*^{a} + \psi_*^{b}\right)|_{\xi_1=0}d\zeta$$

$$= \int_{-1}^{+1}d\zeta\int_{-\infty}^{0}\left(\eta^{2-3s+c}\frac{1}{A_{10}}\frac{\partial\psi_*^{a}}{\partial\xi_1} + \frac{1}{A_{10}}\frac{\partial\psi_*^{b}}{\partial\xi_1}\right)A_{10}d\xi_1$$

$$= -\int_{-1}^{+1} d\zeta \int_{-\infty}^{0} \left(\eta^{2-3s+c} E_{1*}^{a} + E_{1*}^{b}\right) A_{10} d\xi_1$$

$$= \int_{-1}^{+1} d\zeta \int_{-\infty}^{0} \left[\eta^{2-3s+c}\left(t_1 \tau_{13*}^{a} - D_{1*}^{a}\right) + \left(t_1 \tau_{13*}^{b} - D_{1*}^{b}\right)\right] A_{10} d\xi_1$$

$$= -\int_{-1}^{+1} d\zeta \int_{-\infty}^{0} \xi_1 \left[\eta^{2-3s+c}\left(t_1 \frac{\partial \tau_{13*}^{a}}{\partial \xi_1} - \frac{\partial D_{1*}^{a}}{\partial \xi_1}\right)\right.$$

$$\left. + \left(t_1 \frac{\partial \tau_{13*}^{b}}{\partial \xi_1} - \frac{\partial D_{1*}^{b}}{\partial \xi_1}\right)\right] A_{10} d\xi_1$$

$$= \int_{-1}^{+1} d\zeta \int_{-\infty}^{0} A_{10}\xi_1 \left[t_1 \left(\frac{\partial\, \tau_{33*}^{b}}{\partial\ \zeta} + \eta^{2-3s+c}\frac{\partial \tau_{33*}^{a}}{\partial \zeta}\right)\right.$$

$$\left. + \eta^{2-3s+c} t_1 X_3^{a} - \frac{\varepsilon_{33}^{T}}{\varepsilon_{11}^{T}}\left(\frac{\partial D_{3*}^{b}}{\partial \zeta} + \eta^{2-3s+c}\frac{\partial D_{3*}^{a}}{\partial \zeta}\right) - \eta^{2-3s+c} M^{a}\right] A_{10} d\xi_1$$

$$= \eta^{2-3s+c}\frac{1}{A_{20}}\frac{\partial}{\partial \xi_2}\int_{-1}^{+1} d\zeta \int_{-\infty}^{0} A_{10}\xi_1 \left(t_1 \tau_{32*}^{a} - D_{2*}^{a}\right) A_{10} d\xi_1 = 0. \tag{44.9}$$

These transformations allow us to include all the boundary layer quantities in the boundary conditions 44.4 and write them in the form

$$\begin{aligned}
2\tau_{11,0} &= 0\\
2\tau_{21,0} - \eta^{1-s} n_1 \frac{1}{A_{20}}\frac{\partial \psi_{,1}}{\partial \xi_2} &= 0\\
\eta^{1-2s+c}\frac{2}{3}\tau_{11,1} - \eta^{2-3s+c} 2J \frac{1}{A_{20}}\frac{\partial \tau_{21,1}}{\partial \xi_2} &= 0\\
2\tau_{31,0} + \eta^{1-2s+c}\frac{2}{3}\tau_{31,2} + \eta^{s}\frac{R}{R_{2,0}} n_1 \psi_{,1} + \eta^{1-2s+c}\frac{2}{3}\frac{1}{A_{20}}\frac{\partial \tau_{21,1}}{\partial \xi_2} &= 0\\
\psi_{,0} + \eta^{1-2s+c}\frac{1}{3}\psi_{,2} &= V_* .
\end{aligned} \tag{44.10}$$

We pass to the forces and moments in formulas 44.10 as in formulas 36.12. After the transformation, we get the following boundary conditions *at the free electrode-covered edge* $\alpha_1 = \alpha_{10}$:

$$\begin{aligned}
T_1 &= 0\\
S_{21} - h n_1 \frac{d_{31}^2}{2\varepsilon_{33}^{T} s_{11}^{E}}\frac{1}{A_2}\frac{\partial T_2}{\partial \alpha_2} &= 0 \quad [\eta^{1-s}]\\
G_1 + 3J\frac{h}{A_2}\frac{\partial H_{21}}{\partial \alpha_2} &= 0 \quad [\eta^{1-s}]
\end{aligned}$$

$$N_1 - n_1 \frac{d_{31}^2}{2\varepsilon_{33}^T s_{11}^E} \frac{h}{R_2} T_2 - \frac{1}{A_2} \frac{\partial H_{21}}{\partial \alpha_2} = 0 \quad [\eta^s]$$

$$\psi^{(0)} - \frac{d_{31}}{4h\varepsilon_{33}^T} \left(G_2 - 3J \frac{1}{A_2} \frac{\partial H_{21}}{\partial \alpha_2} \right) = V \quad [\eta^{2-3s+c}]. \qquad (44.11)$$

The rest of the boundary conditions can be found similarly. We give them without derivation.

Rigidly fixed electrode-covered edge:

$$u_1 + m_2 h 2a \frac{1}{A_1} \frac{\partial u_1}{\partial \alpha_1} = 0, \quad [\eta^{1-s}]$$

$$u_2 = 0$$

$$\gamma_1 + \rho_3 h a \kappa_1 = 0 \quad [\eta^1]$$

$$w + \rho_4 h^2 a \kappa_1 = 0 \quad [\eta^{1-c}]$$

$$\psi^{(0)} - \frac{1}{4h} \frac{d_{31}}{\varepsilon_{33}^T} (G_1 + G_2) + \frac{h}{12\varepsilon_{33}^T} \left[d_{31} \left(\frac{T_1}{R_1} + \frac{T_2}{R_2} \right) + d_{33} \left(q_3^+ + q_3^- \right) \right] = V[\eta^1] \qquad (44.12)$$

where a is found from 43.7.

The number m_2 is found by solving auxiliary plane problems I and II with the following conditions at the edge $\xi_1 = 0$:

$$\begin{array}{llll} \text{I.} & v_{1*}^b + 1 = 0, & v_{3*}^b = 0, & \psi_*^b = 0 \\ \text{II.} & v_{1*}^b = 0, & v_{3*}^b + \zeta = 0, & \psi_*^b = 0, \qquad m_2 = \Gamma_1/\Gamma_2. \end{array}$$

The numbers ρ_3 and ρ_4 are found by solving auxiliary plane problems III–V with the following conditions at the edge $\xi_1 = 0$:

$$\begin{array}{llll} \text{III.} & v_{1*}^b + \zeta = 0, & v_{3*}^b = 0, & \psi_*^b = 0 \\ \text{IV.} & v_{1*}^b = 0, & v_{3*}^b + 1 = 0, & \psi_*^b = 0 \\ \text{V.} & v_{1*}^b = 0, & v_{3*}^b + \zeta^2 = 0, & \psi_*^b = 0 \end{array}$$

$$\rho_3 = \frac{\Delta_3}{\Delta}, \quad \rho_4 = \frac{\Delta_4}{\Delta}$$

$$\Delta = \begin{vmatrix} N_3 & N_4 \\ M_3 & M_4 \end{vmatrix}, \quad \Delta_3 = \begin{vmatrix} N_4 & N_5 \\ M_4 & M_5 \end{vmatrix}, \quad \Delta_4 = \begin{vmatrix} N_5 & N_3 \\ M_5 & M_3 \end{vmatrix}.$$

Here Γ_i, N_i, and M_i are, as before, the horizontal resultant, the vertical resultant, and the resulting bending moment of the edge stresses respectively.

Hinge supported electrode-covered edge $\alpha_1 = \alpha_{10}$:

$$u_2 = 0, \qquad w = 0, \qquad T_1 + n_3 hk_2 \{T_2\} = 0 \quad [\eta^1]$$

$$\psi^{(0)} - \frac{d_{31}}{4h\varepsilon_{33}^T} G_2 + \frac{hd_{31}}{12\varepsilon_{33}^T} \left\{ \left[-\frac{1}{R_1} + \frac{d_{31}}{\left(d_{31}^2 - \varepsilon_{33}^T s_{11}^E\right)} \right.\right.$$
$$\left.\left. \times \left(2d_{15} - \frac{\varepsilon_{11}^T}{\varepsilon_{33}^T} d_{31}\right) \frac{1}{R_2} \right] T_2 + \frac{d_{33}}{d_{31}} \left(q_3^+ + q_3^-\right) \right\} = V \quad [\eta^1]$$
$$G_1 - \left\{ \frac{h^2}{3} \frac{\varepsilon_{11}^T D_{31}^2}{\varepsilon_{33}^T \left(d_{31}^2 - \varepsilon_{33}^T s_{11}^E\right)} \frac{T_2}{R_2} \right\} = 0 \quad [\eta^1]$$

(44.13)

where

$$n_3 = \int_{-1}^{+1} d\zeta \int_{-\infty}^{0} \tau_{22*}^b A_{10} d\xi_1$$

is defined by the solution of the plane problem with the edge conditions

$$\tau_{11*}^b = 0, \quad v_{3*}^b + \zeta \left(\frac{s_{13}^E}{s_{11}^E} + \frac{d_{33} d_{31}}{s_{11}^E \varepsilon_{33}^T} \right) = 0, \qquad \psi_*^b + \zeta \frac{d_{31}^2}{S_{11}^E \varepsilon_{33}^T} = 0.$$

Hinge-supported edge $\alpha_1 = \alpha_{10}$ without electrodes:

$$u_2 = 0, \quad w = 0, \quad T_1 + n_4 hk_2 \{T_2\} = 0 \quad [\eta^1]$$

$$G_1 - \left\{ \frac{h^2}{3} \frac{\varepsilon_{11}^T d_{31}^2}{\varepsilon_{33}^T \left(d_{31}^2 - \varepsilon_{33}^T s_{11}^E\right)} \frac{T_2}{R_2} \right\} = 0 \quad [\eta^1]$$

$$\frac{1}{A_1} \frac{\partial}{\partial \alpha_1} \left[\psi^{(0)} - \frac{d_{31}}{4h\varepsilon_{33}^T} (G_1 + G_2) \right.$$
$$\left. - \frac{h}{12\varepsilon_{33}^T} \left\{ d_{31} \left(\frac{T_1}{R_2} + \frac{T_2}{R_1} \right) + d_{33} \left(q_3^+ + q_3^-\right) \right\} \right] = 0 \; [\eta^1] \quad (44.14)$$

In order to find

$$n_4 = \int_{-1}^{+1} d\zeta \int_{-\infty}^{0} \tau_{22*}^b A_{10} d\xi_1$$

we should solve auxiliary problem VI with the following conditions at the edge $\xi_1 = 0$:

$$\text{VI.} \qquad \tau_{1*}^b = 0, \qquad v_{3*}^b + \zeta\left(\frac{s_{13}^E}{s_{11}^E} + \frac{d_{33}d_{31}}{s_{11}^E\varepsilon_{33}^T}\right) = 0, \quad D_{1*}^b = 0.$$

Free edge $\alpha_1 = \alpha_{10}$ without electrodes:

$$T_1 = 0, \qquad S_{21} = 0, \qquad G_1 + 3J\frac{h}{A_2}\frac{\partial H_{21}}{\partial \alpha_2} = 0 \quad [\eta^{1-s}]$$

$$N_1 - \frac{1}{A_2}\frac{\partial H_{21}}{\partial \alpha_2} - \left\{\frac{h^2}{3R_2}\frac{d_{31}^2\varepsilon_{11}^T}{\left(d_{31}^2 - s_{11}^E\varepsilon_{33}^T\right)\varepsilon_{33}^T}\frac{1}{A_{10}}\frac{\partial}{\partial \alpha_1}T_2\right\} = 0 \quad [\eta^{1+s}]. \tag{44.15}$$

Rigidly fixed edge $\alpha_1 = \alpha_{10}$ without electrodes:

$$u_1 + m_2 2a\frac{h}{A_1}\frac{\partial u_1}{\partial \alpha_1} = 0 \quad [\eta^{1-s}], \qquad u_2 = 0$$

$$\gamma_1 + \rho_5 ha\kappa_1 = 0 \quad [\eta^1]$$

$$w + \rho_6 ha\kappa_1 = 0 \quad [\eta^{1-c}] \tag{44.16}$$

where $m_3 = \Gamma_7/\Gamma_8$ has been determined from the solution of plane problems VII and VIII with the following conditions at the edge $\xi_1 = 0$:

$$\begin{array}{llll} \text{VII.} & v_{1*}^b + 1 = 0, & v_{3*}^b = 0, & D_{1*}^b = 0 \\ \text{VIII.} & v_{1*}^b = 0, & v_{3*}^b + \zeta = 0, & D_{1*}^b = 0. \end{array}$$

The numbers ρ_5 and ρ_6 are found by solving auxiliary problems IX–XI:

$$\begin{array}{llll} \text{IX.} & v_{1*}^b + \zeta = 0, & v_{3*}^b = 0, & D_{1*}^b = 0 \\ \text{X.} & v_{1*}^b = 0, & v_{3*}^b + 1 = 0, & D_{1*}^b = 0 \\ \text{XI.} & v_{1*}^b = 0, & v_{3*}^b + \zeta^2 = 0, & D_{1*}^b = 0. \end{array}$$

$$\rho_5 = \frac{\Delta_5}{\Delta}, \rho_6 = \frac{\Delta_6}{\Delta}$$

$$\Delta_5 = \begin{vmatrix} N_{10} & N_{11} \\ M_{10} & M_{11} \end{vmatrix}, \qquad \Delta_6 = \begin{vmatrix} N_{11} & N_9 \\ M_{11} & M_9 \end{vmatrix}, \qquad \Delta = \begin{vmatrix} N_9 & N_{10} \\ M_9 & M_{10} \end{vmatrix}.$$

At the free and rigidly fixed edges without electrodes, the electrical condition is in the form typical for hinge-supported edges without electrodes.

45 BOUNDARY CONDITIONS IN THE THEORY OF SHELLS WITH TANGENTIAL POLARIZATION (ELECTRODE-COVERED FACES)

As before, we will assume that the shell is polarized along the α_2-lines. Let the three-dimensional conditions for rigid fixation

$$v_1 = 0, \qquad v_2 = 0, \qquad v_3 = 0, \qquad D_1 = 0$$

be fulfilled at the shell edge $\alpha_1 = \alpha_{10}$. We represent each quantity as a sum of four terms: the principal electroelastic state, the electroelastic states with great variability, and the antiplane and plane boundary layers

$$\begin{aligned}
\eta^{1-s}\left(v_{1,0}^{(p)} + \eta^{1-2s}\zeta v_{1,1}^{(p)}\right) + \eta^{\delta+q}\left(v_{1,0}^{(a)} + \eta^{1-2q}\zeta v_{1,1}^{(a)}\right)& \\
+\eta^{2-s+\alpha} v_{1*}^{a} + \eta^{1+\beta} v_{1*}^{b} &= 0 \\
\left(\eta^{1-s} v_{2,0}^{(p)} + \zeta v_{2,1}^{(p)}\right) + \eta^{\delta+2q-s}\left(v_{2,0}^{(a)} + \eta^{1-2q}\zeta v_{2,1}^{(a)}\right)& \\
+\eta^{1+\alpha} v_{2*}^{a} + \eta^{2-s+\beta} v_{2*}^{b} &= 0 \\
\eta^{1-2s}\left(v_{3,0}^{(p)} + \eta^{1}\zeta v_{3,1}^{(p)} + \eta^{2s}\zeta^2 v_{3,2}^{(p)}\right) + \eta^{\delta}\left(v_{3,0}^{(a)} + \eta^{1}\zeta v_{3,1}^{(a)}\right)& \\
+\eta^{2-s+\alpha} v_{3*}^{a} + \eta^{1+\beta} v_{3*}^{b} &= 0 \\
\left(\eta^{1-2s} D_{1,0}^{(p)} + \zeta D_{1,1}^{(p)}\right) + \eta^{\delta+q-1}\left(D_{1,0}^{(a)} + \zeta D_{1,1}^{(a)}\right)& \\
+\eta^{\alpha} D_{1*}^{a} + \eta^{1-s+\beta} D_{1*}^{b} &= 0.
\end{aligned} \tag{45.1}$$

As in Section 24, the superscripts (p) and (a) denote the quantities of the principal electroelastic state and of the electroelastic state with great variability, respectively. The numbers δ, α, and β are chosen so that

$$\delta = q, \qquad \alpha = -1, \qquad \beta = -s. \tag{45.2}$$

By substituting 45.2 into 45.1, we get

$$\begin{aligned}
v_{1,0} + \eta^{s}\zeta v_{1,1} + R v_{1*} &= 0 \\
\eta^{1-s} v_{2,0} + \zeta v_{2,1} + \eta^{1}\zeta^2 v_{2,2} + v_{2*}^{a} + \eta^{2-2s} v_{2*}^{b} &= 0 \\
\eta^{q-1+s} v_{3,0} + \eta^{q+s}\zeta v_{3,1} + \eta^{s}\zeta^2 v_{3,2} + v_{3*} &= 0 \\
\eta^{2-2s} D_{1,0} + \eta^{1}\zeta D_{11} + D_{1*}^{a} + \eta^{2-2s} D_{1*}^{b} &= 0.
\end{aligned} \tag{45.3}$$

Here, we have used the notation

$$v_{1,0} = v_{1,0}^{(p)} + \eta^{2q-1+s} v_{1,0}^{(a)}, \qquad v_{1,1} = v_{1,1}^{(p)} + \eta^{1-3s} v_{1,1}^{(a)}$$

$$v_{2,0} = v_{2,0}^{(p)} + \eta^{3q-1} v_{2,0}^{(a)}, \quad v_{2,1} = v_{2,1}^{(p)} + \eta^{1+q-s} v_{2,1}^{(a)}$$
$$v_{3,0} = \eta^{1-q-2s} v_{3,0}^{(p)} + v_{3,0}^{(a)}, \quad v_{3,1} = \eta^{1-q-2s} v_{3,1}^{(p)} + v_{3,1}^{(a)}$$
$$v_{3,2} = v_{3,2}^{(p)}, \quad D_{1,0} = D_{1,0}^{(p)} + \eta^{2q-1+s} D_{1,0}^{(a)}$$
$$D_{1,1} = D_{1,1}^{(p)} + \eta^{2q-s} D_{1,1}^{(a)}$$
$$v_{1*} = v_{1*}^{a} + v_{1*}^{b}, \quad v_{3*} = v_{3*}^{a} + v_{3*}^{b}.$$

We can obtain the system of equations for defining the quantities $v_{1*}, v_{3*}, \ldots$ by summing up the equations of the principal subsystem for the plane boundary layer and the auxiliary subsystem for the antiplane boundary layer up to the quantities $O(\eta^1)$:

$$\frac{1}{A_{10}} \frac{\partial \tau_{11*}}{\partial \xi_1} + \frac{\partial \tau_{13*}}{\partial \zeta} = -X_1^a$$
$$\frac{1}{A_{10}} \frac{\partial \tau_{31*}}{\partial \xi_1} + \frac{\partial \tau_{33*}}{\partial \zeta} = -X_3^a$$
$$\frac{1}{A_{10}} \frac{\partial v_{1*}}{\partial \xi_1} - \tau_{11*} + \nu_1 \tau_{22*} + \nu \tau_{33*} = -P_1^a$$
$$\frac{\partial v_{3*}}{\partial \zeta} + \nu \tau_{11*} + \nu_1 \tau_{22*} - \tau_{33*} = -P_3^a$$
$$\frac{\partial v_{1*}}{\partial \zeta} + \frac{1}{A_{10}} \frac{\partial v_{3*}}{\partial \xi_1} - \sigma_1 \tau_{13*} = 0$$
$$-\nu_1(\tau_{11*} + \tau_{33*}) + \sigma_3 \tau_{22*} = P_2^a. \qquad (45.4)$$

The quantities $X_1^a \ldots, P_2^a$ are found from the formulas 38.12.

We will treat the first and third conditions 45.3 as the boundary conditions for the plane boundary layer

$$v_{1*} + v_{1,0} + \eta^s \zeta v_{1,1} = 0$$
$$v_{3*} + \eta^{q-1+s} v_{3,0} + \eta^{q+s} \zeta v_{3,1} + \eta^s \zeta^2 v_{3,2} = 0.$$

We represent the plane boundary layer as a sum of the symmetric (corresponding to the bending of the half-band) and the antisymmetric (corresponding to the extension of the half-band) parts of the plane boundary layer.

For the antisymmetric part the boundary conditions will be written as

$$v_{1*} + v_{1,0} = 0, \qquad v_{3*} + \eta^{q+s} \zeta v_{3,1} = 0. \qquad (45.5)$$

Since the problem is linear, we can represent the solution of equations 45.4 with the edge conditions 45.5 and the homogeneous conditions on the faces as a linear combination of three problems with the following conditions at the edge $\xi_1 = 0$:

I. $\nu_{1*} + 1 = 0, \quad \nu_{3*} = 0$
II. $\nu_{1*} = 0, \quad \nu_{3*} + \zeta = 0$
III $\nu_{1*} = 0, \quad \nu_{3*} = 0.$

We solve problems I and II by integrating the system of equations of the plane problem with the nonhomogeneous conditions I and II. We solve problem III by integrating the nonhomogeneous equations 45.4 with the homogeneous conditions III. We multiply the solutions of I–III by the factors

I. $\nu_{1,0}$
II. $\eta^{q+s}\nu_{3,1}$
III η^1

respectively, and take their sum to obtain the solution to the system of equations 45.4 under conditions I–III. The solution of problem III is multiplied by η^1 because the terms which correspond to the conditions 45.5 in the right-hand sides of equations 45.4 have this order.

The stresses at the infinitely distant edge of the half-band are equal to zero. For the band to be at equilibrium, we should equate the resultant of the horizontal (directed along the lines) stresses at the edge to zero, which corresponds to problems I, II and III ($\Gamma_1, \Gamma_2, \Gamma_3$):

$$\Gamma_1\nu_{1,0} + \Gamma_2\eta^{q+s}\nu_{3,1} + \Gamma_3\eta^1 = 0.$$

By neglecting the quantities $O(\eta^1)$, we get

$$\nu_{1,0} + \mu_1\eta^{q+s}\nu_{3,1} = 0, \qquad \mu_1 = \Gamma_2/\Gamma_1$$

For the symmetric part of the plane boundary layer, the boundary conditions have the form

$$\nu_{1*} + \eta^s\zeta\nu_{1,1} = 0$$
$$\nu_{3*} + \eta^{q-1+s}\nu_{3,0} + \eta^s\zeta^2\nu_{3,2} = 0.$$

We use the linearity of the problem and represent the solution to the system 45.4 as a linear combination of four problems with the following conditions at the edge:

IV. $\nu_{1*} + \zeta = 0, \quad \nu_{3*} = 0$
V. $\nu_{1*} = 0, \quad \nu_{3*} + 1 = 0$
VI. $\nu_{1*} = 0, \quad \nu_{3*} + \zeta^2 = 0$
VII. $\nu_{1*} = 0, \quad \nu_{3*} = 0.$

and with the following factors, respectively:

IV. $\eta^s v_{1,1}$
V. $\eta^{q-1+s} v_{3,0}$
VI. $\eta^s v_{3,2}$
VII. η^s.

In order to solve problems IV, V, and VI, we should integrate the homogeneous system of equations 45.4 with the edge conditions IV, V, and VI. In order to solve problem VII, we should integrate the nonhomogeneous system (equations 45.4) with the homogeneous conditions VII. On the right-hand side of equations 45.4, we have the quantities of the antiplane problem

$$X_1^a = \left(\tau_{12*}^a + \tau_{21*}^a\right), \qquad X_3^a = \tau_{23*}^a, \qquad P_1^a = k_{10} v_2^a$$

found from the homogeneous equations of the antiplane problem with the nonhomogeneous edge conditions ($\xi_1 = 0$)

$$D_1^a = 0, \qquad v_{2*}^a + \zeta v'_{2,1} = 0. \tag{45.6}$$

In order to fulfill the damping conditions, we should require that the resultants of the vertical forces and bending moments in the four problems vanish at the edge. We use $N_4, \ldots, N_7$ and $M_4, \ldots, M_7$ to denote the resultants of the vertical forces and bending moments in problems IV, ..., VII, respectively. We write the equilibrium conditions for the half-band with edge unloaded at infinity:

$$\eta^s v_{1,1} N_4 + \eta^{q-1+s} v_{3,0} N_5 + \eta^s (N_6 v_{3,2} + N_7) = 0$$
$$\eta^s v_{1,1} M_4 + \eta^{q-1+s} v_{3,0} M_5 + \eta^s (M_6 v_{3,2} + M_7) = 0.$$

Solving these equations with respect to $v_{1,1}$ and $v_{3,0}$, we get

$$v_{1,1} = \rho_1, \quad v_{3,0} = \eta^{-q+1} \rho_2, \quad \rho_1 = \frac{\Delta_1}{\Delta}, \quad \rho_2 = \frac{\Delta_2}{\Delta}$$

$$\Delta_1 = \begin{vmatrix} N_5, & N_6 v_{3,2} + N_7 \\ M_5, & M_6 v_{3,2} + M_7 \end{vmatrix}, \qquad \Delta_2 = \begin{vmatrix} N_6 v_{3,2} + N_7, & N_4 \\ M_6 v_{3,2} + M_7, & M_4 \end{vmatrix}$$

$$\Delta = \begin{vmatrix} N_4 & N_5 \\ M_4 & M_5 \end{vmatrix}.$$

The second and fourth conditions in 45.3 are the conditions for the antiplane boundary layer. Consider the symmetric part of the antiplane boundary layer with the edge conditions

$$v_{2*}^a + \eta^{1-s} v_{2,0} + \eta^1 \zeta^2 v_{2,2} = 0$$
$$D_{1*}^a + \eta^{2-2s} D_{1,0} = 0.$$

We represent this problem as a sum of auxiliary problems with the following edge conditions:

VIII. $v_{2*}^{a} + 1 = 0, \quad D_{1*}^{a} = 0$
IX. $v_{2*}^{a} + \zeta^2 = 0, \quad D_{1*}^{a} = 0$
X. $v_{2*}^{a} = 0, \quad D_{1*}^{a} = 0$
XI. $v_{2*}^{a} = 0, \quad D_{1*}^{a} + 1 = 0.$

The solutions of the auxiliary problems should be multiplied by the respective factors

VIII. $\eta^{1-s} v_{2,0}$
IX. $\eta^{1} v_{2,2}$
X. $\eta^{1} R k_{10} v_{2,1}$
XI. $\eta^{2-2s} D_{1,0}.$

The solutions to problems VIII, IX, and XI are found from the homogeneous equations. The solution to problem X is found from the nonhomogeneous equations of the plane problem, the right-hand sides of which are found by solving the antiplane problem with the conditions 45.6 at the edge. In problems VIII, IX, X, and XI, we equate the horizontal resultants of the forces $\Gamma_8, \Gamma_9, \Gamma_{10}, \Gamma_{11}$ directed along the ξ_1-line at the edge to zero and get the condition

$$\Gamma_8 \eta^{1-s} v_{2,0} + \Gamma_9 \eta^{1} v_{2,2} + \Gamma_{10} \eta^{1} v_{2,1}' + \Gamma_{11} \eta^{2-2s} D_{1,0} = 0$$

or

$$v_{2,0} + \eta^{s} \mu_2 v_{2,2} + \eta^{s} \mu_3 v_{2,1}' + \eta^{1-s} \mu_4 D_{1,0} = 0$$

where

$$\mu_2 = \frac{\Gamma_9}{\Gamma_8}, \qquad \mu_3 = \frac{\Gamma_{10}}{\Gamma_8}, \qquad \mu_4 = \frac{\Gamma_{11}}{\Gamma_8}. \tag{45.7}$$

Thus, we have obtained the following conditions for the rigidly fixed edge $\xi_1 = 0$:

$$v_{2,0} + \eta^{s} \mu_2 v_{2,2} + \eta^{s} \mu_3 v_{2,1}' + \eta^{1-s} \mu_4 D_{1,0} = 0$$

$$v_{1,0} + \mu_1 \eta^{q+s} v_{3,1} = 0, \qquad v_{1,1} = \rho_1, \qquad v_{3,0} = \eta^{1-q} \rho_2.$$

In the notation of shell theory, the boundary conditions for a *rigidly fixed edge* $\alpha_1 = \alpha_{10}$ will be

$$u_1 + \frac{\mu_1}{2} \left(s_{12}^{E} T_1 + s_{13}^{E} T_2 \right) = 0 \quad \left[\eta^{q+s} \right]$$

$$u_2 + h\left(\frac{1}{2R_1}\mu_2 - k_1\mu_3\right)d_{15}V + h\mu_4 d_1 q_3\omega = 0 \quad \left[\eta^s\right]$$

$$\gamma_1 + \rho_1 = 0 \quad \left[\eta^0\right]$$

$$w + h\rho_2 = 0 \quad \left[\eta^{1-q}\right]. \tag{45.8}$$

If we have $k_1 = 0$ at the edge (k_1 is the geodesic curvature of the α_1-line), then $\rho_1 = \rho_2 = 0$. The constants $\mu_1, \mu_2, \mu_3, \mu_4, \rho_1$, and ρ_2 are found by solving the auxiliary plane problems I–XI. We will give the remaining boundary conditions without deriving them.

For the *hinge-supported edge* $\alpha_1 = \alpha_{10}$

$$\tau_{11} = v_2 = v_3 = D_1 = 0. \tag{45.9}$$

We write each of these quantities as a sum of the type in equation 45.1 and take $\alpha = 1, \beta = -s$ and $\delta = -1 + 2q$. We get the following boundary conditions for the *hinge-supported edge* $\alpha_1 = \alpha_{10}$

$$T_1 - J_1\frac{h^2}{s_{11}^E}\left[k_1\left(\frac{1}{R_2} - \frac{1}{R_1}\right) + \frac{1}{A_2}\frac{\partial}{\partial\alpha_2}\left(\frac{1}{R_2}\right)\right]d_{15}V = 0$$

$$u_2 + h\left(\frac{\mu_2}{2R_1} - k_1\mu_3\right)d_{15}V + h\mu_4 d_1 q_3\omega = 0 \quad \left[\eta^s\right]$$

$$G_1 = 0, \qquad w = 0. \tag{45.10}$$

The second term in the first condition is $O(\eta^{2s})$ for $k_2(\alpha_{10}) = 0$ and $O(\eta^{2-3q})$ for $k_2(\alpha_{10}) \neq 0$. The numbers μ_2, μ_3, and μ_4 are found by formulas (45.7) from the solutions to problems VIII–XI. The number J_1 is calculated using equation 45.11, where we put $i = 1$:

$$J_i = \int_1^{+1}\zeta d\zeta \int_{-\infty}^{0} S_{12}^a A_{10} d\xi_1 \tag{45.11}$$

where S_{12}^a has been found from the homogeneous equations of the antiplane problem with the following conditions at the edge $\xi_1 = 0$

$$v_{2*}^a + \zeta = 0, \qquad D_{1*}^a = 0. \tag{45.12}$$

For the *free edge* $\alpha_1 = \alpha_{10}$

$$(\tau_{11} = \tau_{21} = \tau_{13} = D_1 = 0, \qquad \alpha = 0, \qquad \beta = 1 - s, \qquad \delta = -1 + 2q)$$

$$T_1 = 0$$

$$G_1 + 3J_2\frac{h}{A_2}\frac{\partial H_{21}}{\partial\alpha_2} = 0 \quad \left[\eta^{1-s}\right]$$

$$S_{21} + \frac{H_{21}}{R_2} = 0$$

$$N_1 - \frac{1}{A_2}\frac{\partial H_{21}}{\partial \alpha_2} = 0. \qquad (45.13)$$

Here, J_2 is found by equation 45.11 ($i = 2$) from the solution of the antiplane problem with the following conditions at the edge $\xi_1 = 0$:

$$\tau^a_{12*} + \zeta = 0, \qquad D^a_{1*} + t_1\zeta = 0. \qquad (45.14)$$

The numbers d_1, q_3, and t_1 were introduced by equation 34.7.

For the rigidly fixed edge $\alpha_2 = \alpha_{20}$

$$v_1 = v_2 = v_3 = D_2 = 0, \qquad \alpha = -1 + q - s,$$

$$\beta = -q, \qquad \delta = -1 + q$$

After transformations, the conditions at the *rigidly fixed edge* $\alpha_2 = \alpha_{20}$ will be written as

$$u_1 = 0, \qquad w = 0$$

$$\gamma_2 + d_{15}\frac{V}{h} - \frac{3\mu}{2h^2}\frac{d_{15}}{\varepsilon^T_{33}}(d_{31}G_1 + d_{33}G_2) = 0 \quad \left[\eta^{1-q}\right]$$

$$u_2 + \frac{\delta_1}{2}\left(s^E_{12}T_1 + s^E_{13}T_2\right) + \frac{\delta_2}{2}\frac{d_{15}}{\varepsilon^T_{33}}(d_{31}T_1 + d_{33}T_2) = 0 \quad \left[\eta^{1-q}\right]. \qquad (45.15)$$

In order to find μ, α_1, and α_2, we should solve the homogeneous system of equations for the plane boundary layer with the following conditions at the edge $\xi_2 = 0$:

XII.	$v_{2*} + \zeta = 0,$	$v_{3*} = 0,$	$D_{2*} = 0$
XIII.	$v_{2*} = 0,$	$v_{3*} + 1 = 0,$	$D_{2*} = 0$
XIV.	$v_{2*} = 0,$	$v_{3*} = 0,$	$D_{2*} + \zeta = 0$
XV.	$v_{2*} + 1 = 0,$	$v_{3*} = 0,$	$D_{2*} = 0$
XVI.	$v_{2*} = 0,$	$v_{3*} + \zeta = 0,$	$D_{2*} = 0$
XVII.	$v_{2*} = 0,$	$v_{3*} = 0,$	$D_{2*} + 1 = 0.$

Then we should find the resultants of the vertical forces N_{12}, N_{13}, and N_{14}; the bending moments M_{12}, M_{13}, and M_{14} in problems XII, XIII, and XIV; and the horizontal forces Γ_{15}, Γ_{16}, and Γ_{17} in problems XV, XVI, and XVII at the edge $\xi_2 = 0$. The numbers μ, δ_1, and δ_2 are found by the formulas

$$\mu = \frac{\Delta_1}{\Delta}, \qquad \delta_1 = \frac{\Gamma_{16}}{\Gamma_{15}}, \qquad \delta_2 = \frac{\Gamma_{17}}{\Gamma_{15}}$$

$$\Delta_1 = \begin{vmatrix} N_{13} & N_{14} \\ M_{13} & M_{14} \end{vmatrix}, \qquad \Delta = \begin{vmatrix} N_{12} & N_{13} \\ M_{12} & M_{13} \end{vmatrix}.$$

For the *hinge-supported edge* $\alpha_2 = \alpha_{20}$

$$T_2 = 0, \qquad G_2 = 0, \qquad u_1 = 0, \qquad w = 0. \tag{45.16}$$

For the *free edge* $\alpha_2 = \alpha_{20}$

$$T_{12} + \mu_5 h k_1 T_1 = 0 \quad \left[\eta^{2-3q}\right], \qquad S_{12} - \mu_5 \frac{h}{2A_1}\frac{\partial T_1}{\partial \alpha_1} = 0$$
$$G_2 + 3hJ_3 \frac{1}{A_1}\frac{\partial H_{12}}{\partial \alpha_1} = 0 \quad \left[\eta^{1-s}\right], \qquad N_2 - \frac{1}{A_1}\frac{\partial H_{12}}{\partial \alpha_1} = 0 \tag{45.17}$$

where

$$\mu_5 = \frac{d_{31}d_{15}}{\varepsilon_{33}^T s_{11}^E}\int_{-1}^{+1} d\zeta \int_{-\infty}^{0} \tau_{11*}^b A_{20} d\xi_2.$$

The number μ_5 is found from the solution of the plane problem with the following conditions at the edge $\xi_2 = 0$:

$$\tau_{22*}^b = 0, \qquad \tau_{23*}^b = 0, \qquad D_{2*}^b + 1 = 0.$$

The number J_3 is found by equation 45.11 from the solution of the homogeneous equations of the antiplane problem (equations 40.6 and 40.7) with the conditions

$$\tau_{12*}^a + \zeta = 0$$

at the edge $\xi_2 = 0$.

46 BOUNDARY CONDITIONS IN THE THEORY OF SHELLS WITH TANGENTIAL POLARIZATION (FACES WITHOUT ELECTRODES)

On faces without electrodes, the electrical boundary condition has the form

$$D_3 \left|_{\zeta=\pm 1}\right. = 0. \tag{46.1}$$

For the *rigidly fixed edge* $\alpha_1 = \alpha_{10}$ *without electrodes*

$$u_1 + \frac{\mu_6}{2}\left[s_{12}^E T_1 + s_{13}^E T_2 + 2hd_{31}E_3^{(0)}\right] = 0 \quad \left[\eta^{1-s}\right]$$
$$u_2 = 0, \qquad w = 0$$
$$\gamma_1 + \frac{3\mu_7}{2h^2}\left[s_{12}^E G_1 + s_{13}^E G_2\right] = 0 \quad \left[\eta^{1-s}\right]$$
$$D_1^{(0)} - \frac{\mu_6}{2}\frac{b_1}{A_2}\frac{\partial}{\partial \alpha_2}\left[s_{12}^E T_1 + s_{13}^E T_2 + 2hd_{31}E_3^{(0)}\right] = 0 \quad \left[\eta^{1-s}\right]$$
$$\mu_6 = \frac{\Gamma_2}{\Gamma_1}, \qquad \mu_7 = \frac{N_3}{N_4}, \qquad b_1 = \frac{d_{31}s_{33}^E - d_{33}s_{13}^E}{s_{12}^E s_{33}^E - \left(s_{13}^E\right)^2} \tag{46.2}$$

where Γ_1 and Γ_2 are the horizontal resultants of the edge stresses found from the solution of the simultaneous homogeneous equations of the plane boundary layer with the following conditions at the edge $\xi_1 = 0$:

$$\text{I.} \quad v^b_{1*} + \zeta = 0, \quad v^b_{3*} = 0$$
$$\text{II.} \quad v^b_{1*} = 0, \quad v^b_{3*} + 1 = 0.$$

N_3 and N_4 are the vertical resultants of the edge stresses found from the plane auxiliary problems III and IV with the following conditions at the edge $\xi_1 = 0$:

$$\text{III.} \quad v^b_{1*} + \zeta = 0, \quad v^b_{3*} = 0$$
$$\text{IV.} \quad v^b_{1*} = 0, \quad v^b_{3*} + \zeta^2 = 0.$$

For the *hinge-supported edge* $\alpha_1 = \alpha_{10}$ *without electrodes*

$$T_1 = 0, \qquad u_2 = 0, \qquad w = 0, \qquad G_1 = 0$$
$$2D_1^{(0)} - \mu_8 \frac{b_2}{2A_2} \frac{\partial}{\partial \alpha_2} \left(s^E_{13} T_2 - 2hd_{31} E_2^{(0)} \right) = 0 \quad \left[\eta^{1-s}\right] \tag{46.3}$$

$$\mu_8 = \int_{-1}^{+1} d\zeta \int_{-\infty}^{0} \tau^b_{33*} A_{10} d\xi_1$$
$$b_2 = d_{31} - d_{33} \frac{s^E_{13}}{s^E_{33}}$$

where μ_8 is found from the solution of the plane problem with the edge conditions ($\xi_1 = 0$)

$$\text{V.} \quad \tau^b_{11*} = 0, \quad v^b_{3*} + \zeta = 0.$$

For the *free edge* $\alpha_1 = \alpha_{10}$ *without electrodes*

$$T_1 = 0, \qquad S_{21} = 0, \qquad G_1 + 3J_4 \frac{h}{A_2} \frac{\partial H_{21}}{\partial \alpha_2} = 0 \quad \left[\eta^{1-s}\right]$$
$$N_1 - \frac{1}{A_2} \frac{\partial H_{21}}{\partial \alpha_2} = 0, \qquad D_1^{(0)} = 0. \tag{46.4}$$

Here, J_4 is found by equation 45.11 ($i = 4$) from the solution of the antiplane problem in equation 39.6 with edge conditions in equation 45.12 and electrical conditions of the type in equation 46.1.

For the *rigidly fixed electrode-covered edge* $\alpha_1 = \alpha_{10}$,

$$u_1 + \frac{\mu_6}{2}\left[s_{12}^E T_1 + s_{13}^E T_2 + 2hd_{31}E_3^{(0)}\right] = 0 \quad \left[\eta^{1-s}\right], \qquad u_2 = 0, \qquad w = 0$$

$$\gamma_1 + \frac{3\mu_7}{2h^2}\left[s_{12}^E G_1 + s_{13}^E G_2\right] = 0 \quad \left[\eta^{1-s}\right] \qquad \psi^{(0)} = V. \qquad (46.5)$$

The numbers μ_6 and μ_7 are the same as for the electrode-covered rigidly fixed edge $\alpha_1 = \alpha_{10}$ without electrode.

For the *hinge-supported electrode-covered edge* $\alpha_1 = \alpha_{10}$

$$u_2 = 0, \qquad w = 0, \qquad T_1 = 0, \qquad G_1 = 0, \qquad \psi^{(0)} = V. \qquad (46.6)$$

For the *free electrode-covered edge* $\alpha_1 = \alpha_{10}$

$$T_1 = 0, \qquad S_{21} = 0, \qquad G_1 + 3J_5 \frac{h}{A_2}\frac{\partial H_{21}}{\partial \alpha_2} = 0 \quad \left[\eta^{1-s}\right]$$

$$N_1 - \frac{1}{A_2}\frac{\partial H_{21}}{\partial \alpha_2} = 0, \qquad \psi^{(0)} = V. \qquad (46.7)$$

The number J_5 is found by equation 45.11 from the solution of the antiplane problem with conditions

$$\tau_{23*}^a = 0, \qquad D_{3*}^a = 0$$

at the faces $\zeta = \pm 1$.

For the *rigidly fixed edge* $\alpha_2 = \alpha_{20}$ *without electrodes*

$$u_1 = 0, \qquad w = 0, \qquad D_2^{(0)} = 0$$

$$u_2 + \frac{\mu_{10}}{2}\left[s_{12}^E T_1 + s_{13}^E T_2 + 2hd_{31}E_2^{(0)}\right] = 0 \quad \left[\eta^{1-s}\right]$$

$$\gamma_2 + \mu_{11}\frac{3}{2h^2}\left[s_{11}^E G_1 + s_{13}^E G_2\right] = 0 \quad \left[\eta^{1-s}\right]. \qquad (46.8)$$

For defining μ_{10} and μ_{11}, we should integrate the homogeneous conditions of the plane boundary layer with the following conditions at the edge $\xi_2 = 0$:

VI.	$v_{2*}^b + 1 = 0$,	$v_{3*}^b = 0$,	$D_{2*}^b = 0$
VII.	$v_{2*}^b = 0$,	$v_{3*}^b + \zeta = 0$,	$D_{2*}^b = 0$
VIII	$v_{2*}^b + \zeta = 0$,	$v_{3*}^b = 0$,	$D_{2*}^b = 0$
IX.	$v_{2*}^b = 0$,	$v_{3*}^b = 0$,	$D_{2*}^b = 0$.

In problems VI and VII, we should find the resultants of the horizontal forces Γ_6 and Γ_7. In problems VIII and IX, we should find the resultants of the vertical edge forces N_8 and N_9 and the resultants of the edge bending moments M_8 and M_9. The numbers μ_{10} and μ_{11} are calculated by

$$\mu_{10} = \frac{\Gamma_6}{\Gamma_7}, \qquad \mu_{11} = \frac{M_8}{M_9}.$$

For the *hinge-supported edge* $\alpha_2 = \alpha_{20}$ *without electrodes*

$$T_2 = 0, \qquad u_1 = 0, \qquad w = 0, \qquad G_2 = 0, \qquad D_2^{(0)} = 0. \tag{46.9}$$

For the *free edge* $\alpha_2 = \alpha_{20}$ *without electrodes*,

$$\begin{aligned} T_2 = 0, \qquad S_{12} = 0, \qquad G_2 + 3J_3 \frac{h}{A_1}\frac{\partial H_{12}}{\partial \alpha_1} = 0 \\ N_2 - \frac{1}{A_1}\frac{\partial H_{12}}{\partial \alpha_1} = 0, \qquad D_2^{(0)} = 0. \end{aligned} \tag{46.10}$$

The number J_3 is defined by equation 45.11.

For the *rigidly fixed electrode-covered edge* $\alpha_2 = \alpha_{20}$

$$u_1 = 0, \qquad w = 0$$

$$\begin{aligned} &\gamma_2 + \frac{3\mu_{12}}{2h^2}\left(s_{11}^E G_1 + S_{13}^E G_2\right) = 0 \quad \left[\eta^{1-s}\right] \\ &u_2 + \mu_{13}\frac{h}{A_2}\frac{\partial}{\partial \alpha_2}\Big[\left(s_{12}^E n_{12} + s_{13}^E n_{22}\right) u_2 \\ &\qquad - \left(d_{31} - s_{12}^E c_1 - s_{13}^E c_2\right)\psi^{(0)}\Big] = 0 \quad \left[\eta^{1-s}\right] \\ &\psi^{(0)} + \frac{\mu_{14}}{d_{15}}\frac{h}{A_2}\frac{\partial}{\partial \alpha_2}\Big[\left(s_{12}^E n_{12} + s_{13}^E n_{22}\right) u_2 \\ &\qquad - \left(d_{31} - s_{12}^E c_1 - s_{13}^E c_2\right)\psi^{(0)}\Big] = V \quad \left[\eta^{1-s}\right]. \end{aligned} \tag{46.11}$$

For defining μ_{13} and μ_{14}, we should solve three auxiliary plane problems

X.	$v_{2*}^b + 1 = 0$,	$v_{3*}^b = 0$,	$\psi_*^b = 0$
XI.	$v_{2*}^b = 0$,	$v_{3*}^b + \zeta = 0$,	$\psi_*^b = 0$
XII.	$v_{2*}^b = 0$,	$v_{3*}^b = 0$,	$\psi_*^b + 1 = 0.$

In problems X, XI, and XII, we calculate the horizontal resultants Γ_{10}, Γ_{11}, and Γ_{12} of the forces at the edge $\xi_2 = 0$ and the resultants D_{10}, D_{11}, and D_{12} of the electric induction vector component normal to the edge surface:

$$\mu_{13} = \frac{\Delta_1}{\Delta}, \qquad \mu_{14} = \frac{\Delta_2}{\Delta}, \qquad \mu_{12} = \frac{N_{13}}{N_{14}}$$

$$\Delta_1 = \begin{vmatrix} \Gamma_{12} & \Gamma_{11} \\ D_{12} & D_{11} \end{vmatrix}, \qquad \Delta_2 = \begin{vmatrix} \Gamma_{11} & \Gamma_{10} \\ D_{11} & D_{10} \end{vmatrix}, \qquad \Delta = \begin{vmatrix} \Gamma_{10} & \Gamma_{12} \\ D_{10} & D_{12} \end{vmatrix}.$$

N_{13} and N_{14} are the vertical resultants at the edge, which have been found from auxiliary problems XIII and XIV with the edge conditions

$$\begin{array}{llll} \text{XIII.} & v^b_{2*} + \zeta = 0, & v^b_{3*} = 0, & \psi^b_* = 0 \\ \text{XIV.} & v^b_{2*} = 0, & v^b_{3*} + \zeta^2 = 0, & \psi^b_* = 0. \end{array}$$

For the *hinge-supported electrode-covered edge* $\alpha_2 = \alpha_{20}$

$$u_1 = 0, \qquad w = 0, \qquad T_2 = 0, \qquad G_2 = 0$$

$$\psi^{(0)} + \frac{\mu_{15}}{d_{15}}\left(\frac{1}{2}s^E_{12}T_1 - \frac{h}{A_2}\frac{\partial \psi^{(0)}}{\partial \alpha_2}\right) = V \quad \left[\eta^{1-s}\right], \qquad \mu_{15} = \frac{N_{15}}{N_{16}}. \tag{46.12}$$

N_{15} and N_{16} are the vertical resultants of problems XV and XVI with the edge conditions ($\xi_2 = 0$)

$$\begin{array}{llll} \text{XV.} & \tau^b_{22*} = 0, & v^a_{3*} = 0, & \psi^b_* + 1 = 0 \\ \text{XVI.} & \tau^b_{22*} = 0, & v^a_{3*} + \zeta = 0, & \psi^b_* = 0. \end{array}$$

For the *free electrode-covered edge* $\alpha_2 = \alpha_{20}$

$$T_2 = 0, \qquad S_{12} = 0, \qquad G_2 + 3J_3\frac{h}{A_1}\frac{\partial H_{12}}{\partial \alpha_2} = 0 \quad \left[\eta^{1-s}\right]$$

$$N_2 - \frac{1}{A_1}\frac{\partial H_{12}}{\partial \alpha_1} = 0, \qquad \psi^{(0)} = V. \tag{46.13}$$

The number J_3 is found by equation 45.11 from the solution of the antiplane problem whose edge condition has the form

$$\tau^a_{12*} + \zeta = 0.$$

Chapter 9

SOME PROBLEMS OF BOUNDARY LAYERS

47 ANTIPLANE BOUNDARY LAYER AT A FREE EDGE OF A SHELL WITH THICKNESS POLARIZATION

We write the dimensionless equations of the antiplane boundary layer given in Section 38 (for the edge $\xi_1 = 0$).

$$\frac{\partial \tau^a_{21*}}{\partial \xi} + \frac{\partial \tau^a_{23*}}{\partial \zeta} = 0, \quad \frac{\partial V^a_{2*}}{\partial \xi} - \sigma_1 \tau^a_{12*} = 0,$$
$$\frac{\partial V^a_{2*}}{\partial \xi} - \sigma_1 \tau^a_{21*} = 0 \quad \frac{\partial V^a_{2*}}{\partial \zeta} - \sigma_2 \tau^a_{23*} = 0. \tag{47.1}$$

Here, we use the notation $A_{10}\xi_1 = \xi$, and σ_1 and σ_2 are defined by formulas 34.7. At the edge $\xi = 0$, we are given the condition

$$\tau^a_{12} + \zeta = 0. \tag{47.2}$$

There are no surface loads at the faces

$$\tau^a_{23*}\Big|_{\zeta=\pm 1} = 0. \tag{47.3}$$

This problem coincides with the antiplane problem of elasticity theory within constant factors. We write its solution, which meets the conditions 47.2 and 47.3 and has been obtained by separating the variables:

$$V^a_{2*} = \Sigma_{n=1}^{\infty} A_n e^{(2n-1)\frac{\pi}{2}k\xi} \sin(2n-1)\frac{\pi}{2}\zeta$$
$$\tau^a_{12*} = \frac{k}{\sigma_1}\frac{\pi}{2}\sum_{n=1}^{\infty}(2n-1)A_n e^{(2n-1)\frac{\pi}{2}k\xi} \sin(2n-1)\frac{\pi}{2}\zeta$$
$$k = \sqrt{s^E_{66}/s^E_{44}}.$$

We take into account that

$$\zeta = \frac{8}{\pi^2}\sum_{n=1}^{\infty}\frac{(-1)^{n+1}}{(2n-1)^2}\sin(2n-1)\frac{\pi}{2}\zeta$$

and use the condition 47.2 to find A_n:

$$A_n = -\frac{\sigma_1}{k}\frac{16}{\pi^3}\frac{(-1)^{n+1}}{(2n-1)^3}.$$

The correction due to the boundary layer, which enters the boundary conditions 43.9, will be

$$J = \int_{-1}^{+1} \zeta d\zeta \int_{-\infty}^{0} \tau_{12*}^{a} d\xi = -0.42\frac{1}{k} = -0.42\sqrt{\frac{s_{44}^{E}}{s_{66}^{E}}}.$$

48 ANTIPLANE BOUNDARY LAYER AT A FREE EDGE OF A SHELL WITH TANGENTIAL POLARIZATION

We will assume that the faces and the edge do not have electrodes. In this case, the simultaneous equations of the antiplane electroelastic problem for the edge $\xi_1 = 0$ have the form (see Section 39)

$$\begin{aligned}
\frac{\partial \tau_{21*}^{a}}{\partial \xi} + \frac{\partial \tau_{23*}^{a}}{\partial \zeta} &= 0 \\
\frac{\partial D_{3*}}{\partial \zeta} + \frac{\partial D_{1*}^{a}}{\partial \xi} &= 0 \\
\frac{\partial V_{2*}^{a}}{\partial \xi} + \frac{\partial \psi_{*}^{a}}{\partial \xi} - \sigma_2 \tau_{21*}^{a} &= 0 \\
\frac{\partial V_{2*}^{a}}{\partial \xi} + \frac{\partial \psi_{*}^{a}}{\partial \xi} - \sigma_2 \tau_{12*}^{a} &= 0 \\
\frac{\partial V_{2*}^{a}}{\partial \zeta} + \frac{\partial \psi_{*}^{a}}{\partial \zeta} - \sigma_1 \tau_{23*}^{a} &= 0 \\
D_{1*}^{a} + \frac{\varepsilon_{11}^{T}}{\varepsilon_{33}^{T}} \frac{\partial \psi_{*}^{a}}{\partial \xi} - t_4 \tau_{21*}^{a} &= 0 \\
D_{3*}^{a} + \frac{\varepsilon_{11}^{T}}{\varepsilon_{33}^{T}} \frac{\partial \psi_{*}^{a}}{\partial \zeta} - t_4 \tau_{23*}^{a} &= 0, \qquad A_1 \xi = \xi.
\end{aligned} \tag{48.1}$$

The homogeneous conditions

$$D_{3*}^{a}\Big|_{\zeta=\pm 1} = 0, \qquad \tau_{23*}^{a}\Big|_{\zeta=\pm 1} = 0 \tag{48.2}$$

should hold on the faces. The load

$$\tau_{12*}^{a} + \zeta = 0, \qquad D_{1*}^{a} + t_4 \zeta = 0 \tag{48.3}$$

is given at the edge $\alpha_1 = \alpha_{10}$.

After some transformations, the system of equations 48.1 can be reduced to a system of two equations with two variables:

$$\frac{\partial^2 \psi_*^a}{\partial \xi^2} + \frac{\partial^2 \psi_*^a}{\partial \zeta^2} = 0$$

$$\frac{\partial V_{2*}^a}{\partial \xi^2} + \frac{1}{k^2}\frac{\partial^2 V_{2*}^a}{\partial \zeta^2} + \frac{\partial^2 \psi_*^a}{\partial \xi^2} + \frac{1}{k^2}\frac{\partial^2 \psi_*^a}{\partial \zeta^2} = 0. \tag{48.4}$$

By separating the variables, we get the solution to equations 48.4 in the form

$$\psi_*^a = \sum_{n=1}^{\infty} A_n e^{(2n-1)\frac{\pi}{2}\xi} \sin(2n-1)\frac{\pi}{2}\zeta$$

$$V_{2*}^a = \sum_{n=1}^{\infty} \left[-A_n e^{(2n-1)\frac{\pi}{2}\xi} + B_n e^{(2n-1)\frac{\pi}{2}\frac{1}{k}\xi} \right] \sin(2n-1)\frac{\pi}{2}\zeta.$$

The constants A_n and B_n are obtained by meeting the conditions 48.3:

$$B_n = -\frac{16}{\pi^3}\frac{\sigma_2}{k}\frac{(-1)^{n+1}}{(2n-1)^3}, \qquad A_n = 0.$$

Whence

$$\tau_{12*}^a = -\frac{8}{\pi^2}\sum_{n=1}^{\infty}\frac{(-1)^{n+1}}{(2n-1)^2} e^{(2n-1)\frac{\pi}{2}\frac{1}{k}} \sin(2n-1)\frac{\pi}{2}\zeta$$

$$J_2 = -0.42k. \tag{48.5}$$

The remaining problems are solved in a similar way. By the definition of $J_1, \ldots, J_5$, we find that

$$J = J_1 = \ldots = J_5 = -0.42k = -0.42\sqrt{\frac{s_{66}^E}{s_{44}^E}}.$$

49 ANTIPLANE BOUNDARY LAYER AT A RIGIDLY FIXED EDGE OF A SHELL WITH TANGENTIAL POLARIZATION

We will assume that the shell's faces are covered with electrodes. We saw in Section 39 that the antiplane boundary layer at a rigidly fixed edge $\alpha_1 = \alpha_{10}$ ($\xi_1 = 0$) (with preliminary polarization being directed along the edge) has a much greater intensity than the inner electroelastic state and is decisive in defining the greatest stresses.

Equations 39.6 of the antiplane boundary layer is given in Section 39 and the boundary conditions have the form

$$\psi_*^a\Big|_{\zeta=\pm 1} = 0, \qquad \tau_{23*}^a\Big|_{\zeta=\pm 1} = 0 \tag{49.1}$$

$$V_{2*}^a + \zeta\Big|_{\xi=0} = 0, \qquad D_{1*}^a\Big|_{\xi=0} = 0. \tag{49.2}$$

In order to solve the problem, we use the Laplace integral transform

$$U = \int_0^\infty V_{2*}^a(\xi,\zeta)e^{-p\xi}d\xi, \qquad \Phi = \int_0^\infty \psi_*^a(\xi,\zeta)e^{-p\xi}d\xi. \tag{49.3}$$

Allowing for 49.3, we write equations 48.4 as

$$\begin{aligned}\frac{d^2U}{d\zeta^2} + k^2p^2U &= k^2\frac{\partial^2 V_{2*}^a}{\partial\xi^2}\Big|_{\xi=0} + k^2pV_{2*}^a\Big|_{\xi=0}\\ &\quad + (1-k^2)\left[p^2\Phi - \frac{\partial^2\psi_*^a}{\partial\xi}\Big|_{\xi=0} - p\psi_*^a\Big|_{\xi=0}\right]\\ \frac{d^2\Phi}{d\zeta^2} + p^2\Phi &= \frac{\partial\psi_*^a}{\partial\xi}\Big|_{\xi=0} + p\psi_*^a\Big|_{\xi=0}.\end{aligned} \tag{49.4}$$

The conditions 49.1 for the quantities we now need assume the form

$$\Phi|_{\zeta=\pm 1} = 0, \qquad \left[\frac{dU}{d\zeta} + \frac{d\Phi}{d\zeta}\right]\Big|_{\zeta=\pm 1} = 0. \tag{49.5}$$

The conditions 49.2 at the edge can be conveniently written as

$$V_{2*}^a\Big|_{\xi=0} = -\zeta, \left[\frac{\varepsilon_{11}^T}{\varepsilon_{33}^T} - \frac{t_4}{\sigma_2}\right]\frac{\partial\psi_*^a}{\partial\xi}\Big|_{\xi=0} - \frac{t_4}{\sigma_2}\frac{\partial V_{2*}^a}{\partial\xi}\Big|_{\xi=0} = 0. \tag{49.6}$$

We integrate the ordinary differential equations in equation 49.4 to obtain

$$\begin{aligned}\Phi_1 &= C_1\cos p\zeta + C_2\sin p\zeta + a(p,\zeta)\cos p\zeta + b(p,\zeta)\sin p\zeta\\ U &= C_3\cos kp\zeta + C_4\sin kp\zeta + \frac{1}{p}\zeta - (a(p,\zeta)\cos p\zeta + b(p,\zeta)\sin p\zeta)\\ &\quad - (C_1\cos p\zeta + C_2\sin p\zeta) + k(c(p,\zeta)\cos kp\zeta + d(p,\zeta)\sin kp\zeta)\\ &\quad + \frac{k}{p}\left(\frac{\varepsilon_{11}^T}{\varepsilon_{33}^T}\frac{\sigma_2}{t_4} - 1\right)[l(p,\zeta)\cos kp\zeta + m(p,\zeta)\sin kp\zeta].\end{aligned}$$

Here, we have used the notation

$$a(p,\zeta) = -\int_0^\zeta \chi\Big|_{\xi=0}\sin p\zeta d\zeta$$

$$b(p,\zeta) = \int_{-1}^{\zeta} \chi\Big|_{\xi=0} \cos p\zeta d\zeta$$

$$c(p,\zeta) = -\int_{0}^{\zeta} \chi\Big|_{\xi=0} \sin kp\zeta d\zeta$$

$$d(p,\zeta) = \int_{-1}^{\zeta} \chi\Big|_{\xi=0} \cos kp\zeta d\zeta$$

$$l(p,\zeta) = -\int_{0}^{\zeta} \frac{\partial\psi_*^a}{\partial\xi}\Big|_{\xi=0} \sin kp\zeta d\zeta$$

$$m(p,\zeta) = \int_{-1}^{\zeta} \frac{\partial\psi_*^a}{\partial\xi}\Big|_{\xi=0} \cos kp\zeta d\zeta$$

$$\chi(p\zeta) = \frac{1}{p}\frac{\partial\psi_*^a}{\partial\xi}|_{\xi=0} + \psi_*^a\Big|_{\xi=0}.$$

By satisfying the conditions 49.6, we find the constants $C_1,\ldots,C_4$:

$$C_1 = C_3 = 0$$

$$C_2 = -\frac{a(p,1)\cos p}{\sin p}$$

$$C_4 = \frac{1}{\cos kp}\left[kc(p,1)\sin kp + \frac{k}{p}\left(\frac{\varepsilon_{11}^T\sigma_2}{\varepsilon_{33}^T t_4} - 1\right) l(p,1)\sin kp - \frac{1}{kp^2}\right].$$

We turn to the initial quantities using the formulas

$$V_{2*}^a = \frac{1}{2\pi i}\int_{\alpha-i\infty}^{\alpha+i\infty} Ue^{p\xi}dp = \sum_{j=1}^{\infty} \text{res } U(p_j,\zeta)e^{p_j\xi}$$

$$\psi_*^a = \frac{1}{2\pi i}\int_{\alpha-i\infty}^{\alpha+i\infty} \Phi e^{p\xi}dp = \sum_{j=1}^{\infty} \text{res } (p_{j'},\zeta)e^{p_j\xi}.$$

The function $U(p_j,\zeta)$ has residues for $p_m = \pm\frac{(2m-1)\pi}{2k}$ and $p_n = \pm n\pi$. The function $\Phi(p_j,\zeta)$ has residues for $p_n = \pm n\pi$.

We omit simple algebra and write the finite formulas

$$\psi_*^a = -\frac{2}{\pi}\sum_{n=1}^{\infty}\frac{B_n}{n}\sin p_n\zeta \cdot e^{-p_n\xi}$$

$$\tau_{12*}^a = \frac{2\varepsilon_{11}^T s_{11}^E}{d_{15}^2}\sum_{m=1}^{\infty} A_m \sin q_m\zeta \cdot e^{-q_m\xi/k}$$

$$\tau_{23*}^a = -\frac{2}{k}\frac{\varepsilon_{11}^T s_{11}^E}{d_{15}^2}\sum_{m=1}^{\infty} A_m \cos q_m\zeta \cdot e^{-q_m\xi/k}$$

$$V_{2*}^{a} = -\frac{2k\varepsilon_{11}^{T}s_{44}^{E}}{d_{15}^{2}}\sum_{m=1}^{\infty}\frac{A_m}{q_m}\sin q_m\zeta\cdot e^{-q_m\xi/k} + \frac{2\varepsilon_{11}^{T}s_{44}^{E}}{d_{15}^{2}}\sum_{n=1}^{\infty}\frac{B_n}{p_n}\sin p_n\zeta\cdot e^{-p_n\xi}$$

$$p_n = n\pi, \qquad q_m = (m-1/2)\pi.$$

The constants B_n are found from the system

$$2\sum_{m=1}^{\infty}\left[(m-1)\frac{d_{15}^{2}}{\varepsilon_{11}^{T}s_{44}^{E}} - kn\right]\frac{(-1)^n B_n}{\left(n-m+\frac{1}{2}\right)\left(n+m-\frac{1}{2}\right)} = -\frac{d_{15}^{2}}{\varepsilon_{11}^{T}s_{44}^{E}}\frac{1}{\left(m-\frac{1}{2}\right)}.$$

Knowing B_n,we can find the constants A_m from the formulas

$$A_m = \frac{2}{\pi}\sum_{n=1}^{N}B_n\frac{(-1)^{n+m}}{\left(n-m+\frac{1}{2}\right)\left(n+m-\frac{1}{2}\right)}, \qquad m = 1, 2, \ldots, N.$$

The larger N, the more accurate is the computation of the electroelastic state. Since the series are sign-alternating, we can, in practice, restrict ourselves to a few terms.

Figure 31 presents the variations of the dimensionless quantities ψ_{*}^{a}, τ_{12*}^{a} and τ_{23*}^{a} versus the thickness coordinate ζ at the edge $\xi = 0$. Figures 32 and 33 give τ_{12*}^{a} and V_{2*}^{a} at the face $\zeta = 1$ as functions of ξ.

50 PLANE BOUNDARY LAYER AT A RIGIDLY FIXED EDGE OF A SHELL WITH TANGENTIAL POLARIZATION

In Section 45, we formulated the problem to be solved for obtaining corrections in the shell's boundary conditions at a rigidly fixed edge $\alpha_1 = \alpha_{10}$. We assume that the shell is polarized along the α_2-lines and its faces are covered with electrodes.

We write the equations and boundary conditions for the plane boundary layer:

$$\frac{\partial\tau_{11*}}{\partial\xi} + \frac{\partial\tau_{13*}}{d\zeta} = X_{1*}^{a}, \qquad \frac{\partial\tau_{13*}}{\partial\xi} + \frac{\partial\tau_{33*}}{\partial\zeta} = X_{3*}^{a} \tag{50.1}$$

$$\begin{aligned}
\tau_{11*} &= \frac{\partial V_{1*}}{\partial\xi} + \nu\frac{\partial V_{3*}}{\partial\xi} - P_{1*}^{a}\\
\tau_{33*} &= \frac{\partial V_{3*}}{\partial\zeta} + \nu\frac{\partial V_{1*}}{\partial\zeta} - \nu P_{1*}^{a}\\
\tau_{13*} &= \sigma\left(\frac{\partial V_{1*}}{\partial\zeta} + \frac{\partial V_{3*}}{\partial\xi}\right)
\end{aligned} \tag{50.2}$$

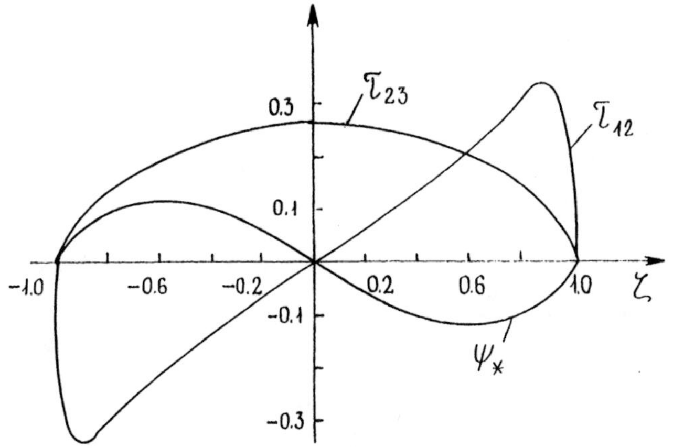

FIGURE 31. Stresses and electrical potential of the boundary layer at the shell edge as functions of the thickness coordinate

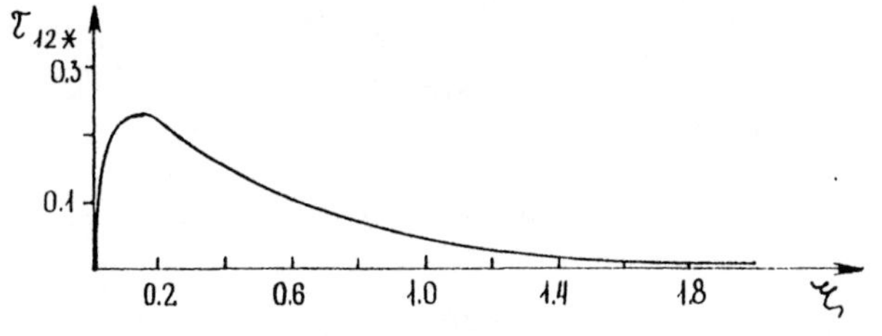

FIGURE 32. The stress τ_{12*} on a shell face as a function of the ξ- coordinate

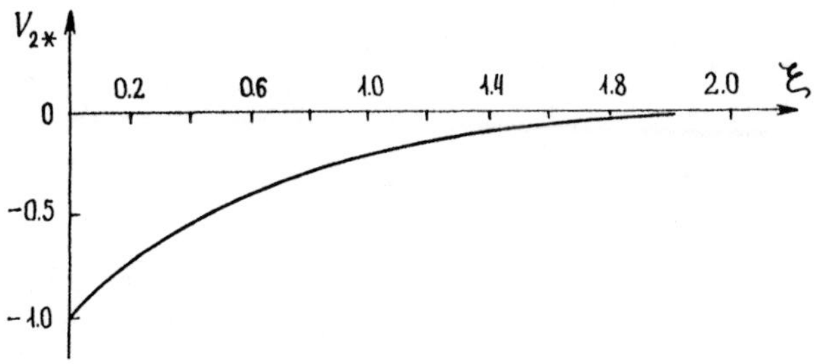

FIGURE 33. The displacement V_{2*} on a shell face as a function of the ξ- coordinate

$$\tau_{33*}\Big|_{\zeta=\pm1} = 0, \qquad \tau_{13*}\Big|_{\zeta=\pm1} = 0 \tag{50.3}$$

$$V_{1*}\Big|_{\xi=0} = -\zeta\rho_{1*}, \qquad V_{3*}\Big|_{\xi=0} = -\rho_{2*} - \zeta^2. \tag{50.4}$$

Here,

$$P^a_{1*} = V^a_{2*}, \qquad X^a_{1*} = \frac{2}{s^E_{11}E_u}\tau^a_{12*}$$

$$X^a_{3*} = \frac{1}{s^E_{11}E_u}\tau^a_{23*}, \qquad E_u = \frac{a}{a^2-b^2},$$

$$\nu = -\frac{b}{a} \qquad \sigma = \frac{1}{E_u s^E_{66}},$$

$$a = s^E_{11} - \frac{(s^E_{13})^2}{s^E_{33}}, \qquad b = s^E_{12} - \frac{(s^E_{13})}{s^E_{33}}$$

$$k_1 = \nu + \sigma, \qquad 1 - \sigma = k.$$

The numbers ρ_{1*} and ρ_{2*} are connected with ρ_1 and ρ_2, in Section 45, by the formulas

$$\rho_1 = k_{1,0}d_{15}E_3^{(0)}\rho_{1*}, \qquad \rho_2 = k_{1,0}d_{15}E_3^{(0)}\rho_{2*} \tag{50.5}$$

where $k_{1,0}$ is the geodesic curvature of the α_1-lines at the edge.

The quantities V^a_{2*}, τ^a_{12*}, and τ^a_{23*} have been found in Section 49 by solving the antiplane problem. We will seek the solution to equations 50.1 to 50.4 using the Laplace integral transform with respect to the variable

$$U(p,\xi) = \int_0^\infty V_{1*}e^{-p\xi}d\xi, \qquad W(p,\xi) = \int_0^\infty V_{3*}e^{-p\xi}d\xi. \tag{50.6}$$

By writing the equilibrium equations in terms of the displacements and allowing for equations 50.6, we get a system of ordinary differential equations

$$\begin{aligned}\frac{d^2U}{d\zeta^2} + \frac{p^2}{\sigma}U + \frac{kp}{\sigma}\frac{dW}{d\zeta} &= \Phi(p,\zeta)\\ \frac{d^2W}{d\zeta^2} + \sigma p^2 W + kp\frac{dU}{d\zeta} &= \Psi(p,\zeta)\end{aligned} \tag{50.7}$$

$$\begin{aligned}\Phi(p,\zeta) &= \frac{1}{\sigma}\frac{\partial V_{3*}}{\partial\xi}\Big|_{\xi=0} + \frac{k}{\sigma}\frac{\partial V_{3*}}{\partial\zeta}\Big|_{\xi=0} + p^a_{1*}\Big|_{\xi=0} + Y_1\\ \Psi(p,\zeta) &= k\frac{\partial V_{3*}}{\partial\zeta}\Big|_{\xi=0} + \sigma\frac{\partial V_{3*}}{\partial\xi}\Big|_{\xi=0} + \sigma p^a_{3*}\Big|_{\xi=0} + Y_3\end{aligned} \tag{50.8}$$

where

$$Y_1 = \int_0^\infty X^a_{1*}e^{-p\xi}d\xi, \qquad Y_3 = \int_0^\infty X^a_{3*}e^{-p\xi}d\xi.$$

After the transformations 50.6, the boundary conditions 50.3 assume the form

$$\left(\frac{dW}{d\zeta} + \sigma p U - V_{1*}\Big|_{\xi=0}\right)\Big|_{\zeta=\pm 1} = Q^{\pm} \qquad \left(\frac{dU}{d\zeta} + pW - V_{3*}\Big|_{\xi=0}\right)\Big|_{\zeta=\pm 1} = 0 \tag{50.9}$$

where

$$Q^{\pm} = \int_0^{\infty} \nu P^a_{1*} e^{-p\xi}\Big|_{\zeta=\pm 1} d\xi.$$

It follows from Section 49 that X^a_{1*} and V^a_{2*} are odd functions of the variable ζ, and X^a_{3*} is an even function of ζ. Therefore, the needed function U will be odd with respect to ζ, and W will be an even function. The general solution to equations 50.7 contains two arbitrary functions dependent on p and has the form

$$\begin{aligned}
U(p\zeta) &= a_1(p)\sin p\zeta + a_2(p)p\zeta\cos p\zeta + b_1(p,\zeta)\cos p\zeta + b_2(p,\zeta)\sin p\zeta \\
W(p,\zeta) &= \left(a_1(p) - \frac{1+\sigma}{k}a_2(p)\right)\cos p\zeta - a_2(p)p\zeta\sin p\zeta \\
&\quad - (b_1(p,\zeta) + k_1 b_2(p,\zeta))\sin p\zeta + b_2(p,\zeta)p\zeta\cos p\zeta
\end{aligned} \tag{50.10}$$

$$\begin{aligned}
b_1(p,\zeta) &= \frac{k}{2p\sigma}\int_0^{\zeta}\Bigg[-\Psi(p,\zeta)p\zeta\cos p\zeta \\
&\qquad -\sigma\Phi(p,\zeta)\left(\frac{1+\sigma}{k}\sin p\zeta + p\zeta\cos p\zeta\right)\Bigg]d\zeta \\
b_2(p,\zeta) &= \frac{k}{2p\sigma}\int_0^{\zeta}[-\Psi(p,\zeta)\cos p\zeta + \sigma\Phi(p,\zeta)\sin p\zeta]\,d\zeta.
\end{aligned} \tag{50.11}$$

We find $a_1(p)$ and $a_2(p)$ from the boundary conditions 50.9:

$$\begin{aligned}
a_1(p) = -\frac{1}{\psi(p)}\Bigg[& b_1(p,\zeta)\left(\cos^2 p - \frac{1}{k}\right) - pb_2(p,1)\left(p + \frac{\sigma}{k^2 p}\right) \\
& - \frac{k-\sigma}{2\sigma}V_{1*}(0,1)\left(\sin p + \frac{\sigma\cos p}{kp}\right) \\
& + \frac{1}{2}V_{3*}(0,1)\left(\cos p - \frac{\sin p}{kp}\right) - \frac{1}{2\alpha}Q^{+}\left(\frac{\sigma}{k}\frac{\cos p}{p} + \sin p\right)\Bigg]
\end{aligned}$$

$$a_2(p) = -\frac{1}{\psi(p)}\Bigg[-b_1(p,1) + pb_2(p,1)\left(-\frac{k}{\sigma p} + \frac{\cos^2 p}{p}\right)$$

$$- \frac{k-\sigma}{2\sigma} V_{1*}(0,1)\frac{\cos p}{p} - \frac{1}{2} V_{3*}(0,1)\frac{\sin p}{p} - \frac{1}{2\sigma}\frac{\cos p}{p} Q^{+} \Bigg] \tag{50.12}$$

$$\psi(p) = \sin p \cos p - p. \tag{50.13}$$

We take an inverse Laplace transform and compute the integral with the help of the theorem on residues. The equation $\psi(p) = 0$ has a three-tuple root $\psi = 0$ and infinitely many four-tuple roots $p_n, \bar{p}_n, -\bar{p}_n, -p_n$. Since the problem implies a decreasing solution, the residues with respect to the roots p_n and $\bar{p}_n$ with positive real parts should be equal to zero. The residues $U(p, \zeta)$ and $W(p, \zeta)$, with respect to the pole p_n, have the form

$$\begin{aligned}
\mathrm{res}_{p_n} U(p,\zeta)e^{p\zeta} &= F(p_n)\frac{k}{2\sigma p_n \psi'(p_n)} \\
&\quad \left[\left(\cos^2 p_n - \frac{1}{k}\right)\sin p\zeta_n - p_n\zeta \cdot \cos p_n\zeta\right] e^{p_n\zeta} \\
\mathrm{res}_{p_n} W(p,\zeta)e^{p\zeta} &= F(p_n)\frac{k}{2\sigma p_n \psi'(p_n)} \\
&\quad \left[\left(\cos^2 p_n + \frac{\sigma}{k}\right)\cos p_n\zeta + p_n\zeta \cdot \sin p_n\zeta\right] e^{p_n\zeta}
\end{aligned}$$

$$\begin{aligned}
F(p_n) = \int_0^1 \Bigg\{ &\Psi(p_n,\zeta)\left[p_n\zeta \cdot \sin p_n\zeta + \cos p_n\zeta \cdot \left(\cos^2 p_n + \frac{\sigma}{k}\right)\right] \\
&+ \sigma\Phi(p_n,\zeta)\left[p_n\zeta \cdot \cos p_n\zeta - \sin p_n\zeta\left(\cos^2 p_n - \frac{1}{k}\right)\right] \\
&- \frac{1}{1-\sigma} Q^{+}\cos p_n - \frac{\sigma}{k} V_{3*}(0,1)\cdot \sin p_n \Bigg\} d\zeta.
\end{aligned} \tag{50.14}$$

For the solution to be nonincreasing as $\xi \longrightarrow \infty$, we should take $F(p_n) = 0$ at the poles with positive real parts.

The residues with respect to the pole $p = 0$ give exponentially increasing terms. Since $U(p, \zeta)$ has a third-order pole at the origin of coordinates and $W(p, \zeta)$ has a fourth-order pole, we will have the following conditions when searching for the residues $U(p, \zeta)e^{p\xi}$ and $W(p, \zeta)e^{p\xi}$ with respect to $p = 0$ and equating them to zero:

$$\int_0^1 \tau_{13*}(0,\zeta)d\zeta = \alpha,$$

$$\int_0^1 \zeta\tau_{11*}(0,\zeta)d\zeta = \beta$$

$$\int_0^1 \zeta^2 \tau_{13*}(0,\zeta)d\zeta + \frac{8\sigma k}{k-\sigma}\int_0^1 \zeta V_{1*}(0,\zeta)d\zeta = \gamma$$

$$\int_0^1 \zeta^2 \tau_{11*}(0,\zeta)d\zeta + \frac{12\sigma k}{2+k-\sigma}\int_0^1 (1-\zeta^2)V_{3*}(0,\zeta)d\zeta = \delta \quad (50.15)$$

where

$$\alpha = \int_0^\infty \left[(k-\sigma)V_{2*}^a(\xi,1) - \int_0^1 \left(X_{3*}^a - (k-\sigma)\frac{\partial V_{2*}^a}{\partial \zeta}\right) d\zeta\right] d\xi$$

$$\beta = \int_0^\infty \left\{ -(k-\sigma)\xi V_{2*}^a(\xi,1) + \int_0^1 \left[\xi\left(X_{3*}^a + (k-\sigma)\frac{\partial V_{2*}^a}{\partial \zeta}\right)\right.\right.$$
$$\left.\left. - \zeta\left(X_{1*}^a + \frac{\partial V_{2*}^a}{\partial \xi}\right)\right] d\zeta\right\} d\xi - \int_0^1 \zeta V_{2*}^a(0,\zeta)d\zeta$$

$$\gamma = \frac{1}{k-\sigma}\int_0^\infty \left\{ \left[(k-\sigma)+\xi^2\right](k-\sigma)V_{2*}^a(\xi,1)\right.$$
$$+ \int_0^1 \left[2\xi\zeta\left(X_{1*}^a + \frac{\partial V_{2*}^a}{\partial \xi}\right)\right.$$
$$\left.\left. -(\xi^2 + (k-\sigma)\zeta^2)\left(X_{3*}^a + (k-\sigma)\frac{\partial V_{2*}^a}{\partial \zeta}\right)\right] d\zeta\right\} d\xi$$

$$\delta = \frac{3}{1+2k}\int_0^\infty \left\{ \left[-(1+2k) + \frac{1}{3}\xi^2\right]\xi(k-\sigma)V_{2*}^a(\xi,1)\right.$$
$$+ \int_0^1 \left[\zeta\left(\xi^2 - \frac{1+2k}{3}\zeta^2\right)\left(X_{1*}^a + \frac{\partial V_{2*}^a}{\partial \xi}\right)\right.$$
$$+ \left(4k - \frac{1}{3}\xi^2 - (k-\sigma)\zeta^2\right)\xi$$
$$\left.\left. \times \left(X_{3*}^a + (k-\sigma)\frac{\partial V_{2*}^a}{\partial \zeta}\right)\right] d\zeta\right\} d\xi$$
$$- \int_0^1 \zeta^3 V_{2*}^a(0,\zeta)d\zeta. \quad (50.16)$$

After some transformations, the conditions in equation 50.14 will be written as

$$\int_0^1 \left[\tau_{11*}(0,\zeta)h(p_n,\zeta) + \tau_{13*}(0,\zeta)g(p_n,\zeta) + 2\sigma V_{1*}(0,\zeta)s(p_n,\zeta)\right.$$

$$+2\sigma V_{3*}(0,\zeta)t(p_n,\zeta)\Big]d\zeta - \frac{k-\sigma}{k}\cos p_n \int_0^\infty V^a_{2*}(\xi,1)e^{-p_n\xi}d\xi$$

$$+\int_0^1 \left[h(p_n,\zeta)\int_0^\infty \left(X^a_{1*} + \frac{\partial V^a_{2*}(\xi,\zeta)}{\partial \xi}e^{-p_n\xi}d\xi\right]d\zeta\right.$$

$$+\int_0^1 h(p_n,\zeta)V^a_{2*}(0,\zeta)d\zeta = 0. \tag{50.17}$$

Here,

$$\begin{aligned}
h(p_n) &= p_n\zeta\cos p_n\zeta + \sin p_n\zeta \cdot \left(\sin^2 p_n + \frac{\sigma}{k}\right)\\
g(p_n,\zeta) &= p_n\zeta\sin p_n\zeta - \cos p_n\zeta \cdot \left(\cos^2 p_n + \frac{\sigma}{k}\right)\\
s(p_n,\zeta) &= p_n\Big[p_n\zeta\cos p_n\zeta + \left(\sin^2 p_n + 1\right)\cdot \sin p_n\zeta\Big]\\
t(p_n,\zeta) &= p_n\Big[p_n\zeta\sin p_n\zeta - \cos p_n\zeta \cdot \sin^2 p_n\Big].
\end{aligned} \tag{50.18}$$

Equations 50.15 and 50.17 are sufficient for defining τ_{11*} and τ_{13*} at the edge and the constants ρ_1 and ρ_2 in the boundary conditions of the inner problem. As in [27–28], we will solve the problem of defining ρ_1 and ρ_2 in two stages. In the first stage, we find $\tau_{11*}(0,\zeta), \tau_{13*}(0.\zeta)$ from the first two formulas of equations 50.15 and equation 50.17. In the second stage, we compute ρ_{1*} and ρ_{2*} from the third and fourth formulas in equations 50.15.

We write equations 50.15, 50.17, and 50.18 in matrix form and add the conditions $F(\bar{p}_n) = 0$ to the system:

$$T(\zeta) = \begin{pmatrix} \tau_{11*}(0,\zeta) \\ \tau_{13*}(0,\zeta) \end{pmatrix}$$

$$\Phi_{n+1} = \begin{pmatrix} h(p_n,\zeta), g(p_n,\zeta) \\ \bar{h}(p_n,\zeta), \bar{g}(p_n,\zeta) \end{pmatrix}$$

$$\Phi_1(\zeta) = \begin{pmatrix} 0 & 1 \\ \zeta & 0 \end{pmatrix}$$

$$A_{n+1} = \begin{pmatrix} a_{n+1} \\ \bar{a}_{n+1} \end{pmatrix}$$

$$A_1 = \begin{pmatrix} \alpha \\ \beta \end{pmatrix}$$

$$a_{n+1} = -2\sigma\int_0^1 \Big[V_{1*}(0,\zeta)s(p_n,\zeta) + V_{3*}(0,\zeta)t(p_n,\zeta)\Big]d\zeta$$

$$+\frac{k-\sigma}{k}\cos p_n \int_0^\infty V^a_{2*}(\xi,1)e^{-p_n\xi}d\xi$$

$$- \int_0^1 \left[h(p_n, \zeta) \int_0^\infty \left(X_{1*} + \frac{\partial V_{2*}^a(\xi, \zeta)}{\partial \xi} \right) e^{-p_n \xi} d\xi \right.$$

$$\left. + g(p_n, \zeta) \int_0^\zeta \left(X_{3*} + (k - \sigma) \frac{\partial V_{2*}^a(\xi, \zeta)}{\partial \zeta} \right) e^{-p_n \xi} d\xi \right] d\zeta$$

$$- \int_0^1 h(p_n, \zeta) V_{2*}^a(0, \zeta) d\zeta, \quad n = 1, 2, 3, \ldots$$

In matrix form, the system in equations 50.15 and 50.17 will be written as

$$J(\Phi_n, T) = A_n, \quad n = 1, 2, 3. \tag{50.19}$$

Here,

$$J(\Phi_n, T) = \int_0^1 \Phi_n T d\zeta. \tag{50.20}$$

We will define the matrix $T(\zeta)$ by the conditions in equation 50.19 just as a function is defined by its expansion in a nonorthogonal system of functions.

From the matrix system $\Phi_n(\zeta)(n = 1, 2, \ldots)$, we go to an orthoganal matrix system $\Psi_n(\zeta)$, where

$$\int_0^1 \psi_n(\zeta) . \psi_k^*(\zeta) d\zeta = 0, \quad n \neq k. \tag{50.21}$$

Here, and below, the asterisk denotes a transposed matrix. The matrices $\Psi_n(\zeta)$ will be defined as

$$\Psi_1 = \Phi_1$$

$$\Psi_L = \sum_{M=1}^{L-1} C_M^L \Psi_M + \Phi_L, \quad L = 2, 3, \ldots \tag{50.22}$$

The square matrix C_M^L in equation 50.22 is chosen so that the matrix Ψ_n will be orthogonal to all the matrices $\Psi_{n-1}, \ldots, \Psi_1$. We write the formulas for the matrix

$$C_M^L = J(\Phi_L, \Psi_M^*) \cdot J^{-1}(\Phi_M, \Psi_M^*) \quad M = 1, 2, 3, \ldots, \quad L = 1, 2, 3, \ldots \tag{50.23}$$

The matrix $T(\zeta)$ is expanded in the orthogonal matrix system $\Psi_n(\zeta)$ as

$$T(\zeta) = \sum_{n=1}^{\infty} \Psi_*^n(\zeta) \cdot \beta_n. \tag{50.24}$$

The matrices β_n are obtained by premultiplying both sides of equation 50.24 by $\Psi_n(\zeta)$ and then integrating them with respect to ζ from 0 to 1 taking into account equation 50.21:

$$\beta_n = J^{-1}(\Psi_n, \Psi_n^*) \cdot J(\Psi_n, T). \tag{50.25}$$

In turn, $J(\Psi_n, T)$ can be found using equations 50.20 and 50.22:

$$J(\Psi_1, T) = A_1 = B_1$$
$$J(\Psi_L, T) = \sum_{M-1}^{L-1} C_M^{L-1} B_M + A_L, \quad L = 2, 3, 4, \ldots . \tag{50.26}$$

We substitute equations 50.26 into equation 50.25, then we find the required expansion for the matrix $T(\zeta)$ in equation 50.24. Knowing the expansion for $T(\zeta)$, we apply the still unused conditions 50.15 for calculating ρ_{1*} and ρ_{2*}.

BIBLIOGRAPHY

[1] Adelman, N. T. and Stavsky, J., Flexural-extensional behavior of composite piezoelectric circular plates, *J. Acoust. Soc. Am.*, 67, (3), 819, 1980.

[2] Adelman, N. T. and Stavsky, J., Vibrations of radially polarized composite piezoelectric cylinders and disks, *J. Sound and Vibration*, 43, (1), 37, 1975.

[3] Adelman, N. T., Stavsky, J., and Segal, E., Axisymmetric vibrations of radially polarized piezoelectric ceramic cylinder, *J. Sound and Vibration*, 38, (2), 245, 1975.

[4] Alike, H., Webman, K., and Hunt, J. T., Vibration response of sonar transducer using piezoelectric finite elements, *J. Acoust. Soc. Am.*, 56, No 6, 1782, 1984.

[5] Armstrong, G. A. and Crampin, S., Piezoelectric surface-wave calculations in multilayered anisotropic media, *Electron. Lett.*, 8, (21), 521, 1972.

[6] Berlincourt, D. A., Current developments in piezoelectric applications of ferroelectrics, *Ferroelectrics*, No 10, 111, 1976.

[7] Berlincourt, D. A., Piezoelectric and ferroelectric energy conversion, *IEEE Trans. Sonics and Ultrasonics*, 15, (2), 89, 1968.

[8] Berlincourt, D. A., Curran, D. R., and Jaffe, H., Piezoelectric and piezomagnetic materials and their function as transducers, in *Physical Acoustics,* Mason, W. P., Ed., 1A, Academic Press, New York, 1964.

[9] Boriseiko, V. A., Martynenko, V. S., and Ulitko, A. F., On the theory of vibrations in piezoceramic shells, *Mathematical PHysics*, No 21, 71, Kiev, 1977 (in Russian).

[10] Boriseiko, V. A., Martynenko, V. S., and Ulitko, A. F., Electroelasticity relations in piezoceramic shells polarized along a meridional coordinate, *Soviet Appl. Mech.*, 15, (12), 1155, 1979.

[11] Boriseiko, V. A., and Ulitko, A. F., Axisymmetric vibrations of a thin piezoceramic spherical shell, *Soviet Appl. Mech.*, 10, (10), 1041, 1974.

[12] Burfoot, J. C., *Ferroelectrics. An Introduction to the Physical Principles*, Princeton, New Jersey, Toronto, 1967.

[13] Cady, W. G., *Piezoelectricity*, Dover, New York, 1964.

[14] Cheng, N. C. and Sun, C.T., Wave propagation in two-layered piezoelectric plates, *J. Acoust. Soc. Am.*, 57, (3), 632, 1975.

[15] Chernykh, K. F., *Linear Theory of Shells*, Vol 1,2, NASA -TT-F-11-562, 1968.

[16] Cole, J. D., *Perturbation Method in Applied Mathematics*, Blaisdell Publishing Company, Toronto-London, 1968.

[17] Dieulesant, E. and Royer, D., *On des Elastiques Dans les Solides: Application au Traitment Dusignales*, Maisson, Paris, 1974.

[18] Eringen, A. C. and Maugen, G. A., *Electrodynamics of Continia*, Vol. 1, Springer-Verlag, New York, 1990, chap. 6.

[19] Drumheller, D.S. and Kalnins, G.L., Dynamic shell theory for ferroelectric ceramics, *J. Acoust. Soc. Am.*, 47, (5), 1343, 1970.

[20] Getman, I. P. and Ustinov, Iu. A., On the theory of inhomogeneous electroelastic plate, *J. Appl. Math. and Mech.*, 43, (5), 995, 1979.

[21] Gol'denveizer, A. L., *The Theory of Elastic Thin Shells*, Pergamon Press, New York, 1961.

[22] Gol'denveizer, A. L., *The Theory of Thin Elastic Shells*, 2nd ed, Nauka, Moscow, 1976 (in Russian).

[23] Gol'denveizer, A. L., Tovstik, P. E., and Lidsky, V. B., *The Free Vibrations of Thin Elastic Shells*, Nauka, Moscow, 1979 (in Russian).

[24] Gordon, E. Martin, Vibrations of longitudinally polarized ferroelectric cylindrical tubes, *J. Acoust. Soc. Am.*, 35, (4), 510, 1963.

[25] Graham, F. McDearmon, The addition of piezoelectric properties to structural finite element programs by matrix manipulations, *J. Acoust. Soc. Am.*, 76, (3), 666, 1984.
[26] Grinchenko, V. T., Ulitko, A. F., and Shul'ga, N. A., *Ser. Mechanics of Coupled Fields in Structural Elements, Electroelasticity,* 5, Naukova Dumka, Kiev, 1989 (in Russian).
[27] Gusein- Zade, M. I., On necessary and sufficient conditions for the existence of decaying solutions of the plane problems of the theory of elasticity for a semistrip, *J. Appl. Math. and Mech.*, 29, (4), 892, 1965.
[28] Gusein- Zade, M. I., On a plane problem of elasticity theory for a half-strip, *J. Appl. Math. and Mech.*, 41, (1), 115, 1977.
[29] Donnel, L. H., *Beams, Plates and Shells*, McGraw-Hill, New-York, 1976.
[30] Haskins, G. F. and Walsh, J. L., Vibrations of ferroelectric cylindrical shells with transverse isotropy, *J. Acoust. Soc. Am.*, 29, (6), 729, 1957.
[31] Holden, A., Longitudinal modes of elastic waves in isotropic cylinders and bars, *Bell Syst. Techn. J.*, 30, (4), 956, 1951.
[32] Holland, R., Countour extensional resonant properties of rectangular piezoelectric plates, *IEEE Trans. Sonics and Ultrasonics*, 15, (2), 97, 1968.
[33] Holland, R. and Eer Nisse, E. P., *Design of Resonant Piezoelectric Devices*, MIT Press, Cambridge, 1969.
[34] Holland, R. and Eer Nisse, E. P., Variational evaluation of admittances of multielectroded three-dimensional piezoelectric structures, *IEEE Trans. Sonics and Ultrasonics*, 15, (2), 119, 1968.
[35] Horgan, C. O. and Knowles, J. K., Recent developments concerning Saint-Venant's principle, *Advances in Appl. Mech.*, 23, 179, 1983.
[36] Infimovskaya, A. A., Chernyshev, G. N., and Rogacheva, N. N., Use of the thin piezoceramic gauges to measure of small dynamic strains, *Mechanics of Solids*, 24, (2), 518, 1989.
[37] Jaffe, B., Cook, W. R., and Jaffe, H., *Piezoelectric Ceramics,* Academic Press, London and New York, 1971.
[38] Jan Soderkvist, Dynamic behavior of a piezoelectric beam, *J. Acoust. Soc. Am.*, 90, (2), 692, Pt. 1, 1991.
[39] Kagawa, V., A new approach to analysis and design of electromechanical filters by finite-element technique, *J. Acoust. Soc. Am.*, 49, (5), pt 1., 1348, 1971.
[40] Kagawa, V. and Samabuch, T., Finite element simulation of electromechanical problems with an application to energy trapped and surface-wave devices, *IEEE Trans. Sonics and Ultrasonics*, 23, (6), 376, 1976.
[41] Kagawa, Y. and Yamabuchi, T., Finite element approach for piezoelectric circular rod, in *IEEE Trans. Sonics and Ultrasonics*, SU-23, (6), 1976.
[42] Karlash, V. L., Electroelastic oscillations of a compound hollow piezoceramic cylinder with radial polarization, *Soviet Appl. Mech.*, 26, (5), 440, 1990.
[43] Koiter, W. T. and Simmonds, J. G., Foundation of shell theory, in *Proc. of 13th Congr. Theor. and Appl. Mech.*, Springer, Berlin, 1973.
[44] Kosmodamianskii, A. S. and Lozhkin, V. N., Electroelastic equilibrium of a thin anisotropic layer with piezoelectric effects taken into account, *J. Appl. Math. and Mech.*, 42, (4), 781, 1978.
[45] Landau, L. D. and Lifshitz, E. M., *Electrodynamics of Continius Media*, Pergamon Press, New York, 1960, 337.
[46] Le Khan Chau, Basic relations of the theory of piezoelectric shells, *Vestnik MGU (Matem. Mekh.)*, No 6, 77, 1982.
[47] Le Khan Chau, The theory of piezoelectric shells, *J. Appl. Math. and Mech.*, 50, (1), 98, 1986.
[48] Lee, C. Y., Syngellakis, S., and Hou, J. P., A two-dimensional theory for high-frequency

vibrations of piezoelectric crystal plates with and without electrodes, *J. Appl. Phys.*, 61, (4), 1987.

[49] Loza, I. A., Medvedev, K. V., and Shul'ga, N. A., Propogation of acoustoelectric waves in a planar layer made of piezoceramics of hexagonal syngony, *Soviet Appl. Mech.*, 23, (3), 611, 1987.

[50] Loza, I. A. and Shul'ga, N. A., Effect of electrical boundary conditions on the propagation on axisymmetric acoustoelectric waves in hollow cylinder with axial polarization,, *Soviet Appl. Mech.*, 23, (9), 832, 1987.

[51] Lypacewich, V. and Filipipezyncki, J., Vibrations of piezoelectric ceramic transducer loaded mechanically, in *Proc. Vibr. Prob. Acad. Sci.*, 11, (3), 285, 1972.

[52] Madorskii, V. V. and Ustinov, Yu. A., Symmetric vibrations in piezoelectric plates, *Izv. Arm. Akad. Nauk, Mechanika*, 39, (4), 51, 1976 (in Russian).

[53] Mason, W. P., *Piezoelectric Crystals and Their Application to Ultrasonics*, Princeton, New Jersey, 1950.

[54] Mason, W. P., Fifty years of ferroelectricity, *J. Acoust. Soc. Am.*, 50, (5), pt 2, 1281, 1972.

[55] Mason, W. P., *Electromechanical Transducer and Wave Filters*, Princeton, New York - London, 1948.

[56] Martin, G. E., Vibrations of longitudinally polarized ferroelectric cylindrical tubes, *J. Acoust. Soc. Am.*, 35, (4), 510, 1963.

[57] Maugin, G. A., *Continium Mechanics of Electromagnetic Solids*, North-Holland, Amsterdam, 1988, chap.4.

[58] Mindlin, R. D., Coupled piezoelectric vibrations of quarts plates, *Inter. J. Solids and Struct.*, 10, (4), 543, 1974.

[59] Mindlin, R. D., High frequency vibrations of piezoelectric crystal plate, *Inter. J. Solids and Struct.*, 8, (7), 895, 1972.

[60] Mindlin, R. D., Forced thickness-shear and flexural vibrations of piezoelectric crystal plates, *J. Appl. Phys.*, 23, (1), 83, 1952.

[61] Mindlin, R. D. and Lee, P. C., Thickness-shear and flexural vibrations of piezoelectric crystal plates, *J. Appl. Phys.*, 23, (1), 83, 1952.

[62] Nye, J. F., *Physical Properties of Crystals*, Oxford, 1957.

[63] Mohammed, A., Expression for electromechanical coupling factor in term of critical frequencies, *J. Acoust. Am.*, 39, (2), 289, 1966.

[64] Munk, E. C., The equivalent electrical circuit for radial modes of a piezoelectic ceramic disk with concentric electrodes, *Philips Res. Repts.*, 20, (2), 170, 1965.

[65] Naghdi, P. M., On the theory of thin elastic shell, *Quart. Appl. Math.*, 14, (4), 369, 1957.

[66] N-Nagy, F. L. and Joice, G. C., *Solid Stage Control Elements Operating on Piezoelectric Principles, Physical Acoustics*, Mason, W. P. and Thursten, R. N., Eds., Acad. Press, New York, 9, 129, 1972.

[67] Novacki, W., *Elecrtomagnetic Interactions in Elastic Solids*, Parkus, H., Ed., Springer-Verlag, Wien, 1979, 105.

[68] Novozhilov, V. V., *The Theory of Thin Shells*, P. Noordoff, Groningen, 1959.

[69] Okazaki K. Recent developments in piezoelectric ceramics in Japan, *Ferroelectrics*, 35, (1-4), 173, 1981.

[70] Onoe, M., The contour vibrations of thin rectangular plates, *J. Acoust. Soc. Am.*, 30, (11), 1159, 1958.

[71] Onoe, M. and Pao, Y. H., Edge mode of thin rectangular plate of barium titanate, *J. Acoust. Soc. Am.*, 33, (11), 1628, 1961.

[72] Onoe, M., Tiersten, H. F., and Meitzler, A. H., Shift in the location of resonant frequencies caused by large electromechanical coupling in thickness-mode resonators, *J. Acoust. Soc. Am.*, 35, (1), 36, 1963.

[73] Parton, V. Z. and Kudryavtsev, B. A., *Electromagnetoelasticity*, Gordon and Breach, New York, 1988.

[74] Paul, H. S., Vibration of circular cylindrical shells of piezoelectric silver jodide crystals, *J. Acoust. Soc. Am.*, 40, (5), 1077, 1960.

[75] Paul, H. S. and Raman, G. V., Vibrations of pyroelectric plates, *J. Acoust. Soc. Am.*, 90, Pt. 1, (4), 1729, 1991.

[76] Pawell, R. Z. and Stetson, K. A., Interferometry vibration analysis by wave-front reconstructions, J. Opt. Soc. Am., 55, (12), 1593, 1965.

[77] Reissner, E., On bending of elastic plates, *Quart. Appl. Math.*, 5, (1), 55, 1947.

[78] Rogacheva, N. N., Equations of electroelasticity for free piezoceramic shells, *Mechanics of Solids*, 15, (6), 122, 1980.

[79] Rogacheva, N. N., The refined theory of piezoceramic shells, *Izv. Akad. Nauk Arm. SSR, Mechanika*, 34, (1), 55, 1981 (in Russian).

[80] Rogacheva, N. N., Equations of state of the piezoceramic shells, *J. Appl. Math. and Mech.*, 45, (5), 677, 1981.81.

[81] Rogacheva, N. N., On boundary conditions in the theory of piezoceramic shells polarized along coordinate lines, *J. Appl. Math. and Mech.*, 47, (2), 220, 1983.

[82] Rogacheva, N. N., Boundary conditions in the theory of piezoceramic shells with thickness polarization, *Izv. Acad. Nauk Arm. SSR, Mechanika*, 36, (6), 50, 1983 (in Russian).

[83] Rogacheva, N. N., On Saint Venant type conditions in the theory of piezoelastic shells, *J. Appl. Math. and Mech.*, 48, (2), 213, 1984.

[84] Rogacheva, N. N., Classification of free piezoceramic shell vibrations, *J. Appl. Math. and Mech.*, 50, (1), 106, 1986.

[85] Rogacheva, N. N., Applicability of general theorems of electroelasticity to the theory of piezoceramic shells, *Mechanics of Solids*, 21, (5), 172, 1986.

[86] Rogacheva, N. N., Influence of fluid viscosity and pressure on shell vibrations in a fluid, *J. Appl. Math. and Mech.*, 51, (4), 504, 1987.

[87] Rogacheva, N. N., Forced vibrations of a piezoceramic cylindrical shells with longitudinal polarization, *J. Appl. Math. and Mech.*, 52, (5), 641, 1988.

[88] Rogacheva, N. N., Refined theory of membrane electroelastic shells, *J. Appl. Math. and Mech.*, 54, (4), 518, 1990.

[89] Rogacheva, N. N., Asymptotic theory of piezoelectric shells,in *Proc. of the IUTAM Symp. on the Mechanical Modellings of New Electromagnetic Materials*, Hsieh, R. K. T., Ed., Elsevier, Amsterdam, 1990, 348.

[90] Rogacheva, N. N., Piezoceramic gage of dynamic strains, in *Proc. of the Third Int. ISEM Symp. on the Application of Electromagnetic Forces*, Tani, J. and Tagaki, T., Eds., Elsevier, 1991, 211.

[91] Rogacheva, N. N., Multilayered elastic, viscoelastic, electroelastic shells, in *Proc. of the 7th All-Union Congress on Applied and Theoretical Mechanics*, Nauka, Moscow, 1991 (in Russian).

[92] Rogacheva, N. N., Resonance method of measuring fluid viscosity with the use of piezoeffect, in *Proc. of the Fourth Int. Symp. on Nonlinear Phenomena in Electromagnetic Fields*, Furuhashi, T. and Uchikawa, Y., Eds, Elsevier, 1992, 581.

[93] Rogacheva, N. N., Multilayered piezoelectric shells, in *Proc. of the Fourth Int. Symp. on Nonlinear Phenomena in Electromagnetic Fields*, Furuhashi, T. and Uchikawa, Y., Eds, Elsevier, 1992, 577.

[94] Rogacheva, N. N., The use of piezoeffect for vibration suppression, in *Proc. of the Fourth Int. Symp. on Nonlinear Phenomena in Electromagnetic Fields*, Furuhashi, T. and Uchikawa, Y., Eds, Elsevier, 1992, 573.

[95] Schnessler, H. H., Filters and resonators, *IEEE Trans. Sonics and Ultrasonics*, 21, (4), 257,

1974.

[96] Sinka, B. K. and Tiersten, H. F., Elastic and piezoelectric surface waves guided by tin film, *J. Appl. Phys.*, 44, (11), 4831, 1973.

[97] Stavsky, J. and Adellman, N. T., Radial vibrations of composite piezoelectric ceramic disks, *Appl. Sci. Res.*, 31, (2), 128, 1975.

[98] Stephenson, C. V., Higher modes of radial vibrations in short, hollow cylinders of barium titanate, *J. Acoust. Soc. Am.*, 28, (5), 928, 1956.

[99] Stephenson, C. V., Radial vibrations on short hollow cylinders of barium titanate, *J. Acoust. Soc. Am.*, 28, (1), 51, 1956.

[100] Takashi Maeno, Takayuki Tsukimoto, and Akita Miyake, Finite-element analysis of the rotor/stator contact in a ring-type ultrasonics motor, *IEEE Trans. Ultrason., Ferroelec. and Freq. contr.*, 39, (6), 668, 1992.

[101] Tiersten, H. F., *Linear piezoelectric plate vibrations*, Plenum Press, New York, 1969.

[102] Tiexrsten, H. F., Wave propagation in an infinite piezoelectric plate, *J. Acoust. Soc. Am.*, 35, (2), 234, 1963.

[103] Tiersten, H. F., Thickness vibrations of piezoelectric plates, *J. Acoust. Soc. Am.*, 35, (1), 53, 1963.

[104] Tiersten, H. F. and Mindlin, R. D., Forced vibrations on piezoelectric crystal plates, *Quart. Appl. Math.*, 20, (2), 107, 1962.

[105] Timoshenko, S. P., *Theory of Elasticity*, McGraw-Hill, New York, 1970.

[106] Timoshenko, S. P. and Woinowsky-Krieger, S., *Theory of Plates and Shells*, McGraw-Hill, New York, 1959.

[107] Toulis, W. J., Electromechanical coupling and composite transducers, *J. Acoust. Soc. Am.*, 35, (1), 74, 1963.

[108] Venkateswara Sarma K., Torsional oscillations of a finite inhomogeneous piezoelectric cylindrical shell, *Inter. J. Eng. Sci.*, 21, (12), 1483, 1983.

[109] Wager, R. S. and Kino, G. S., Analysis of the elastic potential of an acoustic surface wave propagating on a layered halfspace, *IEEE Trans. Sonics and Ultrasonics*, 21. (3), 209, 1974.

[110] Wellkins, C. J., Vibrations of backed piezoceramic disk-transducers with annular electrodes and matching layer, *IEEE Trans. Sonics and Ultrasonics.*, 29, (1), pt. 2. 37, 1982.

[111] Whittaker, E. T. and Watson, G. N., *A Course of Modern Analysis*, University Press, Cambridge, 1927.

[112] Wilson, O. B., *Introduction to the Theory and Design of Sonar Transducers*, Peninsula, Los Atlos, 1988.

[113] Woollett, R. S., Comments on "Electromechanical coupling and composite transducers", *J. Acoust. Soc. Am.*, 35, (12), 1837, 1963.

[114] Woollett, R. S., Effective coupling factor of single-degree-of-freedom transducer, *J. Acoust. Soc. Am.*, 40, (5), 1112, 1966.

[115] Yong-Kong Yong, James, T. Stewart, Lacques Détaint, Albert Zarka Bernard Capelle, and Yunlin Zheng, Thickness-shear mode shapers and mass-frequency influence surface of a circular and electroded AT-cut quartz resonator, *IEEE Trans. on Ultrason., Ferroelec. and Freq. Contr.*, 39, (5), 609, 1992.

[116] Yuan Yiquan, Investigation of a new-wide-band transducer with piezoelectric layered ring, *IEEE Trans. on Ultrason., Ferroelec. and Freq. Contr.*, 39, (1), 39, 1992.

INDEX

A

B

C

R

S

T

V

Z